Integrated Science

Volume 39

Editor-in-Chief
Nima Rezaei, Tehran University of Medical Sciences, Tehran, Iran

The **Integrated Science** Series aims to publish the most relevant and novel research in all areas of Formal Sciences, Physical and Chemical Sciences, Biological Sciences, Medical Sciences, and Social Sciences. We are especially focused on the research involving the integration of two of more academic fields offering an innovative view, which is one of the main focuses of Universal Scientific Education and Research Network (USERN), science without borders.

Integrated Science is committed to upholding the integrity of the scientific record and will follow the Committee on Publication Ethics (COPE) guidelines on how to deal with potential acts of misconduct and correcting the literature.

Joaquín Vial

Sustainable Development and Natural Resources

A View from the South

EDICIONES UC

Springer

Joaquín Vial
Instituto de Economía and CLAPES UC
Pontificia Universidad Católica de Chile
Santiago, Chile

ISSN 2662-9461 ISSN 2662-947X (electronic)
Integrated Science
ISBN 978-3-032-12262-9 ISBN 978-3-032-12263-6 (eBook)
https://doi.org/10.1007/978-3-032-12263-6

The original submitted manuscript has been translated into English. The translation was done using artificial intelligence. A subsequent revision was performed by the author(s) to further refine the work and to ensure that the translation is appropriate concerning content and scientific correctness. It may, however, read stylistically different from a conventional translation.

Translation from the Spanish language edition: "DESAFÍOS PARA UN DESARROLLO SUSTENTABLE DE CHILE EN EL SIGLO XXI" by Joaquín Vial, © Pontificia Universidad Católica de Chile 2022. Published by Ediciones UC. All Rights Reserved.

Jointly published with EdicionesUCIntegrated Science

This Springer imprint is published by the registered company Springer Nature Switzerland AG
The registered company address is: Gewerbestrasse 11, 6330 Cham, Switzerland

Preface

These notes originate from courses taught at various universities in Chile over the past 20 years, but especially from those we have taught—first with José Pablo Arellano and later individually—at the Pontifical Catholic University of Chile.

This is a substantially revised second edition of the book *Desafíos para el desarrollo sustentable de Chile en el Siglo XXI*, published in 2023 by Ediciones UC. The focus of the first edition was to help Chilean university students better understand the economic and social context in which they will develop their professional careers, as well as some of the major challenges they will face within the framework of these trends, which, while global in nature, manifest in each country in particular ways. This second edition shifts the focus from Chile to emerging countries in general, especially to those rich in natural resources. The latter face specific challenges because these resources interact directly with the environment, are heterogeneous, and are available in limited quantities. These characteristics make it very difficult for them to replicate the successful experiences of the Asian Tigers or China—experiences that, moreover, are no longer available—and require a reflection that addresses the particularities of natural resources.

A common theme throughout these notes is the realization that the era we are living through is unprecedented and presents unique challenges for our species. Never in human history has the planet experienced changes of the magnitude we are witnessing now, in such a short period of time and, most notably, because of the actions of a single species.

It is true that cosmic accidents, such as collisions with asteroids, have caused traumatic, planet-wide changes over very short periods, dramatically altering the course of the evolution of life on Earth. However, what is remarkable about our era is that, precisely because of that evolution, a species has emerged with the capacity to take life to other parts of the cosmos, as well as to destroy it on our own planet. From my parents' generation to the present, humanity has had to contend with the risk of nuclear holocaust—something created and managed exclusively by humans. Current generations will also have to face the consequences of human progress on the environment, something that, unlike nuclear self-annihilation, does not depend solely on human action and is subject to great uncertainty. Our generation and those to come are defining the fate of our species, and even of life itself on our planet, whether by action or omission.

Another recurring theme in these notes is the need to meet the demands for progress in material living conditions, especially for those living in poverty, while doing so in a manner compatible with the global and local constraints imposed by the environment. The so-called economic problem, which economists learn about in the early years of university, is summarized as addressing the scarcity arising from ever-increasing material needs when resources are limited. Today, this problem presents itself on a global scale, as over the past century humanity has begun to exceed the limits for safe operation within a closed ecosystem.

We have reached a point where the availability of certain finite resources is being tested globally, but even more so, the planet's capacity to absorb the impacts of human activity. This is happening at a time when instantaneous information flows make it clear that a large portion of the population still lives in miserable conditions, facing hunger and hardship that much of the world left behind long ago.

In line with the above, the main theme of this document is the challenge of reconciling long-term economic progress in the poorest countries with the global and national constraints arising from various environmental limits, which are now under intense pressure due to the exponential growth of population and income at both global and local levels. For small countries rich in natural resources, the prospects for sustained improvements in living standards necessarily depend on expanding exports of goods and services, as the size of the domestic market is insufficient for this purpose. In the analysis and projection of the country's economic integration into the global economy, specialization in the production and export of such products is key. This type of specialization makes these economies particularly sensitive both to the physical impacts of environmental constraints at the global and local levels and to the indirect effects that adjustments by the world's major economies to address these constraints may have.

The book is divided into four parts: the first discusses the major global trends that frame the possibilities for sustainable development at the global level, outlining the challenges of improving the quality of life for most of the world's population living in middle- or low-income countries, while at the same time avoiding a global environmental catastrophe. Several challenges arise from this, which differ depending on the level of development of each country: in the poorest countries, the priority must be inclusive economic development that opens opportunities for women and rapidly increases productivity, in order to accelerate the demographic transition and reduce pressures on biodiversity from land use change and deforestation, which is their main contribution to global warming. In middle-income countries, which concentrate the largest share of the population, economic activity, and greenhouse gas emissions, the main challenge is threefold: to accelerate the energy transition and reduce land use change to avoid environmental catastrophes, to accelerate the demographic transition, and to develop more environmentally friendly consumption patterns. Finally, developed countries must accelerate their energy transition, make profound adjustments to their consumption patterns to make them more sustainable, and increase resource transfers to enable the necessary changes in middle- and low-income countries.

The solution to these challenges requires significant adjustments in three variables underlying the environmental impact of human activity: the evolution of population size, the resource requirements to meet per capita consumption levels, and the efficiency in the use of nature, understood as the environmental impact per unit of consumption. The following parts of the book are devoted to examining these three major topics.

The second part focuses on demographic trends and their relationship with economic development, as well as the challenges posed by demographic adjustment compatible with a sustainable trajectory of global economic development.

The third part analyzes in greater depth the impact of economic activity on natural resources, identifying the causes of overexploitation and policies to address it. It also examines the growing impact and tensions between the exploitation of natural resources and the preservation of ecosystem capacity to maintain their integrity. This part also includes a chapter on the macroeconomic challenges faced by resource-rich countries.

The fourth and final part is devoted to analyzing the determinants of two components of the global environmental crisis: the loss of biological diversity and climate change. Both are consequences of abuses of nature that are putting at risk the continuity of many environmental services that, while essential for life in general, are also vital for our species, and which until recently were taken for granted. Avoiding a global catastrophe requires a substantial improvement in the efficiency with which we interact with the environment.

The future awaiting today's younger generations will not be easy, regardless of the professional field they choose. As E. O. Wilson said, in the coming decades humanity must pass through a demographic and economic "bottleneck" to transition to a future of sustainable material living conditions and unimaginable opportunities for progress. Achieving this successfully will require the cooperation of all nations, even the smallest, but a prerequisite is to understand these challenges as thoroughly as possible.

I hope these notes help to better understand some of the problems that these young people will have to face in their professional future, and spark interest in studying how to address them. Advancing toward improvements in the well-being of current and future generations will be very difficult and leaves no room for decisions based on prejudice, ideology, or unfounded intuition. On the contrary, it will require ingenuity, creativity, and significant advances in scientific and technical knowledge, as well as institutions and individuals capable of cooperating to design and implement the actions needed to move forward.

Santiago, Chile Joaquín Vial

Acknowledgments

This book gradually took shape over the past two decades, based on courses taught at Adolfo Ibáñez, Católica, and the University of Chile. It is difficult to pinpoint the specific contributions of teaching assistants and students over these years, but without a doubt, they were fundamental, both through their questions in class and their critiques in evaluations and comments on the early drafts. The inspiration behind these courses was to familiarize students with some of the major challenges they would encounter in their professional lives, because of global trends that were already emerging as decisive for humanity.

As for many economists raised and living in developing economies, the issue of economic development and its linkages with natural resources were ever present in all my formative years, so I have a great debt with my teachers and colleagues at the University of Chile and the University of Pennsylvania. The same happens with many colleagues both in Government and Central Banks in Chile and many Latin American countries. However, what drove me to get involved in a project like this was my work at Center for International Development (CID) in Harvard and at the Earth Institute in Columbia under the guidance of Jeff Sachs, as well as—on the opposite side to the institutional map—BBVA Research, where, under the leadership of José Luis Escrivá, I was put in charge of a team to study medium- and long-term trends and their impact on the financial sector. The final push to write a book on these issues started with conversations with my friend, co-teacher, and former colleague at CIEPLAN, José Pablo Arellano, as well as with Elías Albagli, at the Central Bank of Chile.

The final impulse for this project was made possible by the support of CLAPES UC, led by Felipe Larraín, and the Faculty of Economics and Management at Universidad Católica de Chile, and especially to the Dean José Miguel Sanchez who sponsored the publication of a first edition focused on Chile by Ediciones UC.

The materialization of this project was possible thanks to the excellent technical support given by Francisca Cuadros, a junior researcher at CLAPES UC, and the interest and dedication of Gonzalo Córdova and Alexis Rivas, the editors at Springer Nature.

To all of them, as well as to the many colleagues and friends impossible to name here, my most sincere thanks.

Finally, I want to thank María Consuelo and my children, who gave me the space, support, and motivation to persist in this endeavor. At the end of the day, this book was written thinking about the world in which Magdalena, Joaquín, Agustín, and my grandchildren will live.

Contents

Part II The Demographic Lever: The Transition Toward a Sustainable Path for the World Population

Part III Inefficiencies and Excesses in the Use of Natural Resources

Part I

Impacts of Economic Development on Nature. A Historical Overview

Introduction 1

Abstract

This chapter provides motivation and an overview of the main issues discussed in the book. The Industrial Revolution unleashed an era of economic and population growth, fueled by innovation, capital accumulation and international trade. This process opened unlimited opportunities for the progress of Humanity. However, the environmental consequences of this progress are already testing the capacity of the planet to absorb its impacts. In the most recent decades, the trade off between economic progress and sustainability has become evident, while there is still a large fraction of humanity that lives in unacceptable poverty, or aspires to achieve consumption levels like those of more advanced countries. To accommodate both goals, major changes must take place: population growth must slow down and eventually become negative, as is already happening in most of the world, except in the poorest areas. Consumption of physical goods and energy must decline in more advanced nations to leave room for expansion in low-income countries and the environmental impact of human activities per unit of output must decline significantly. Development economics must bring back Nature into the design and application of Economic and Social policies.

Keywords

Sustainable development · Natural capital · Ecosystem services · Comprehensive wealth · Planetary boundaries · Global income disparities

This era has been marked by change, to the extent that today we see it as something natural. A 3-year-old computer, perfectly suitable for writing notes, performing routine calculations, and communicating with the world, is now considered obsolete. A company whose results fail to surprise analysts loses value and may become a target for acquisition by a more successful competitor. Workers are challenged daily by changes in their tools, communication systems, and even new ways to organize

J. Vial, *Sustainable Development and Natural Resources*, Integrated Science 39,
https://doi.org/10.1007/978-3-032-12263-6_1

work. More generally, an economy whose output does not grow, or grows only very slowly, is considered "stagnant" and may be labeled "sclerotic."

Change dominates life, dazzling us with the new opportunities it creates, but also raising concerns about the threats it poses. It can also be painful, as many of its benefits come with costs and the loss of much of what shaped our upbringing and personalities.

Although current generations live in a rapidly changing world, very few are aware that this is a recent phenomenon, dating back little more than a couple of centuries. In other words, a mere blink of an eye in the history of a species that has roamed the planet for about 100,000 years[1].

Looking back in time, at the beginning of the seventeenth century, for example, humanity lived in a very different reality. Most people lived in rural areas, engaged in agriculture, using techniques very similar to those of four or five millennia earlier. The main sources of energy were the same as those used since prehistoric times, and almost all were renewable: human and animal power, ultimately based on photosynthesis; wind, which moved ships and goods around the world; firewood, also a product of photosynthesis; and, in very limited applications, hydropower. All of these, except for coal and to a lesser extent firewood, were renewable and virtually inexhaustible sources, but with low energy density. Moreover, when overused, for example, causing severe deforestation, the consequences were quickly felt, curbing excesses at the cost of increased mortality that adjusted population size.

Nevertheless, even this limited economic progress required major changes in human societies. As populations grew and encountered other human groups, economic and social relations based on trust among lifelong acquaintances, following the tacit rules of the tribe or clan, were no longer sufficient.

The domestication of plants and animals enabled the development of agriculture, which in turn led to the establishment of permanent settlements that, together with population growth, provided an economic foundation for specialization and commercial exchange.

All of this resulted in an exponential leap in the development of increasingly complex norms and institutions, which in turn had to be enforced and defended. The provision of public goods (those that are not depleted by individual use, making it impossible to exclude non-payers) led to the introduction of taxes and treasuries, as well as bureaucracy and public forces, both to ensure internal order and to defend against external aggression.

Money emerged in what is now the Middle East more than 5000 years ago to facilitate exchange, an essential element for specialization and increased productivity, but which also conferred enormous power and wealth on those who produced or validated it. Over time, once the use of standardized metal objects to represent the value of goods became widespread, credit appeared in the city-states of Italy, which were the center of commercial exchange with the East.

[1] To put human history in perspective, it is worth remembering that the first forms of life appeared on Earth about 500 billion years ago.

By the end of the Middle Ages, many of the institutions that form the basis of modern states had already been established. As the relationships within human societies became more complex, so too did the institutions that supported social and economic interactions within and between them[2].

Until then, changes in material living conditions were largely cyclical, with good and bad periods, very different from the exponential progress experienced in recent centuries. People were at the mercy of nature, and many of the tensions—personal, political, and social—experienced up to the seventeenth century were heavily influenced by natural cycles (Fagan 2000; Brooke 2014). Rain or daily temperature—now something that occupies perhaps 5 min of our attention, just to decide what to wear or plan weekend activities—was once the difference between abundance and hunger, or even between life and death for the most vulnerable.

This long period could be characterized as one of limited development, but without major global constraints. Migration helped to alleviate natural limitations in overexploited territories, although it often led to violent conflict with local populations.

For a long period of 20,000 years, humanity lived in a stationary equilibrium, in which all economic progress translated into population growth, but without substantial improvements in living conditions (Albagli and Vial 2022). Malthus described the dynamics of this equilibrium without progress based on two postulates considered indisputable in his time:

> First, that food is necessary for human existence.
>
> Second, that the passion between the sexes is necessary and will remain close to its current state (Malthus 1798, p. 70).

From this, he concluded that the exponential growth potential of the population is much greater than the linear capacity of the land to produce more food, concluding that "this implies a strong and constantly operating check on population from the difficulty of subsistence" (Malthus 1798, p. 71).

In this context, with an uneducated and low-skilled workforce, any innovation leading to productivity increases would not result in permanent wage gains, as these would be offset by competition from new entrants to the labor force. This is what classical economists called the "Iron Law of Wages," a view that prevailed among economists until the late nineteenth century, leading Carlyle to dub economics the "dismal science," a term still in use today (albeit for different reasons).

The Industrial Revolution broke this stagnation. It occurred when knowledge accumulated over a long period began to be applied to the development of new technologies, resulting in new products and services[3]. These changes led to new ways of producing goods and to the development of activities, products, and services that gave rise to new forms of division of labor, greater specialization, and

---

[2] See Ferguson (2008).

[3] This point is explored in great depth for the United States in the book by Robert J. Gordon (2016). See also Gordon's TED talk: https://www.ted.com/talks/robert_gordon_the_death_of_innovation_the_end_of_growth

the development of new skills (and thus, heterogeneity). These changes granted greater bargaining power, both to individual workers and to those organized in unions, enabling them to negotiate wage increases that allowed them to capture part of the returns from innovation, as identified by Alfred Marshall in his "Principles of Economics" (1890), a work that laid the foundations of modern economics.

The economic development that followed the Industrial Revolution was a comprehensive process, resulting in an exponential increase in the production of goods and services and in the use of natural resources. The new levels of income and wealth fostered improvements in nutrition and health conditions, leading to increases in both life expectancy and population size, which in turn drove productivity growth and the expansion of demand.

In this environment, the workforce became increasingly educated and specialized, with qualifications unimaginable just a few decades earlier. All this took place amid an extraordinary increase in the complexity of social and economic relations and, consequently, of the norms and institutions that support them.

This has been possible because material progress generated surpluses to develop institutions, build infrastructure, and create new services, which in turn accelerated progress. Just one example: modern societies could not have achieved current levels of economic progress without the prior existence of mass education systems, both at the school level and in vocational and postgraduate training. The same can be said for health, legislative, governmental, security and law enforcement institutions, among many others.

Economic and social development multiplied both the population and its average income tenfold between the nineteenth and twenty-first centuries. Figure 1.1, based

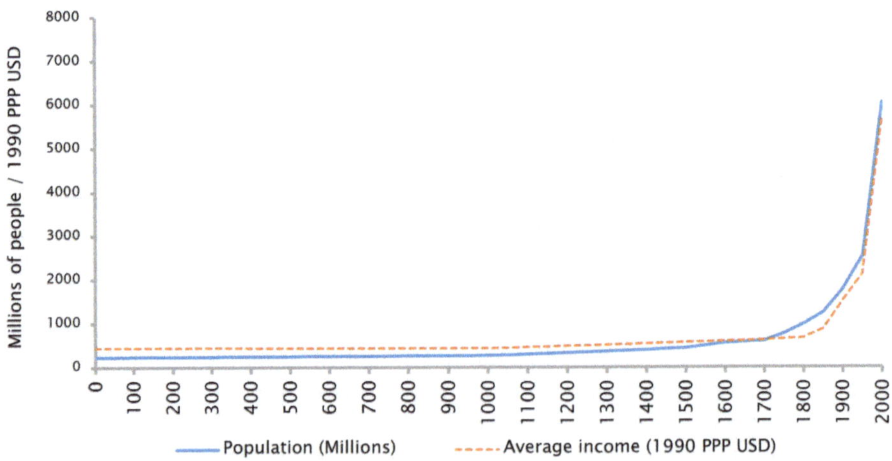

Fig. 1.1 World population and per capita income 0–2000. (Source: Maddison (2000) and United Nations (2019))

on J. Sachs (2008), describes the characteristics of the so-called "Anthropocene"[4], showing the increase in per capita income and the size of the world population[5].

1.1 Why Has Nature Disappeared from the Analysis of Economic Development?

A look at the past must acknowledge the tremendous success of humanity, reflected both in the increase in its population and in improvements in material living conditions.

Alongside demographic and economic trends, there have been enormous advances in the development of knowledge and skills across a wide range of fields: from the arts and culture to greater freedom to enjoy the new phenomenon of "leisure time." This was unthinkable when the main concern was survival for oneself and one's family. All this has been possible because the conditions and incentives for innovation were created.

The Industrial Revolution broke the Malthusian chains thanks to the massive incorporation of new territories and technologies, as well as much more powerful energy sources (fossil fuels) into production processes and transportation. Although this did not initially translate into immediate improvements in people's living conditions, as productivity gains began to result in higher wages, it started to provide access to goods and services previously unimaginable.

Access to water, sewage, and electricity networks in cities, and their gradual expansion to rural areas, freed people from tasks that consumed much of their time and energy, with a very large—though difficult to quantify—impact on well-being[6]. The replacement of horses in urban freight and passenger transport not only improved speed and capacity but also brought about an incalculable improvement in the sanitary conditions of urban life.

The development of the railway drastically shortened distances between people, between resources, and between productive enterprises goods that used them, as well as between suppliers of goods and consumers. This, in turn, opened business opportunities that were previously unimaginable without these technologies. In the field of maritime transport, the replacement of wind power by new mechanical means expanded the possibilities of international trade, creating opportunities for consumption and production on a scale that extended beyond the continents. Air

---

[4] The term was originally coined by Crutzen and Stoermer (2000) to refer to the period beginning around 1800, marking the onset of a planet completely transformed by human activity. Sachs (2008).

[5] This chart may underestimate changes in living conditions, as GDP does not capture welfare gains that are not reflected in market values. As Gordon (2016) illustrates, the connection of households to urban networks of gas, water, electricity, and sewage represented a revolution in the time and effort devoted to heavy domestic chores, along with a drastic improvement in the sanitary conditions of homes and urban areas. Very little of this was likely captured in much later GDP measurements.

[6] See Gordon (2016) for a comprehensive analysis of this process in the United States.

transport further amplified this effect from the second half of the twentieth century onward.

In broader terms, the impact of the development of human activities on both the capacity to manage the environment and the use of its resources was enormous. After 100,000 years of life dependent on the climate, adjusting all activities to the cycle of sunlight and the seasons, humanity was suddenly freed from these constraints[7]. In just over a century, society moved from Malthusian pessimism to an optimism that leads many to believe that almost anything is possible, from leaving the planet to colonize distant worlds, to overcoming physical limitations and transcending time and space (Harari 2015).

Economic thought captured these changes, summarizing them in an aggregate production function, known as the Solow model, using only two factors: capital and labor. Everything not attributable to the growth of these two factors, in both quantity and quality, falls under the term Total Factor Productivity, which is generally associated with all types of innovations and overall efficiency gains.

In this context, total and per capita GDP can grow indefinitely, provided that human capital, physical capital, and productivity continue to increase. Economic science, overwhelmed by the weight of evidence from more than a century of accelerated development, left behind the classical notions of scarcity of Nature's services.

In the early introductory economics classes, at least those from the early 1970s, the functioning of the economy was represented by what was known as the "circular flow of income": the physical and monetary exchanges that occurred between the different "economic agents." In its simplest version, there were two: households, which consumed and worked, and firms, which produced consumer goods and employed workers. The physical transactions of consumer goods and labor services had their counterpart in monetary flows: income and expenditure, which in this simple framework had to be equal. In this static economy, there are no stocks (capital) and time does not matter. There is also no growth, and nature does not enter into the equation.

A formal representation of these relationships forms the basis of the national accounting systems currently used worldwide[8]:

$$Y = C \tag{1.1}$$

where Y refers to household income[9], which is the product of their labor:

$$Y = w^{*}L \tag{1.2}$$

[7] Louise Leakey puts these issues in perspective in a brief and dynamic TED talk from February 2008. Available at: http://www.ted.com/index.php/talks/louise_leakey_digs_for_humanity_s_origins.html?utm_source=SubscriberMail&utm_medium=email&utm_campaign=This%20week%3A%20Stories%20of%20our%20shared%20humanity&utm_term=&utm_content=5abd78e9662048bc8cd187e96237120b

[8] Those not interested in the formal foundations of the National Accounts system may skip the following without much loss.

[9] In this ultra-simplified static world (without a temporal dimension), all households are single-parent or working families.

where w is the wage rate and L is the amount of labor performed by households.

C is the value of the consumer goods produced by firms:

$$Y = p^{*}Q \tag{1.3}$$

where p is the price of the goods produced by firms and Q is the quantity of goods (produced and consumed).

In equilibrium, the following identity holds:

$$w^{*}L = p^{*}Q \tag{1.4}$$

This process is schematically represented in Fig. 1.2, which shows the real flows of goods and services between the two economic agents (households and firms) and their monetary counterparts.

The analysis of economic growth requires the introduction of the temporal dimension and the inclusion of "stocks" or various forms of capital that evolve over time. Typically, this initial approach includes two forms of capital: produced physical capital, which corresponds to the stock of machines and structures that contribute to production, and human capital, which recognizes that, beyond physical strength, workers use knowledge and experience accumulated over time, which also contribute to production.

These two forms of capital provide services to production and are not exhausted in the process. Over time, they age and wear out, which is usually summarized by the concept of depreciation, although in the case of human capital, skills are also acquired. These forms of capital can also increase, either through the addition of new machines and structures or by increasing the number of workers.

Both produced capital and human capital can incorporate improvements that make them more productive, whether through new designs and technological

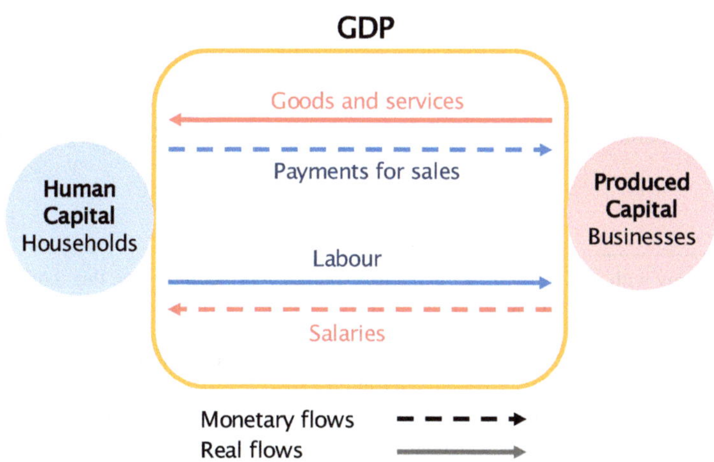

Fig. 1.2 Simplified representation of the real and monetary flows between firms and households. (Source: Own elaboration)

advances or through the acquisition of new knowledge and skills. Moreover, over time, new goods and services are created, advances are made in the organization of production processes, and new technologies or more efficient inputs increase the overall productivity of the various forms of capital. In traditional economic terminology, this is referred to as Total Factor Productivity.

The different forms of capital alter the flow relationships described above: the accumulation of physical capital requires investments (the purchase of machinery) that will last for a certain period while providing their services to production. To purchase these machines, consumption must be postponed and savings accumulated[10].

A similar situation arises with human capital: households now have children to support and educate, which requires resources beyond those used to maintain the working population. The upbringing and care of children are traditionally recorded as household consumption, although they contribute to the accumulation of human capital.

Traditional accounting includes labor services but does not recognize that behind them lies human capital, among other reasons because it is very difficult to disentangle the underlying transactions. This is relevant because, among other things, it underestimates the contribution of parental work to the accumulation of human capital in children. In societies where most of this work is performed by women, this results in a significant underestimation of their contribution to capital accumulation and future income generation[11].

Despite these simplifications, formal representation becomes somewhat more complex. Income, viewed from the perspective of how families spend it, transforms Eq. 1.1 into 1.5:

$$\mathrm{Y} = \mathrm{C} + \mathrm{S} \tag{1.5}$$

where S is household savings.

The origin of household income comes not only from labor, but also from the returns generated by capital[12].

$$\mathrm{Y} = \mathrm{w}^{*}\mathrm{L} + \mathrm{r}^{*}\mathrm{K} \tag{1.6}$$

where r represents the flow of returns generated by capital (K) during the period.

[10] In an economy with division of labor and a minimum level of complexity, this requires specialized intermediaries to collect these savings and channel them to those who invest, which in turn leads to the operation of a specialized market for financial services. This in turn requires the development of specialized institutions for the provision of health and education services, among others, aimed at strengthening and increasing human capital.

[11] Not to mention its contribution to what has come to be known as social capital, related to values and behavioral norms, which are essential for the proper functioning of increasingly complex societies.

[12] In an economy with financial markets, these may derive directly from profits, or as interest payments on savings in these markets.

On the production side, firms produce not only consumer goods but also capital goods, which are sold during the period. This is what is known as investment (I), so Eq. 1.3 becomes:

$$Y = p^{*}Q = C + I = C + S \tag{1.7}$$

Thus, Eq. 1.7 represents macroeconomic equilibrium and allows us to identify a new relationship: aggregate savings must now equal investment, so that there is neither surplus nor shortage.

$$S = I \tag{1.8}$$

Equilibrium is achieved through the functioning of markets and the adjustment of goods prices (p) and the remuneration of labor (w) and capital (r).

The model is completed with a description of the economy's growth path, based on the growth of the different types of capital and technological progress, reflected in what is known as Total Factor Productivity (TFP). Using "g" as a proxy for the growth rates of the various variables and assuming a Cobb-Douglas production function with constant returns to scale[13], GDP growth is determined by the growth of the factors that drive it[14]:

$$Y = TFP^{*}K^{a*}L^{(1-a)} \tag{1.9}$$

Economic growth (gY) can thus be described by the trajectory of the three variables involved in the production function:

The growth of human capital (gL), which is determined by demographic variables, household behavior in relation to the labor market, and the accumulation of knowledge and skills by workers.

In turn, capital growth depends on the acquisition of capital goods by firms, that is, investment, minus the depreciation (d) of existing assets.

The final element is TFP growth, which is very difficult to identify, so economists, following Solow's (1956) approach from the mid-twentieth century, still calculate it ex post as a residual. Incidentally, this requires knowing the value of "a". When there are constant returns to scale and competitive factor markets, "a" can be approximated from the observed shares of capital and labor in National Income.

National Accounts adopts these concepts as a basic framework, adding many other variables to reflect a much more complex reality.

Economic equilibrium is achieved when all household and business plans satisfy the (ex ante) conditions reflected in Eqs. 1.7 and 1.8. In the real economy at any

---

[13] This means that if capital and labor inputs increase by a given proportion, output will grow by that same proportion.

[14] The chosen functional form, known as the Cobb-Douglas function, implies a constant elasticity of substitution between capital and labor and, in this specification where the sum of the coefficients of K and T equals 1, assumes constant returns to scale. These are common assumptions in economic analysis, but they are far from innocuous, especially the latter, as will be seen below.

given moment, the analysis must include inventories (which may be higher or lower than planned by firms), as well as the possibility that not all workers are employed and that the capacity of machinery and structures is not utilized at its optimum level. Therefore, the National Accounts measures summarized in Eqs. 1.7 and 1.8 are ex post accounting identities.

1.2 The Reintegration of Nature into Economic Analysis

The economic representation implicit in the Solow model is usually the first "model" that students encounter to describe an economy and as previously mentioned, it provides the conceptual foundation for National Accounting systems. It is a simple yet highly useful representation.

However, it suffers from a critical omission: it does not incorporate the interactions between the economy and nature, which are substantial—unless one assumes that the services provided by nature are unlimited. In the intellectual climate of the early and mid-twentieth century, when the limits imposed by nature over millennia of Malthusian equilibrium were broken daily, this omission may have been understandable and even excusable.

A century later, with overwhelming evidence that human activity is exceeding the biosphere's capacity for recovery, the role of natural capital in the analysis of economic processes can no longer be ignored. In practice, this omission results in an optimistic bias in all long-term economic projections, as well as other errors stemming from the exclusion of key variables for economic policy analysis.

Indeed, from the second half of the last century, a vigorous debate emerged regarding the sustainability of global economic growth. The tremendous economic progress of the past two centuries is increasingly being questioned due to its impact on the environment and the availability of the resources on which it depends[15].

To illustrate this, it is worth examining the consequences of changes in energy use and production. The Industrial Revolution and subsequent progress were largely based on the application of new forms of energy which are far more powerful than those previously available. Coal began to replace animal and human power with that of machines, but new applications of mechanical power required new energy sources, such as oil and natural gas. At the same time, electricity became essential to daily life and was initially generated using the same fossil fuels and, where available, hydropower. Atomic energy became an important energy source, as did new renewable energies, which are helping to use electricity to replace human intellectual power through electronic and, more recently, digital means. Other advances in knowledge, communications, information transfer and storage, and transportation also depend on increased energy use.

These new uses of energy come at a cost: the growing use of fossil fuels, with a significant impact on local pollution at first and, more recently, through the accumulation of greenhouse gases in the atmosphere, with global consequences for the

[15] See Dasgupta (2021)

climate. Although there are alternatives to fossil fuels for electricity generation, they have not yet been able to fully replace them, while in other areas, such as transportation, fossil fuels remain indispensable.

This phenomenon is just one of many manifestations of environmental pressure. The increase in goods production has also altered land use. Initially, this was due to the rapid expansion of areas devoted to agriculture, and later to the need for raw materials for various industrial activities. Added to this is the growing use of land for human habitation, which, as prosperity advances, demands more living space for both productive and recreational purposes.

All these activities have been accompanied by increasing waste generation and ecosystem destruction, altering the environmental conditions that support life on the planet, with significant losses of biodiversity and declines in the biomass of all types of wild species.

By the end of the twentieth century, it was already clear that human activity was exceeding the safe operating limits of the processes that sustain life in terrestrial systems.

Initially, concern arose over the availability of resources to feed the growing population, as well as to provide energy and raw materials for industrial activity, as reflected in the so-called Club of Rome Report (Meadows et al. 1972). The authors of the report, in a revised version of Malthusian arguments, emphasized the consequences of the dual exponential growth of population and per capita income in a closed system. The first oil crisis in 1973, when the newly created OPEC cartel cut production to force up prices, gave this concern a tremendous sense of legitimacy and urgency.

Soon, however, the perspective had to be broadened to consider the effects of pollution, habitat destruction, and biodiversity loss, once it was recognized that these phenomena were global in scope and affected life across the planet. The so-called Brundtland Report (Brundtland 1987) embodied this new perspective. A significant contribution of this report was the introduction of the concept of sustainable development, defining it as "development that meets the needs of the present without compromising the ability of future generations to meet their own needs."

While market forces promoted substitution and efficiency in the use of oil, alleviating some of the concerns raised by the Club of Rome, economic theory incorporated these concerns in a relatively straightforward manner (Nordhaus 1992), expanding the Solow model to integrate exhaustible and renewable natural resources within a more flexible framework than the traditional Cobb-Douglas function, in which the coefficients vary according to the availability of natural resources. In practice, this meant abandoning the assumption of constant returns to scale and, therefore, opened the possibility that economic growth could slow down or even become negative if the effect of natural resource depletion more than offset increases in traditional production factors and Total Factor Productivity (TFP). It is worth noting that Nordhaus also introduced the distinction between GDP in its traditional measure and a "net" measure that includes welfare losses associated with "bads" or negative externalities such as pollution.

By the late 1980s, several authors began to develop methodologies to measure the value of natural capital associated with the supply of natural inputs for the production of goods and services, or "provisioning services" in current terms (Peskin 1976; Repetto 1991; World Bank 1997)[16]. Paradoxically, this new version of the neoclassical growth model has not taken hold in the professional practice of economists, and the Solow model remains the workhorse in courses on economic growth and in the daily practice of governments, central banks, and multilateral organizations when making medium- and long-term economic projections, even in resource-rich countries such as those in Latin America (Loayza and Fajnzylber 2005; DIPRES 2022). More recently, Dasgupta (2021) proposed reorienting the definition of sustainable development, emphasizing ends rather than means. In his version, development is sustainable when the well-being of current generations does not compromise the well-being of future generations. This distinction is important because well-being depends not only on the material means that support the consumption of goods, but also on the quality of the environment in which we live, including biodiversity. An abbreviated version of his definition is that development is sustainable when a broad measure of wealth (Comprehensive Wealth, in his terminology) increases or remains constant. Comprehensive Wealth includes natural capital and all its services. This new definition goes beyond a mere semantic distinction, as it represents a step toward a less anthropocentric measure of economic development, in which humanity cannot be considered apart from what happens to the ecosystems that support life in general and our species in particular. A more complete and realistic representation appears in Fig. 1.3, derived from Dasgupta's (2021) analysis, which overcomes the omission of Nature in the analysis of economic development. It shows at the bottom the relationships between households and firms with natural

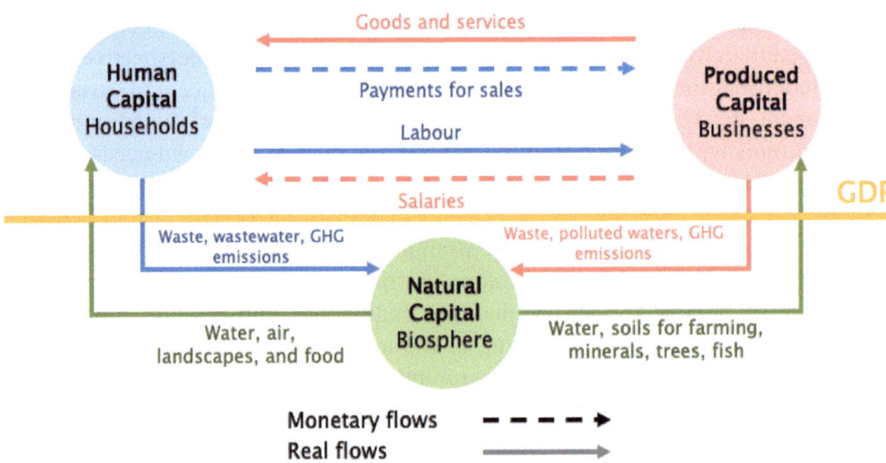

Fig. 1.3 Real and monetary flows between economic agents and nature. (Source: Own elaboration)

[16] Costanza et al. (1997) made a bold first attempt to estimate the value of ecosystem services on a global scale.

capital. Unlike the relationships depicted in Fig. 1.2, many of these transactions do not give rise to financial counterparts, because nature does not charge for its services. Only in those cases where natural resources are appropriable, whether by private parties (firms or households) or by public entities, is it possible to charge, but only for those services that can be traded in markets. Many of these services appear as externalities (positive and negative) for which no markets exist.

The upper part is the traditional representation of the economy shown in Fig. 1.2, corresponding to what is normally measured in GDP when natural resources are involved. Production and the associated payments, such as when fuel is purchased, land is leased for agricultural cultivation, or properties with mineral reserves are acquired for exploitation, give rise to economic transactions that are recorded in national accounts and appear in GDP. However, these accounts do not show payments for the use of the atmosphere to absorb carbon emissions from fuel consumption, nor the loss of biodiversity caused by pesticides to protect crops, nor the environmental degradation due to the disposal of mining waste. In all these cases, natural capital is being used and depreciated. One way to organize these interactions between human activity (households and firms, in this highly simplified world) is to distinguish between the services of nature used directly as inputs in production processes (resources) and those that support the functioning of the biosphere, that is, the continuity of life of different species, including our own. The former[17] are relatively easy to identify (when one is missing, it is noticeable. For example, rainfall, pollinating insects, oil to produce energy, etc.). The latter are more subtle and, in many cases, scientific knowledge about different ecosystems is not yet advanced enough to clearly identify all these interrelationships. This is why efforts led by the World Bank[18] to measure Comprehensive Wealth are still far from providing results that offer a comprehensive view of the value of natural capital. In their current stage of development, they include little more than a valuation of the capital that provides provisioning services. However, work is already underway on the design of methodologies to value these services[19], as well as on the creation of data to support these calculations.

Historical data show that the biosphere's capacity to regenerate lost natural capital is limited, especially when losses occur over short periods (on the timescale of life's evolution and geology, not human lifespans). There is also evidence that, once certain thresholds are crossed, changes become irreversible and, although life on the planet may continue, it will no longer do so under previous conditions, and species that once thrived in those conditions will cease to exist.

In summary, the numerous and diverse indicators on the state of Nature show that, at the beginning of the twenty-first century, the planet is becoming too small to

[17] These are commonly referred to as Provisioning Services in technical literature.

[18] For more information, see the World Bank site The Changing Wealth of Nations (C WON), as well as the currently available data at Changing Wealth of Nations (CWON) (worldbank.org).

[19] There are precedents on which to base these methodologies. In addition to the study conducted by Costanza and coauthors on the global scale, extensive work has been done at the microeconomic level, following the seminal work of Gretchen Daily and her coauthors (Daily, G. 1977).

meet humanity's needs. The costs of the excessive growth of human activity are currently being borne by other forms of life, as well as by present and, above all, future human generations. In this context, it cannot be ruled out that a comprehensive measurement of global wealth would show that wealth has begun to decline, insofar as losses of natural capital may be outpacing increases in produced and human capital, when measured at prices that fully reflect their present and future opportunity costs.

1.3 Planetary Limits to Economic Growth

Historically, the problem of limits to economic growth was a local issue, such as a region or country facing poor harvests, droughts, or the depletion of a key resource, such as water. Pollution problems were also viewed in this context. At the end of the twentieth century was emerging evidence that global limits were beginning to appear, as the impact of human activity was being felt on a planetary scale.

The biosphere is an extremely complex system that has been evolving for hundreds of millions of years and has survived tremendous shocks that caused the mass extinction of species in at least five major events. Life on Earth has demonstrated an admirable capacity for adaptation, while also changing the course of evolution. For example, the mass extinction of the dinosaurs, which occurred just over 60 million years ago, created opportunities for the emergence of mammals, previously relegated to the background in a world dominated by dinosaurs for more than 150 million years. Among mammals, and taking advantage of especially favorable climatic conditions, the first ancestors of modern humans appeared a few hundred thousand years ago. Most striking (and alarming) is that these newcomers, in a tiny period of just over 200 years, have radically transformed the planet, causing a mass extinction of other species[20].

What were once local problems have, over the last century, reached a global scale. The problem of sustainability can no longer s be solved within countries alone but requires policies and actions at the global level as well. For example, in the 1980s, scientists observed the appearance of a large hole in the ozone layer covering much of Antarctica and its surroundings. By reducing protection against ultraviolet radiation, this phenomenon threatened various forms of life, including humans themselves, who would be affected by a higher incidence of skin cancer and other diseases. In this case, governments responded with agreements to phase out the use of certain gases (CFCs) known to damage the ozone layer, although they may have been only one of several causes of the problem. These agreements have been largely

[20] See Brannen (2017). According to this author, it is very likely that some of the earliest—and most lethal—extinctions occurred because of the explosive development of animal species that altered the oxygen content of the oceans and the carbon balance in the atmosphere. Unlike today, these changes took place over millions, not tens, of years.

fulfilled, and CFCs have been mostly eliminated[21]. Although the ozone hole has somewhat diminished, it persists[22], which also demonstrates that the timescales of geological and atmospheric processes can be very different from those of humans.

The ozone layer problem was a wake-up call, but it has been followed by others that are only now beginning to be recognized. One way to represent the risks humanity faces due to its impact on the biosphere is the concept of "planetary boundaries," defined by the limits for the safe functioning of the Earth's critical systems. Rockström et al. (2009) identified nine main processes that are essential for maintaining the integrity and stability of the systems that support life on Earth and made a pioneering effort to assess whether the impact of human activity is endangering the continuity of these processes. The Stockholm Resilience Centre regularly updates this work, incorporating new sources of information and evidence on the evolution of these indicators. These boundaries compare the status of the nine key processes for the continuity of the conditions that have governed the climate and ecosystem during the Holocene. Exceeding them poses a risk of changes that could radically alter the environmental conditions that enabled the life and development of the species that thrived during this epoch.

The planetary boundaries report published in 2023 showed a widespread deterioration of the indicators assessed, and in eight of the nine processes, there was evidence that the level of risk had increased compared to previous measurements. Thus, six of them are already operating with risks above predefined safety thresholds. In fact, humanity continues to operate within the safe zones for only three processes: the state of the atmospheric ozone layer, the only one showing slight improvement, the atmospheric aerosol load, where problems have already emerged in specific regions, and the level of ocean acidification, which is already approaching the safety limit.

In the remaining six processes, human activity is affecting their continuity beyond prudent limits. Among these, climate change and biosphere integrity stand out, both for the speed of their deterioration and the multiple interactions between them. Further setbacks increase the likelihood of an accelerated shift in the environmental conditions that have supported the forms of life that have thrived over the past 15,000 years. The other four processes in which humanity is exceeding safety boundaries are biogeochemical flows[23], land-use change, water availability, and novel entities.[24]

It is important to highlight the interactions among these dimensions. For example, changes in land use result in greenhouse gas (GHG) emissions, which reinforce global warming and biodiversity loss due to habitat destruction. Global warming, in

[21] There have recently been reports of massive violations of these agreements by some companies in China.

[22] Only very recently has scientific evidence emerged of significant reductions in the hole.

[23] This refers to phosphorus and nitrogen flows in streams as a result of the massive use of fertilizers in agriculture.

[24] This refers to synthetic chemicals released into ecosystems without adequately assessing their potential environmental effects.

turn, will accelerate biodiversity loss and forest degradation, making forests more vulnerable to fires and the sudden release of the CO_2 stored within them.

How do these transgressions of boundaries manifest? There are many dimensions, but only two will be examined in greater detail in this book. They refer to two of the processes identified in Fig. 1.4, in which the safety limits have already been greatly exceeded: climate change and biosphere integrity.

Climate change is caused by the accumulation in the atmosphere of gases that produce the so-called greenhouse effect, the most important of which is carbon dioxide (CO_2). Figure 1.5a shows the evolution of the concentration of these gases over the past 50 years, and Figure b shows the correlation between this process and the evolution of global GDP.

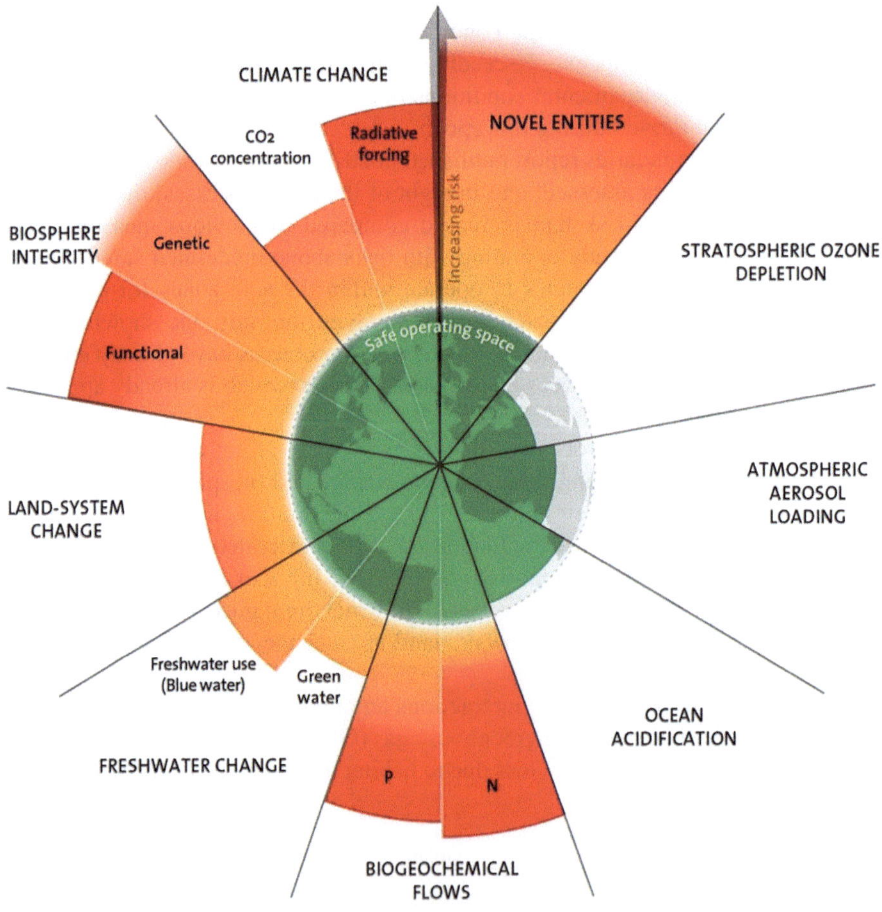

Fig. 1.4 Planetary boundaries for maintaining biosphere integrity. (Source: Resilience Centre, Stockholm University. https://stockholmuniversity.app.box.com/s/sr0nfknm95oydnnsm1zj0c526qzjn1vs/file/1305811057353)

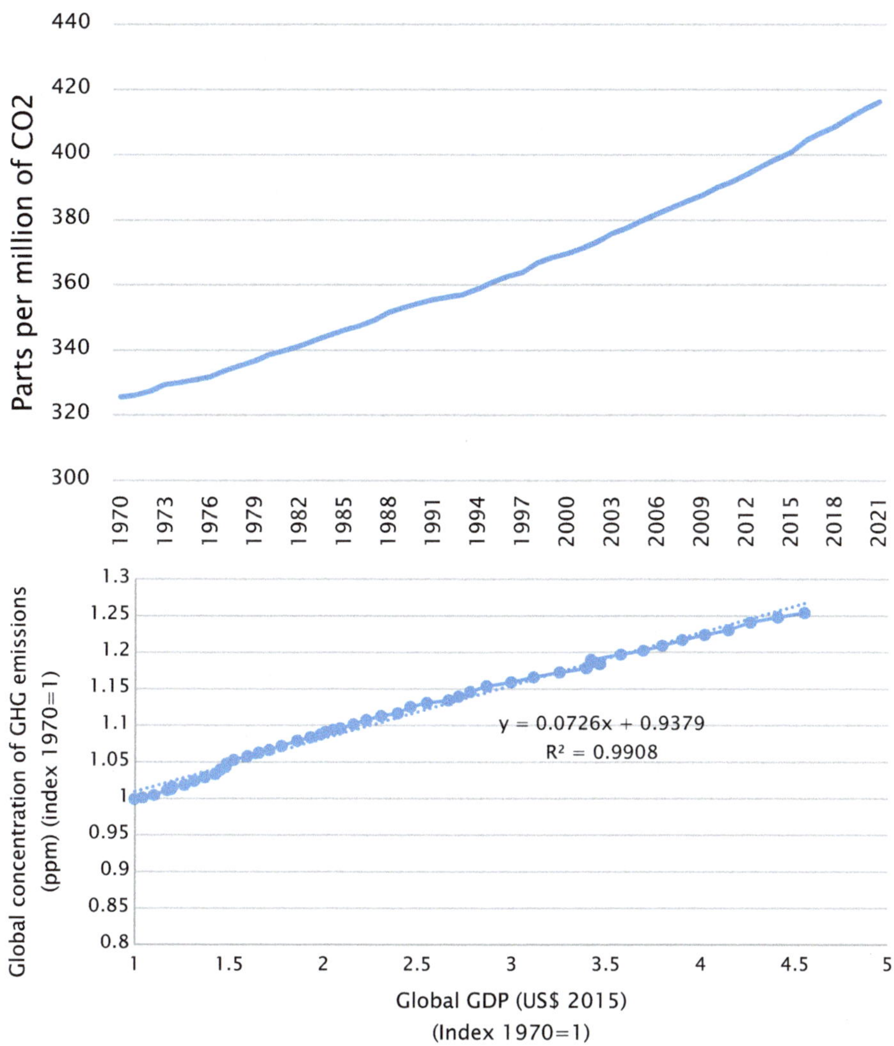

Fig. 1.5 (**a**) Evolution of atmospheric CO_2 concentration. (**b**) Correlation between CO_2 concentration and real GDP trends. (Regarding this graph, the observation that a flow is being compared to a stock is also valid. Source: EPA Indicators on Climate Change in the United States: www.epa.gov/climate-indicators and World Bank: World Development Indicators. World Development Indicators | DataBank (worldbank.org))

Correlation does not imply causation, especially when many other factors are at play, but GDP incorporates at least two key elements: per capita income (consumption capacity) multiplied by the number of inhabitants. In any case, the association is strong enough to place humanity on the list of suspects for the problem.

The scientific evidence compiled in the latest reports of the United Nations expert panel (IPCC), using all the accumulated data from different fields, has confirmed what this graph suggests: if the previous trend between CO_2 emissions and human activity continues, a global-scale climate disaster would occur by the end of this century.

Regarding biosphere integrity, one of the general indicators (although very incomplete due to the complexity of what it aims to measure) is the Living Planet Index (LPI) developed by WWF. Its objective is to show the change in the biomass of wild species over the past 50 years. The evolution of this index and its relationship with global GDP are shown in Fig. 1.6a, b.

In this case, unlike the previous one, we observe a gradual deceleration in the decline of the LPI, after indicating a 70% loss of wildlife in just 50 years (what some refer to as the "Sixth Mass Extinction of Life" on the planet in its entire history), which is not so surprising: the remaining wildlife is confined to areas relatively isolated from human activity, or to species adapted to coexist with humans (such as raptors and rodents in landfills, for example).

This, along with much more available evidence, suggests that the planet is becoming too small to withstand the impact of human activity, at least with current technologies and levels of population and consumption. The great challenge facing humanity in the coming decades is to halt and reverse the deterioration of the processes that regulate the functioning of Earth's systems, and to prepare to face the consequences of those irreversible damages.

Unfortunately, this is complicated by the insatiable human appetite for consumption and novelty, further pressured by the vast disparities in the satisfaction of material needs. While nearly 1 billion inhabitants of wealthy countries enjoy extraordinary levels of consumption, there are about 6 billion people living in low- and middle-income countries who aspire to emulate them, and yet another 1 billion who struggle daily to meet their basic survival needs.

This is the Great Economic Problem of the twenty-first century and, depending on how it is addressed in the coming decades, it will determine the fate of our species.

1.4 Economic Development and Income Disparities: Environmental Implications of Convergence

Before the Industrial Revolution trade between cities and countries was determined primarily by their relative natural advantages, framed by the possibilities offered by transportation technologies. Climate, soil conditions, proximity to food sources or to scarce resources such as salt, timber, or minerals, for example, were key factors in determining specialization in trade, both locally and over long distances.

Improvements in navigation techniques enabled the colonization of territories far from Europe in the Americas and Asia and also laid the groundwork for the development of trade with distant regions, but on a small scale and focused on high-value, easily transported goods such as sugar and spices. With the Industrial Revolution at

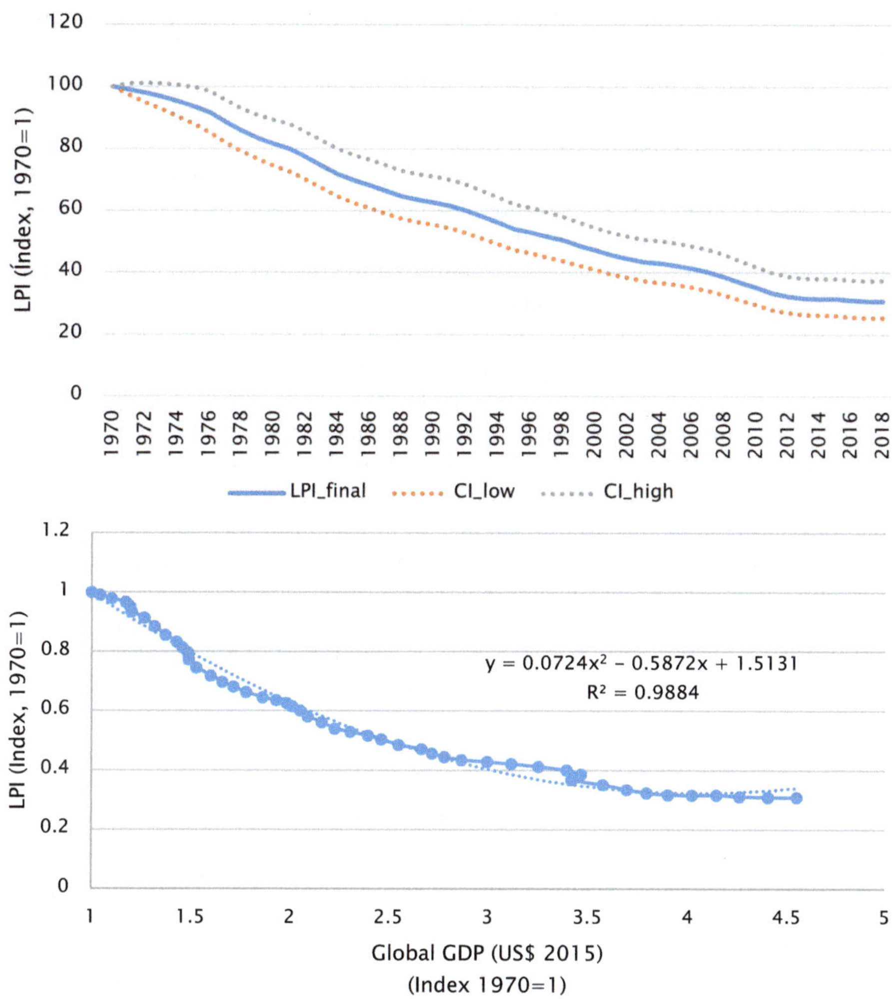

Fig. 1.6 (**a**) Evolution of the Living Planet Index and its confidence bands over the past 50 years. (**b**) Correlation between the Living Planet Index and human activity. (Source: livingplanetindex.org/latest_results and World Bank: World Development Indicators. World Development Indicators | DataBank (worldbank.org))

the end of the eighteenth century, a process of rapid material progress was unleashed, unprecedented in human history. Trade, investment, and information spread this revolution, first to the European continent and North America and, to a lesser extent, to Latin America.

In this new economic order, trade exchanges began to diverge from natural advantages, especially in intra-European trade, where specialization became more closely related to the skills developed in different localities, although some of these had their origins in natural advantages or proximity to energy sources, especially coal.

The United States quickly integrated into these circuits by supplying cotton to the British textile industry, taking advantage of synergies from its proximity to the Caribbean, a region that rapidly specialized in the production of one of the most valued commodities of the time: sugar[25]. Later, and at a slower pace, global trade networks expanded to the rest of the world.

In the early colonial period, Latin America specialized in the export of precious metals and raw materials whose value justified the transport of such heavy goods over long distances. Its trade with the rest of the world was monopolized by Spain, one of the European regions where the Industrial Revolution arrived later. This hindered the early acquisition of capital goods and new technologies. Later, when the independence of these countries ended this monopoly, they opened to trade with North America and Europe, but maintained a strong specialization based on the abundance of natural resources, with very limited development of manufacturing capabilities beyond basic processing of these resources.

The Industrial Revolution (IR) began in Great Britain but quickly spread to Europe, leading to an explosive improvement in material living conditions. The countries and regions that were open and well positioned for trade in goods and financial resources benefited the most. This also coincided with a sharp increase in population, following a rapid decline in mortality rates, especially among children[26].

Available data show that, after the Industrial Revolution, per capita income growth[27] accelerated, first in Great Britain, quickly followed by neighboring countries on the continent, although at different rates.

These differences gave rise to an income gap between the United Kingdom and the rest. Later, Germany and the United States began to close this gap, with the latter taking the lead in the early twentieth century. Spain and Sweden took much longer to join this process and did so with different outcome. Sweden accelerated and began to lead the rest of Europe, taking advantage of natural resources such as iron and timber, key inputs for the industrial activity of its neighbors. Spain, on the other hand, fell behind.

India, despite being part of the British Empire and having a vast domestic market, did not play a very active role in international trade flows of raw materials and consumer goods, partly due to the great distances involved. However, there was limited integration in textile production chains with Great Britain, as well as in the triangular trade of the British Empire with China, supplying one of the few products for which China lacked its own sources: opium.

---

[25] It is important to note that Figure 5B compares a "stock" variable, such as the accumulated level of CO2 in the atmosphere, with a flow variable, such as annual GDP, which reflects human activity each year. A more accurate comparison would be between the flow of emissions of the year with the GDP level, which is presented below and shows a similar result.

[26] Possibly the most decisive changes were improvements in sanitary conditions, such as access to drinking water and the gradual replacement of animal-drawn transport (with the resulting waste on public roads) by mechanized transport.

[27] A similar phenomenon occurs within countries, with a significant increase in inequality. In some way, it seems that rapid progress in material living conditions generates income disparities, at least in its initial stages. See Deaton (2013) for an analysis of this point.

Japan, for its part, joined much later, only after the country adopted a deliberate industrialization strategy, taking advantage of the high natural protection of its domestic market resulting from the vast distances separating it from the industrial powers of the late nineteenth century. As the country lacked exportable natural resources and energy, it was able to capitalize on the abundance of cheap labor to develop labor-intensive manufacturing.

International trade and the exchange of resources and technologies facilitated convergence between Europe and some of its former colonies in North America, driven by increasing trade and financial flows among these countries, in what can be termed "globalization 1.0"[28]. This first wave of globalization ended with World War I and the subsequent period of intense protectionism, exacerbated by the Great Depression. As shown in Fig. 1.7, Spain and Germany were particularly affected during this period, as well as by World War II. However, more recent data (Fig. 1.8) reveal significant convergence in income levels among the most advanced countries.

As can be seen in this last chart, this group of countries experienced sustained economic progress over the past 70 years, as well as a gradual reduction in income disparities, at least until the Global Financial Crisis of 2008. This synchronicity is not surprising considering that the post-World War II period has been characterized by the increasing integration of national economies into a global economic system.

While this integration was initially limited to trade in manufactured goods and capital flows between countries, in recent decades it has also led to the globalization of production chains, extending beyond the borders of the most advanced countries.

Globalization was made possible by significant reductions in transportation costs and advances in real-time communication and information processing technologies. These enabled the fragmentation of industrial production processes, allowing their distribution among plants in different countries and creating new opportunities for value chain integration, in a world where goods, capital, and increasingly, services and people, move or interact freely and in real time.

It is interesting to note that this accelerated process of economic growth in the West, accompanied by population growth, considerably expanded the "human footprint" in these regions and in their former associated colonies[29].

However, this can hardly be taken to mean that economic growth during this phase, and limited to these countries, was endangering the planet's capacity to sustain life, partly because the beneficiaries of this development represented only a small proportion of humanity[30].

In the rest of the world, the diffusion of the Industrial Revolution was much slower, and large income gaps emerged between these countries and the leaders of the Industrial Revolution. Most countries remained on the sidelines, either due to their remoteness—physical or economic isolation—or the absence of resources

[28] This phenomenon is very well described in Ferguson (2006).

[29] Accounts from chroniclers of colonial life at the time already reported the destruction of habitats and species, both due to the expansion of the agricultural frontier and to "sport hunting." See, among many others, Pérez Rosales (1882) and Markham (1942).

[30] In 1960, developed countries accounted for approximately one quarter of the world's population.

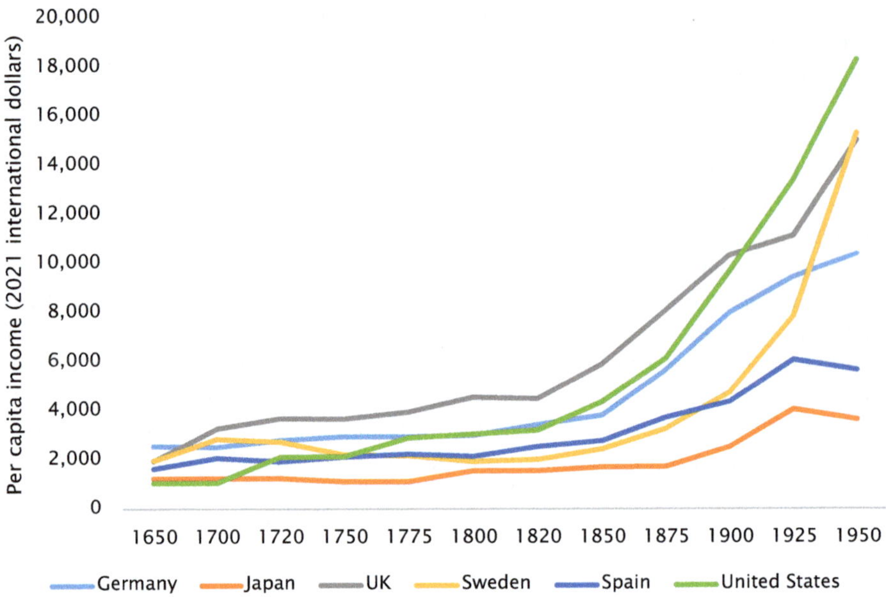

Fig. 1.7 Historical per capita income in advanced countries (in 2021 international dollars). (Source: Maddison, World Bank, and IMF)

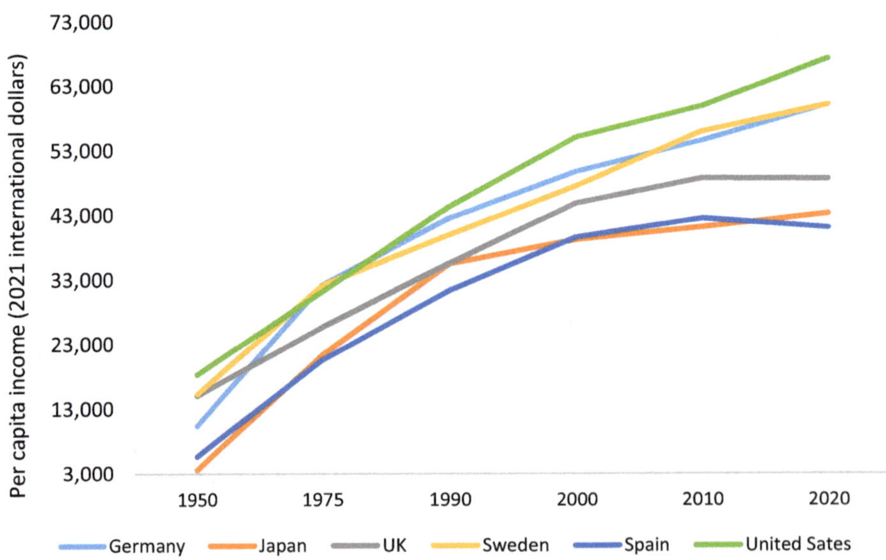

Fig. 1.8 Per capita income in advanced countries (in 2021 international dollars). (Source: Maddison, World Bank, and IMF)

valuable to the developed world. In the words of Pomeranz (2000), this marks the beginning of the Great Divergence in material living conditions, which only a handful of countries such as Japan, the Asian Tigers, and some on the European periphery, like Ireland and Israel, have managed to reverse during the twentieth century[31]. This is shown in more detail in Fig. 1.9.

The economic performance of these three continents could be summed up in a single word for each: Latin America—frustration; Asia—hope; Africa—failure.

In fact, Latin America as a whole, over time, has been unable to narrow the income gap with developed countries, and even those countries that managed to do so for a period (Argentina and Chile at the end of the nineteenth century, Venezuela in the mid-twentieth century) later regressed.

The abundance of natural resources and land suitable for crops and pasture created commercial opportunities in which Latin American countries had advantages in exporting food and minerals, while more developed countries supplied them with capital and consumer goods. The region reached upper-middle income levels at the end of the nineteenth and beginning of the twentieth centuries, at the height of Globalization 1.0, and seemed poised to make the leap to the developed world. However, this process came to a halt with the Great Depression of the early 1930s. Exports collapsed, and the main countries lost access to financial markets. Most countries in the region chose to raise barriers to foreign trade to promote industrialization. This new drive was short-lived due to the small size of domestic markets. Intra-regional trade never took off, partly due to geography and inadequate infrastructure, but also because of highly cumbersome trade barriers (Cornelius and Vial 2002).

In the last decade of the past century, market-oriented economic reforms and a renewed push for liberalization allowed some countries to accelerate their growth, initiating an incipient process of convergence. However, this has only enabled them to recover part of the ground lost during the twentieth century. After the end of the commodity boom around 2014, growth slowed in most of the region and populist policies re-emerged, with serious consequences. The most dramatic case is Venezuela, the richest country in the region in the early 1980s with a per capita income close to $20,000 (PPP, 2017), which has steadily declined and is now approaching $6000. The impoverishment of the country triggered the region's most severe migration crisis, and the country has already lost more than four million people (13% of the population)[32].

Asia, for its part, displays a diversity of experiences, with several countries managing to reduce the per capita income gap with the developed world. The first success story was Japan, which belatedly adopted Western technologies and, after a tragic imperial adventure, closed the gap in the second half of the twentieth century. Shortly thereafter, South Korea, Taiwan, and Singapore followed in Japan's

---

[31] This has been well documented in Pritchett (1997).

[32] In the graphs, missing data between observations were extrapolated using the geometric rate of annual variation between contiguous observed points.

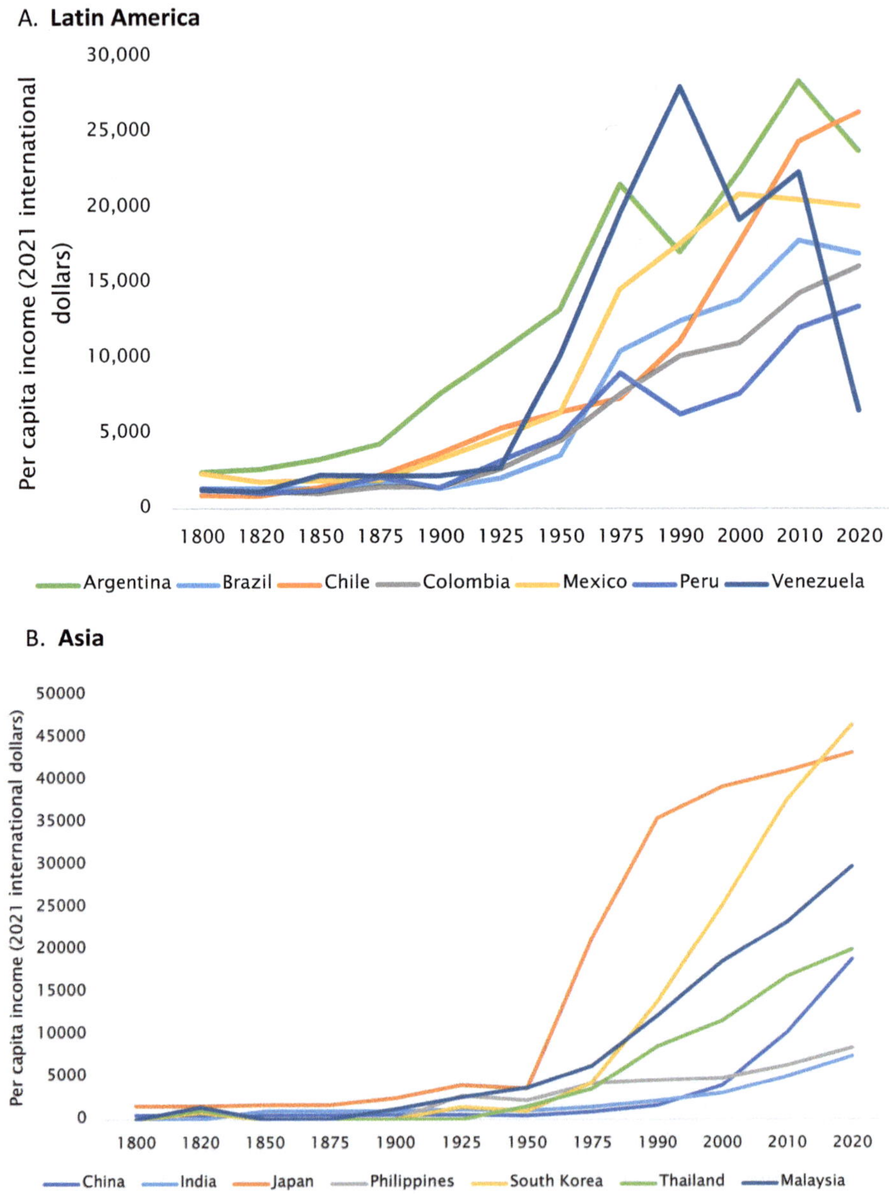

Fig. 1.9 Convergence of peripheral countries to the Industrial Revolution (GDP per capita in 2021 international dollars). (**a**) Latin America, (**b**) Asia, (**c**) Africa. (IMF. World Economic Outlook Database. October 2023. https://www.imf.org/en/Publications/WEO/weo-database/2023/October/weo-report?c=299,&s=NGDP_R,NGDPRPPPPC,LP,&sy=1980&ey=2028&ssm=0&scsm=1&scc=0&ssd=1&ssc=0&sic=0&sort=country&ds=,&br=1. (Source: Maddison, World Bank, and IMF)

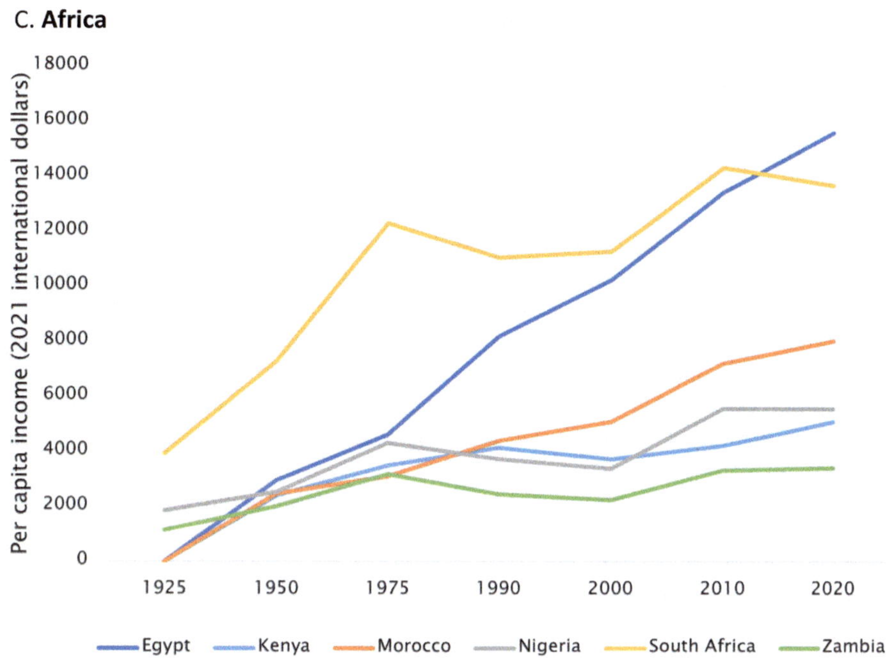

Fig. 1.9 (continued)

footsteps, implementing similar policies. Others are still halfway, such as Malaysia, Indonesia, and Thailand.

China and India deserve special mention due to their size, as they are home to nearly a third of the world's population. Both lagged behind the Industrial Revolution and their income gaps with the developed world widened until the last decades of the twentieth century: only at the end of the twentieth century did China take off strongly and, although it is still far from achieving per capita income levels comparable to advanced countries, by size it is the world's second largest economy in terms of total GDP.

Before the Industrial Revolution, China was a major economic power and the cradle of numerous inventions, as well as the origin of the domestication of many animal and plant species. At that time, China had a per capita income similar to the European average at the dawn of the Industrial Revolution. However, with its arrival, China fell behind, partly due to its isolation and also to the marked deterioration of its institutions.

When Western powers came into conflict with China, it failed to repel them and entered a period of anarchy that led to a severe weakening of central authority, followed by civil war and the autarky of the Maoist period[33]. Only at the end of the twentieth century, with the economic reforms and gradual opening promoted by

[33] One factor that contributed to the weakness of China's response was the opium addiction epidemic which, unlike the conquest of the Americas, was a deliberate process to develop Chinese demand for imported goods.

Deng Xiaoping and his followers, did China begin to recover the ground lost over nearly a century and a half.

India started later and its pace of expansion has been slower than China's, despite having been partially integrated into the production and trade chains of the British Empire. However, there is an incipient process of convergence that gives hope for consolidation, although it is still too early to be certain.

Africa exhibits the greatest income disparities with the developed world and is the region where poverty is most prevalent and severe. Economic progress has been limited, and even South Africa, the only country initially classified as middle income, has experienced relative decline compared to the advanced countries of the West. This country initially benefited from an abundance of natural resources and highly valuable minerals. In addition, its more temperate climate favored the settlement of European colonists who built sophisticated institutions as well as transport and communications infrastructure, enabling it to reach per capita income levels like those of Latin America.

Africa has several dramatic cases of countries that, after achieving political independence in the second half of the twentieth century, saw their per capita income levels decline steadily over a long period, in some cases for decades. At the beginning of the current century, most of the region began to grow and some progress has been made.

The globalization of economic relations has created the conditions for most emerging countries to embark on processes of rapid economic progress, taking advantage of the opportunities offered by international trade and access to capital and technology. Above all, a giant like China accelerated its growth and narrowed income gaps with the developed world. This economic progress, which has materialized in all regions, albeit to varying degrees, has resulted in better jobs and incomes for a large proportion of the world's population. Most people's incomes have surpassed poverty thresholds in Asia, the Americas, and even, to a nascent extent, in African countries, and new middle classes are emerging. More importantly, these populations are gaining access to new levels of consumption of goods, as well as all kinds of services, including financial, health, educational, and even leisure services. Given the scale of these advances and the large population involved, it seems difficult to reverse this process, at least not without conflict and significant losses for all. The main threat to this process comes from rising nationalist and protectionist sentiments in the most advanced countries, particularly in the United States, which has shifted from being the main promoter of free trade and globalization to leading the protectionist wave. This emerging process of income convergence between poor and rich countries, which began to take off at the end of the twentieth century, poses significant challenges to the sustainability of global economic development. First, it means that consumption is increasing for a large part of the world's population. Although their relative per capita consumption levels are much lower than those of the most advanced countries, they are much higher than in the past. Before China's rise, the "converging" countries were few and, at most, small or medium in size. At the same time, in these countries there is still a very large fraction of the population living in poverty, and their economies depend on

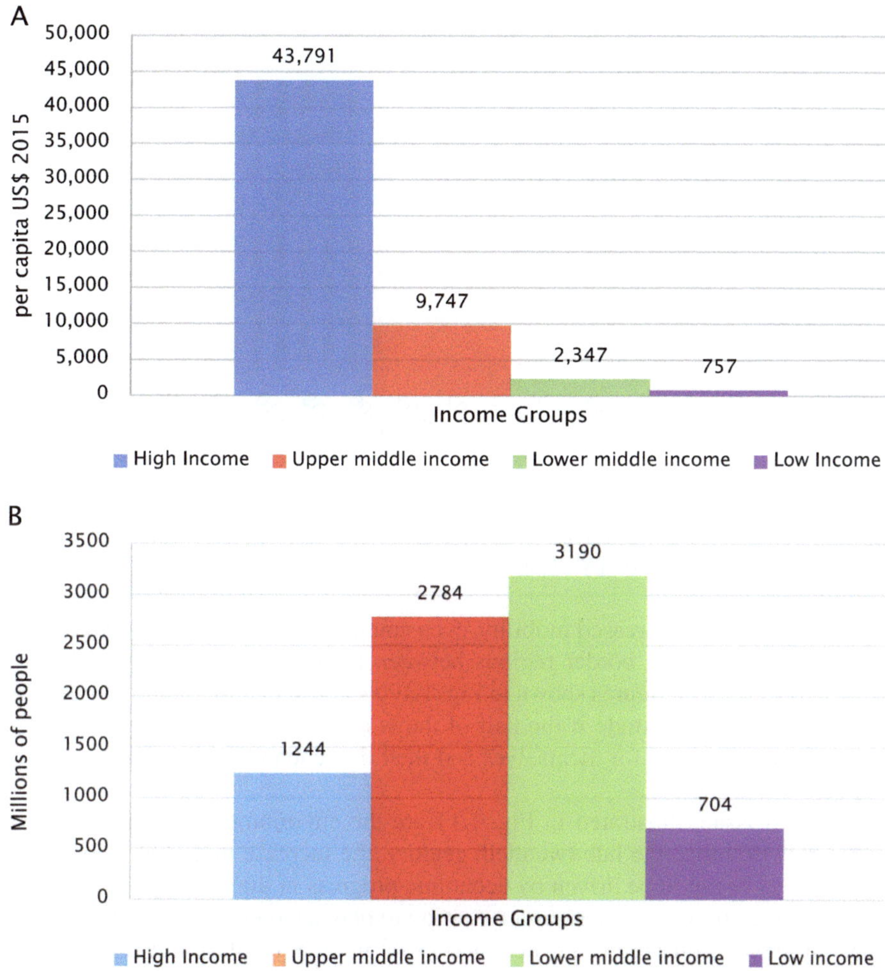

Fig. 1.10 (**a**) Average per capita income in countries with different levels of development in 2022. (**b**) Population size in each country group in 2022. (Source: World Bank: World Development Indicators. World Development Indicators | DataBank (worldbank.org))

growth to create jobs and opportunities for incomes to rise above poverty levels. The magnitude of these problems is illustrated in Figs. 1.10a, b, which show the composition of the world's population grouped by the per capita income level of their countries, on the one hand, and the number of people living in each of these country groups, on the other.

In today's hyperconnected world, residents of the poorest neighborhoods in Nairobi, Caracas, or Delhi can compare their standard of living and amenities with those enjoyed by middle-class families in London, Los Angeles, or Rome. So, it is unthinkable that income disparities such as those shown in Fig. 1.10a can be maintained indefinitely without triggering serious political, economic, and social

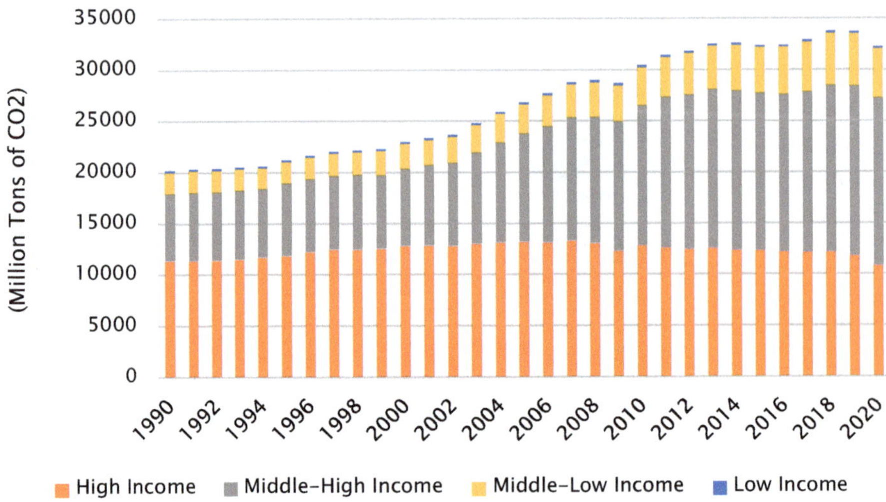

Fig. 1.11 Greenhouse gas emissions by income group of origin. (Source: World Bank: World Development Indicators. World Development Indicators | DataBank (worldbank.org))

conflicts. Moreover, increased mobility, even among the poorest, is already generating migration crises in border regions between the developed regions and those living in poverty. The figures shown in Fig. 1.10b suggest that this could increase by several orders of magnitude if the half of the world's population living in poverty loses hope for progress for themselves and their descendants within their countries of origin[34].

The third issue, illustrated in Fig. 1.11 are the environmental impacts of economic growth. Since the late twentieth century, the increase in global income and consumption began to be driven by economic progress in the developing world. At the same time, these same regions are starting to play a larger role in the increase of greenhouse gas emissions, the destruction of natural habitats due to land use changes, and the demand for natural resources. With limited global budgets for GHG emissions worldwide, and with the dramatic deterioration in the state of nature already reported in this section, humanity and the world as a whole face an unprecedented challenge in the coming decades.

1.5 The Great Global Challenge of Sustainable Development in the Twenty-First Century

When comparing the growth trends of the global economy, both in terms of per capita consumption and demographic growth in different regions, with the deterioration of nature at local and global scales, the sustainability of these trends comes to

[34] This was anticipated by Ferguson (2006).

the fore. Below are the results of some arithmetic exercises projecting historical trends in the global evolution of per capita income, population, and CO_2 emissions, grouped by countries according to their current income levels.

These exercises are not predictions, but they help to gauge the magnitude of the changes that would be necessary to achieve sustainability goals in just one of the dimensions discussed above: climate change. This dimension is important, but it is far from being the only challenge posed by humanity's pressure on the planet. Although there are synergies and many of the adjustments needed to reduce GHG emissions could also help alleviate pressures on the biosphere—such as land use change and water use—this does not mean that all problems would be solved if humanity manages to avoid a climate catastrophe. The main reason for choosing this dimension of the global environmental crisis is simple: unlike the others, here there is a well-defined indicator to measure the progress of the problem (the concentration of GHGs in the atmosphere), with a well-established causal links between this variable and likely climate changes, as a considerable accumulation of knowledge about the human determinants of these emissions. None of this is remotely available in the case of biodiversity loss or other disruptions to the processes that maintain the integrity of the planet's life-support systems[35].

The variables projected for each group of countries are three: population growth, per capita income growth, and a factor (variable) that reflects the intensity of CO_2 emissions relative to GDP, a measure that will later be related to a more general concept of efficiency in the use of nature.

Population growth uses data from World Population Prospects for 2024 published by the United Nations, using the medium variant presented therein.

Per capita GDP growth in the baseline scenario is calculated using trends from the past decade, with slight adjustments to incorporate marginally faster progress in developing regions, focusing on the poorest areas.

For emissions projections, historical trends in the relationship between total real GDP and CO_2 emissions are estimated, which are then extrapolated into the future, assuming a gradual convergence of trends in emerging countries toward those of developed countries[36]. However, for the poorest regions, an intermediate period of increased emissions is assumed, given a possible rise in the intensity of energy use[37].

The first exercise, called "Inertia," is basically an extrapolation of current trends, which include, among other things, a decline in population size in the upper-middle and high-income groups, and a systematic decrease in the emissions-to-GDP ratio following the observed historical trend. It is an inertial world, but different from the present, even if the effects of climate change are excluded from the analysis (Table 1.1).

---

[35] The only exception is the deterioration of the ozone layer, which is a much more limited problem and has already been largely successfully addressed (so far) through international agreements to limit emissions of the gases responsible for this phenomenon.

[36] Appendix 1 provides further background on these projections and assumptions, as well as various numerical exercises.

[37] The main assumptions and results are detailed in Appendix 1.

Table 1.1 Global emissions projections assuming inertia

		Inertia		
		2020	2040	2100
Per capita income	(US$ 2015/person)	11,315	20,031	73,019
Income inequality	(Ratio IPHHigh/IPHLow)	58	65	87
Population	(thousands)	8,021,407	9,177,190	10,180,161
Total CO_2 emissions	(million KT/year)	33.9	57.8	150.9

Key Inputs for the Inertia Scenario					
	High-Income Countries	Upper-Middle-Income Countries	Lower-Middle-Income Countries	Low-Income Countries	WORLD
Annual % change 2022–2100					
Real GDP per capita	1.0%	2.6%	3.3%	0.5%	2.4%
Population	0.00%	−0.42%	0.49%	1.42%	0.31%
Emissions-to-GDP ratios					
2020	0.642	0.776	1.083	0.839	
2040	0.417	0.649	1.076	0.847	
2100	0.177	0.369	0.740	0.787	

This highly simplified calculation is informative in at least two respects: not only do emission levels not fall to zero by mid-century, but they continue to rise throughout the projection horizon, while the income gap between rich and poor countries continues to widen.

An obvious conclusion is that this scenario is unviable. Despite the sharp decline in the emissions-to-GDP ratio in high- and upper-middle-income countries, and the moderation in global population and per capita GDP growth, this is not enough to contain the worsening of the problem, let alone achieve carbon neutrality targets within 30 years[38]. The main message of this exercise is the need for much deeper and more drastic changes in human behavior and technology to address the climate crisis.

In view of this, the question arises as to the magnitude of the changes required to achieve the goal. Given the simplicity of the calculations, it is very difficult to provide clear answers, but at least it is possible to infer an order of magnitude for the necessary adjustments in the three control variables for each group of countries included: per capita GDP growth rates, population growth rates, and emissions intensity relative to GDP. Appendix 1 presents exercises with independent adjustments for each of these three variables and as should be obvious, they are not sufficient. Significant adjustments are required for all of them.

To give an idea of the orders of magnitude involved, an exercise will be presented with strong adjustments in all dimensions but incorporating a significant improvement in income levels for the poorest groups worldwide. In summary, it is assumed that the world population decreases in absolute terms by almost 1.1 billion people by the end of the century, according to the "low variant" of the UN demographic

[38] This was already addressed in Albagli and Vial (2022).

Table 1.2 Projections of global emissions assuming significant adjustments in population and emission efficiency, but with rapid growth among the poorest countries

		COMBINATION + TE MIC&LIC		
		2020	2040	2100
Per capita income	(US$ 2015/person)	11,315	21,143	99,458
Income inequality	(Ratio IPHHigh/IPHLow)	58	42	7
Population	(thousands)	8,021,407	8,784,545	6,986,814
Total CO_2 emissions	(million KT/year)	33.8	13.4	−10.5

Key inputs for the scenario of drastic adjustment in all variables					
	High-income countries	Upper-Middle	Lower-Middle	Low-income	WORLD
Annual % change 2022–2100					
Real GDP per capita	1.0%	2.6%	3.3%	3.8%	2.8%
Population	−0.42%	−0.95%	−0.01%	0.98%	−0.18%
Emissions-to-GDP ratios					
2020	0.640	0.776	1.083	0.839	
2040	0.348	0.202	0.503	0.639	
2100	0.068	0.042	0.103	0.039	

projections. In addition, global growth slows everywhere except in the poorest countries, so that per capita GDP only doubles by the end of the century, with most of the increase concentrated among the poorest countries, instead of quadrupling as in recent decades. Finally, the emissions intensity relative to GDP decreases at an accelerated rate, especially in upper-middle- and high-income countries. The main aggregate results and assumptions are presented in Table 1.2.

This exercise, which is undoubtedly very demanding, shows that even when carbon neutrality is imposed as an assumption for upper-middle- and high-income countries by 2040, this is not sufficient and global emissions not only fail to meet the Paris Agreement, but increase slightly above current levels. As shown in Annex 1, if the same exercise is performed assuming that per capita income in the poorest countries grows as in the baseline scenario, the impact on carbon emission reduction is insignificant (less than one million tons in 2040). Therefore, from the perspective of climate change mitigation, allowing for higher growth in the poorest countries to accelerate global poverty reduction is relatively "cheap" in terms of emissions, provided it is accompanied by greater energy efficiency, leveraging improved access to advanced technologies.

There are many possible pathways to achieve net zero emissions around 2050, with different combinations of the three control variables, but all of them require drastic changes from previous trends. The tool used is too simplistic to address a dynamic optimization exercise that incorporates all the interactions among these variables and their environmental impacts, but even so, the simplicity of these exercises allows us to draw some general conclusions of value for opinion leaders and the public:

(i) Accelerating the demographic transition is very helpful, but its effects are too distant in time to prevent the most severe impacts of climate change. As will be discussed later, this is already happening in middle- and high-income countries, but at a much slower pace in the poorest countries. In any case, the faster progress is made on this front, the more leeway there will be in the long term to ease the pressure to adjust economic growth and to advance rapidly in the energy transition and land use change.
(ii) Effective control of GHG emissions requires a significant adjustment in the levels and composition of per capita consumption, at least in middle- and high-income countries, which currently account for most of the global consumption.
(iii) Expectations worldwide must be adjusted regarding the levels of material goods consumption that can be achieved, at least over the coming decades.
(iv) Progress in the energy transition, land use change, and deforestation reduction, as well as the development of carbon capture, should be greatly accelerated to achieve carbon neutrality within the proposed timeframe.

None of this is easy, but it is not impossible either. There are currently processes underway that point in the right direction. If these are intensified and accelerated, humanity can set itself on a path of sustainable progress in the coming decades, but to do so, it must pass through the "bottleneck" that E. O. Wilson (2002) aptly identified two decades ago, referring to the deterioration of biodiversity and the state of nature in general.

1.6 The Three Levers to Address Global Environmental Degradation

Following the same approach used to construct the exercises described in the previous section, it is possible to identify the actions needed to address the deterioration of life-support systems by different groups of countries.

These actions can be classified into three categories, following the so-called IPAT identity, where the Environmental Impact of Human Activity (I) is the result of the size of the Population (P) multiplied by its Affluence (A)—which in this context can be defined as the level of material consumption per person—and Technology (T), which is essentially analogous to the Solow residual applied to another domain: T encompasses everything related to the efficiency with which humanity uses nature's services. For example, when coal energy is replaced by natural gas, GHG emissions per unit of energy generated are reduced, or when vegetables are grown hydroponically, the amount of land and water used for the same amount of food produced is reduced[39].

[39] A more widely known version is the application of this concept to Greenhouse Gas (GHG) emissions, known as the Kaya Identity, which states that these emissions are the result of multiplying the population by GDP per capita, further multiplied by the energy intensity per unit of GDP, and, in turn, the GHG emission intensity per unit of energy Kaya (1990).

Technology is undoubtedly a key factor, but in many cases significant progress can be made in reducing the impact on nature with the same technology by changing dietary habits, or by recycling or reusing materials from obsolete products, in other words, by adopting behaviors that are more mindful of environmental impacts. The concept of the circular economy aims to this goal.

The first lever to be pulled to achieve an orderly adjustment to the constraints imposed by nature is the size of the human population. After centuries of very moderate growth, it "exploded" around the advent of the Industrial Revolution. Following a long period in which high birth rates coexisted with high mortality, the change began with the rapid reduction in infant mortality rates. This was made possible by improvements in health and sanitation, combined with increased income, which helped improve access to food and housing. As more girls survived, the number of potential mothers increased, and so on. In parallel with these processes, the Industrial Revolution led to a greater demand for increasingly skilled and educated workers, resulting in higher wages, as previously explained.

Over time, and partly due to the massive migration from rural areas to cities, conditions changed and new opportunities for personal development emerged, especially for women, both in terms of education and employment outside the home. These processes, combined with safe and affordable access to birth control options, led to a decline in the number of births per woman in the twentieth century. For the first time in the history of the species, women achieved almost complete control over the number of children they wished to have.

As a result, there has been a sharp decline in birth rates and fertility rates (the number of live births over a woman's reproductive lifetime), first in the most advanced countries and later spreading to middle-income countries at a faster pace in the second half of the last century.

Today, almost all developed countries have fertility rates below 2.1, which is the threshold for stabilizing population size in the long term (in the absence of migration). The same has occurred, with some delays but at a faster rate, in middle-income countries. If current trends continue, many countries in this latter group will begin to experience absolute population declines by the middle of this century, as is currently happening in China.

The United Nations regularly publishes projections of world and country-level populations up to the year 2100[40]. In its most recent "medium scenario" (2024), the world population peaks at 10.3 billion around 2085 and then begins to decline gradually. In this scenario, fertility rates fall from a current global level of 2.4 children per woman to 1.84 in 2100. However, available data suggest that in countries where fertility has dropped well below 2.1 children per woman, it remains low with no signs of recovery, even in middle-income countries. The United Nations also publishes a "low variant," which assumes that the fertility rate falls more rapidly, reaching 1.36 children per woman in 2100. This figure, which is well below the so-called "replacement rate" (2.1 children per woman) needed to stabilize population size, is quite like those currently prevailing in high-income countries and even in many

[40] "World Population Prospects" UN. https://population.un.org/wpp/

upper-middle-income countries. In this variant, the world population peaks at 8.95 billion around 2050 (35 years earlier than in the medium scenario) and then declines in absolute terms to 7 billion by 2100. As can be seen, the results are extremely sensitive to this assumption.

While this would not be enough to halt the environmental crisis, it would make it more feasible. If the fertility rate were to remain stable for a long period at current levels in high- and upper-middle-income countries (half the world's population) and, at the same time, in low- and lower-middle-income countries were to converge relatively quickly (in 2 or 3 decades) to the replacement rate that stabilizes the population in the long term (2.1 children per woman), this process would accelerate and the world population could fall to around 6 billion people by the middle of the next century.

This would be a very different world from today, in which economic growth would be based primarily on technological progress rather than the size of the labor force. It is even possible that global GDP could decrease while per capita income increases.

A world with these characteristics might leave more room for the regeneration of nature, but it would pose enormous challenges for human societies to adjust their functioning after two centuries in which growth was the norm. Adapting to this new environment would not be easy.

There are also variables that operate in the opposite direction. The most important thing is increased longevity. The fact that people live longer means that, for the same number of births, there are more people alive, at least for a time. The dynamics of the population are not altered, as they depend mainly on the evolution of the size of the female population of reproductive age and their fertility decisions.

The transition to a society with a higher proportion of older people also presents challenges, as seen in countries that have aged rapidly, such as Japan, South Korea, some Mediterranean countries in southern Europe, and more recently, China.

Public policies have tended to address these issues from a national than global perspective, with a prevailing concern about the transition to a smaller and older population. Thus, they have focused on promoting higher birth rates. So far, these policies have proven ineffective and quite expensive to finance.

The second lever is global consumption and its composition.

This is directly related to the materials and services provided by nature and used to support human societies. Smil (2022) identifies four main materials that are essential for human civilization in this era, based on the scale of their demand, accessibility, and their "indispensability." These are cement, steel, plastics, and ammonia (for its use in fertilizers).

Unlike other materials, such as oil or copper, which have close substitutes, these four are very difficult to replace. In turn, they require large amounts of energy for their production, especially cement and steel. Plastics and ammonia are derived from hydrocarbons, which, combined with their demand in steel and cement production, make virtually impossible to replace oil and natural gas in all their uses within a reasonably short timeframe.

Ammonia is crucial for producing fertilizers, which in turn are the foundation for growing grains and vegetables. Without them, it would have been impossible to achieve the yields necessary to go from feeding (poorly) 1 billion people to the current 8 billion.

Plastics are now indispensable in all types of packaging and durable consumer goods, as well as in countless single-use applications, due to their ease of handling, lightness, insulation, flexibility, and durability. A similar situation exists with steel and cement, which are present in virtually all existing structures and for which there are no materials that can replace them on a large scale, at least with current technologies.

All these materials have a significant environmental impact, either due to the difficulty of disposing of them at the end of their useful life (plastics, cement), the polluting effects on soil and water (ammonia, plastics), or the large amounts of energy required for their production (all of them). Other materials are also very important (freshwater, copper, and aluminum, for example), but they are relatively abundant (alumina, seawater) and there are good substitutes (seawater desalination for freshwater, aluminum for copper).

It is easy to see that as the population and per capita material consumption levels increase, whether food, housing, transportation, household equipment, clothing, etc. the demand for all underlying materials will also rise. In addition, other goods provided by nature, such as water and land, will be needed for direct urban consumption or use, as well as for food production.

Undoubtedly, there are components of human consumption, such as education, health, and many leisure and tourism support services, that require less energy and physical materials. These have become increasingly important in household consumption expenditures as income surpasses what is needed to meet basic needs for food, clothing, and shelter. However, most of the population in middle- and low-income countries remain below these thresholds and aspire to achieve the material well-being of more advanced societies.

Moreover, worldwide, but particularly within many emerging countries, there are large disparities in income, wealth, and power. While elites and middle classes attain levels of material well-being, comfort, and quality of services like those in developed countries, a large portion of the population lives in much poorer conditions.

The tragic reality is that everyone aspires to replicate the material consumption patterns of those higher up the development ladder. The transition to a sustainable global economy necessarily involves reducing the environmental footprint of consumption in absolute terms. As the global population surpasses poverty thresholds, it is expected that the growing importance of services will contribute to this process.

The materials mentioned earlier, and all the production processes in which they are used, are intended to meet people's material consumption needs. Environmental activism often focuses on the companies that produce these goods, as they are the direct sources of pollution. In this section, the focus has been placed on household consumption because, if people continue to demand the same goods, shutting down some companies will only shift production to others that may be less efficient and, therefore, have a greater adverse impact on the environment.

The third lever consists of improving efficiency in the use of nature's services. This is the primary tool for addressing scarcity issues, but markets do not signal this scarcity when it comes to free or non-appropriable goods, such as air and water.

For 15,000 years or more, since the first human "societies" formed, nature and its services were free and non-appropriable goods. There were often scarcity problems with these free goods, but they were usually resolved through appropriation and conflict, or simply by migration. Only since the second half of the twentieth century have been signs that humanity is exceeding the limits of nature's absorptive capacity: Nature is becoming scarce and, therefore, the planet's life-support systems could fail.

The greatest obstacle to inducing changes in the level and composition of consumption is that many of the impacts on nature are not reflected in the prices of goods and services. This means that the cost perceived by consumers is not the true cost of those goods due to the presence of externalities: the farmer does not perceive the cost of the pollution generated by the fertilizers he uses, the teacher who drives to school is not paying for the carbon emissions and congestion he causes, and so on.

Societies have increasingly imposed restrictions to limit detected environmental damage. This points in the right direction when there are competitive technologies to produce harmful goods, but it is not enough. For example, if the product is a disposable plastic bottle, no matter how cleanly it is produced, it will still require hydrocarbons and energy to make the plastic, and when discarded, it will end up somewhere if not recycled. Greater efficiency can reduce environmental impacts, but in many cases, this is not sufficient.

One way to address these problems is to mobilize markets to help cope with the growing scarcity of nature's services, and the simplest way is to ensure that prices (through taxes, for example) reflect the true costs of consuming and producing various goods and services.

In fact, this has worked in practice: for example, an unintended consequence of the increase in oil prices in the 1970s and in the second decade of this century was a strong impetus to reduce the energy intensity of the use of many goods (cars, for example), as well as to develop new technologies to improve efficiency in the production of goods. But it also led to innovations aimed at increasing the supply of hydrocarbons.

Although governments can act by imposing pollution taxes, their political capacity to do so is limited. On the one hand, this is because the problems have surpassed the local and national spheres and require coordination with other countries, but also because such measures are highly unpopular, as was evident in the United States in the 1970s, in Caracas, Venezuela, in the late 1980s, and more recently in France.

The root of environmental problems lies in household material consumption, with pollution caused by companies being a consequence of responding to that demand. Focusing exclusively on corporate pollution is easier and more popular, but it has the serious drawback of diverting attention from the underlying issue.

Since those affected by pollution can exert pressure and vote, while companies cannot, this can create biases in public policy that end up exacerbating problems rather than solving them.

Nevertheless, there are signs of a global trend toward improved energy efficiency, for example, as measured by the ratio of energy consumption to GDP, as well as a slowdown in wildlife loss, which is also reflected in the relative stagnation of the Living Planet Index. However, much faster progress is needed in these areas if a global environmental tragedy is to be avoided.

1.7 Pathways to Sustainable Development in the Global South

One of the messages developed in the previous sections is that environmental degradation leaves little room for global economic growth in the coming decades.

If the carrying capacities of nature are exceeded, environmental degradation will lead to environmental catastrophes that will generate unmanageable tensions for most societies, resulting in social conflicts and economic crises of unimaginable magnitude. It is enough to observe the current social and political tensions in Europe and North America caused by migration, because of the crisis affecting emerging peripheral countries, to realize the crises that environmental catastrophes could generate in these regions.

In other words, if different societies do not adjust to achieve a relatively orderly transition toward an environmentally sustainable trajectory, it will be the effects on the natural environment that force a chaotic adjustment, which would be especially dangerous in an era when there are means to destroy all of humanity several times over, along with much of the other species that share the planet.

This is bad news for emerging countries that aspire to close the gaps in material well-being within their societies and with the more economically advanced countries. The group of low- and middle-income countries went from contributing about 1/5 of global GDP growth between 1960 and 1980 to more than 2/3 recently. The combination of a large and still rapidly growing population, and rising per capita incomes in these regions, has led to a drastic change in the arithmetic of global growth, as shown in Fig. 1.12.

The demographic dynamics of these regions and recent growth trends in Asia and even Latin America virtually guarantee that these regions will continue to contribute at least 2 percentage points to real global GDP growth, which exceeds the levels required to contain the global environmental crisis.

A first conclusion that can be drawn from this analysis is that expectations of economic progress in developing countries need to be moderated. Although it is still possible to aspire to improved consumption levels, especially among the poorest, it is virtually unfeasible for the emerging middle classes of the developing world to reach per capita consumption levels comparable to those of their counterparts in the developed world, at least for the next several generations.

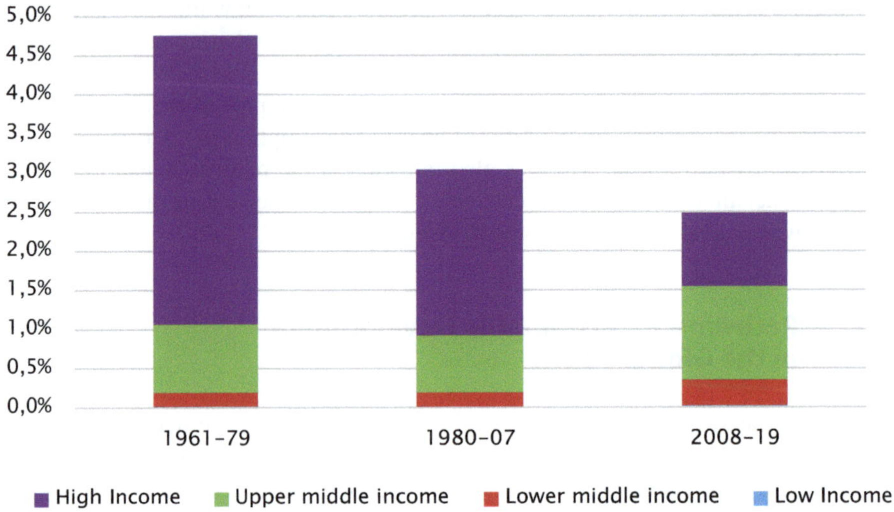

Fig. 1.12 Contribution to global GDP growth by level of development of country groups. (Source: World Bank. World Development Indicators)

If humanity has exceeded the limits imposed by nature, largely due to excessive global consumption, it is inevitable that part of the adjustment to this imbalance will involve a restriction of global consumption, including in emerging countries.

Although it is unfair that part of the restriction should be borne by those who have not enjoyed excess, the size of the population in emerging regions, as well as the progress made in recent decades, make this conclusion unavoidable. It is simply unfeasible for the entire adjustment to be achieved through a reduction in consumption in developed countries: domestic politics in advanced countries would not tolerate it, and, moreover, the distribution of political, military, and economic power on a global scale leaves no room to force such an adjustment.

The major question, then, is how to create opportunities for economic progress in emerging countries, with the means at their disposal. It is possible, but it will not be easy.

A first step is to make use of the valuable assets of these regions, including their natural resources, both those that provide provisioning services (raw materials, food, etc.) and those that support biodiversity. Although the anticipated scarcity of the former has not materialized thanks to technological advances and greater capital intensity in their extraction, this is not guaranteed for the future, and a potential decline in productivity in food production[41] and other resources could increase their value in the future.

[41] As an example, see the series of reports by The New York Times at the end of 2023 on the decline of the major aquifers that have supported high crop yields in the United States. Uncharted Waters - The New York Times (nytimes.com).

It is in the case of other ecosystem services where there is greater potential for development. The protection of natural habitats, with their reserves of wildlife and the carbon stored in their soils and trees, is possibly the most undervalued asset in the wealth of emerging countries. If these services could be valued and, in some way, incorporated as backing for financial transactions, it would open the possibility of income transfers from the more developed world.

The clearest example is the financing of natural habitat preservation through carbon offsets using markets such as the European Union Emissions Trading System (EU ETS), or even with resources from the countries themselves. For Indonesia, Brazil, and other emerging countries in the Americas, Asia, and Africa, the main source of GHG emissions is not the burning of fossil fuels, but rather the release of carbon due to deforestation and land-use change. Investing in their protection would allow these countries to make progress toward meeting their emission reduction targets without having to invest large sums in subsidies to accelerate the energy transition, for example.

The first step in this direction is to advance the valuation of countries' natural capital, beyond the measurement of provisioning services that the World Bank has promoted so far. In fact, this is already on the agenda of various United Nations agencies, including the World Bank itself, as well as the agencies responsible for compiling statistics in several emerging countries.

All of this should be part of a global strategy aimed at improving the efficiency with which these countries use their natural wealth, recognizing that it is increasingly scarce and necessary, both globally and locally. In many cases, this will require creating or strengthening institutions and regulations, but that alone is not enough.

Success will also require engaging markets to guide the allocation of resources that were previously free and are now scarce. For this, it is necessary that the prices paid by companies and consumers reflect the costs of the services provided by nature, which are currently absent due to externalities and failures in their appropriability.

The same applies to the energy transition, which is particularly challenging today in hydrocarbon-producing countries that have historically tended to subsidize the consumption of energy from fossil fuels.

Prices that reflect real costs would not only help achieve a more efficient allocation of economic resources but would also support the much-needed public finances of these countries. One of the paradoxes in many Latin American countries, even when they are not producers of oil, gas, or coal, is the willingness of political leaders to subsidize fuel prices, even though this is very costly and ties up significant public resources, while the main beneficiaries are the wealthier sectors who consume more energy.

In demographic terms, the transition toward lower fertility rates is advancing rapidly in middle-income countries, and the challenge is to recognize this change and incorporate it into the design of public policies in education, health, and social security, among others, rather than trying to delay it.

Experience shows that, in addition to being counterproductive in addressing the environmental crisis, such efforts are ineffective and expensive. In these countries, the focus should be on adapting to the new demographic reality.

In low-income countries, where fertility rates still far exceed the replacement rate (2.1 children per woman), as discussed later, the challenge is to accelerate this transition intelligently. Evidence suggests that this involves expanding economic and self-fulfillment opportunities for women, who are increasingly in control of their reproductive behavior. Nutrition, health, and education policies for women contribute to improving population well-being in all dimensions, accelerating the transition to sustainable fertility rates and, very importantly, increasing women's human capital and income-generating capacity, precisely where they have been most disadvantaged.

All these changes will require significant resources, which in many cases exceed the capacities of the countries themselves. This is where international cooperation from developed countries plays a fundamental role.

Cooperation should range from opening domestic markets and facilitating access to capital and new technologies, to traditional aid focused on supporting the poorest, where increased consumption from current levels has a limited impact on nature. In fact, in many cases, it could even help relieve the pressure to destroy natural areas to expand land for livestock and low-yield crops that are nonetheless necessary for survival.

References

Albagli E, Vial J (2022) Crecimiento económico y biodiversidad: Algo tiene que ceder (Documento de Política Económica No. 70), Banco Central de Chile

Brannen P (2017) The ends of the world: volcanic apocalypses, lethal oceans, and our quest to understand Earth's past mass extinctions. HarperCollins Publishers

Brooke J (2014) Climate change and the course of global history. Cambridge University Press

Brundtland GH (1987) Our common future: Report of the World Commission on Environment and Development. United Nations

Cornelius P, Vial J (eds) (2002) The Latin American competitiveness report 2001–2002. CID – WEF

Costanza R, d'Arge R, de Groot R et al (1997) The value of the world's ecosystem services and natural capital. Nature 387:253–260. https://doi.org/10.1038/387253a0

Crutzen PJ, Stoermer EF (2000) The "Anthropocene". Glob Change Newsl 41:17–18

Dasgupta P (2021) The economics of biodiversity: the Dasgupta review. HM Treasury

Deaton A (2013) The great escape: health, wealth, and the origins of inequality. Princeton University Press

DIPRES (2022) Indicador del balance cíclicamente ajustado: Metodología y resultados 2021

Fagan B (2000) La pequeña edad del hielo. Gedisa

Ferguson N (2006) The War of the World: Twentieth century conflict and the decline of the West. Penguin Press

Ferguson N (2008) The ascent of money: A financial history of the world. The Penguin Press

Gordon RJ (2016) The rise and fall of American growth. Princeton University Press

Harari YN (2015) Homo Deus: A brief history of the future. Penguin Random House

Loayza N, Fajnzylber P (2005) Economic growth in Latin America and the Caribbean: Stylized facts, explanations, and forecasts. World Bank Publications

Maddison A (2000) The world economy: A millennial perspective. OECD Development Centre Studies
Malthus TR (1798) An essay on the principle of population
Markham B (1942) West with the night. Open Road
Meadows D, Meadows D, Randers J (1972) The limits to growth. Potomac Associates
Nordhaus W (1992) Lethal model 2: The limits to growth revisited. Brookings Pap Econ Act 1992(2):1–59
Pérez Rosales V (1882) Recuerdos del pasado (Ed. 1970). Biblioteca Francisco de Aguirre
Peskin HM (1976) A national accounting framework for environmental assets. J Environ Econ Manag 2:255–262
Pomeranz K (2000) The great divergence: China, Europe, and the making of the modern world economy. Princeton University Press
Pritchett L (1997) Divergence, big time. J Econ Perspect 11(3):3–17
Repetto R (1991) La erosión en el balance general: Cómo contabilizar la pérdida en recursos naturales. In: Vial J (ed) Desarrollo y medio ambiente: Hacia un enfoque integrador. CIEPLAN
Rockström J, Steffen W, Noone K, Persson Å, Chapin FS III, Lambin E et al (2009) Planetary boundaries: Exploring the safe operating space for humanity. Ecol Soc 14(2)
Sachs JD (2008) Common wealth: economics for a crowded planet. The Penguin Press
Smil V (2022) How the world really works: the science behind how we got here and where we're going. Viking
Solow RM (1956) A contribution to the theory of economic growth. Q J Econ 70(1):65–94
United Nations (2019) World population prospects: The 2019 revision
Wilson EO (2002) The bottleneck. Sci Am 286(2):82–91
World Bank (1997) Expanding the measure of wealth: Indicators of environmentally sustainable development. World Bank. https://elibrary.worldbank.org/doi/epdf/10.1596/0-8213-3956-7

International Trade, Patterns of Specialization, and Environmental Effects of Globalization

2

Abstract

International trade played a pivotal role in the diffusion of economic progress throughout the world. For many countries in the periphery of the developed world, trade specialization has been determined by their natural resources. For many countries they were instrumental to achieve middle income status, but only a few have been able to advance beyond that. On the other hand, countries with few natural resources in Asia moved ahead exporting industrial goods based initially in cheap labor, and later by the buildup of considerable human and productive capital. The shift of industrial production from advanced countries to emerging ones, as well as the fast expansion of global consumption demanding raw materials and food translated into more environmental degradation in emerging countries, making this problem a truly global crisis. The final section addresses the issue of why we still need economic growth, despite its large impact on global and local sustainability. The answer is that without economic growth, a large fraction of humanity will remain living under unacceptable conditions, so the rest of the world will need to leave space for their progress.

Keywords

International trade · Patterns of specialization · Economic convergence · Natural resources · Trade and growth · Trade and environment

2.1 Trends in the Economic Performance of Different Regions

The nineteenth and twentieth centuries were marked by unprecedented global economic development in human history. In this context, differences in the pace at which countries and regions grew have determined their current position in terms of development level.

J. Vial, *Sustainable Development and Natural Resources*, Integrated Science 39,
https://doi.org/10.1007/978-3-032-12263-6_2

Official economic data on the main comparable national account aggregates is only available from the second half of the twentieth century. However, thanks to the work of Angus Maddison (2000) and his team at the Maddison Project at the University of Groningen, who constructed historical data on per capita income and population, it is now possible to identify the main trends by country and region, as shown in Figs. 2.1 and 2.2.

As shown in Fig. 2.1, all regions experienced positive growth rates in the twentieth century, something that had not occurred previously. In fact, during the nineteenth century, there was virtually no economic growth in Asia and Africa—regions that, at the beginning of the twentieth century, were home to more than half of the world's population. This period marks the "Great Divergence" described by Pomeranz (2000) and Pritchett (1997). Latin America, by contrast, did grow, but at a much slower pace than the economies of Europe and North America, its major trading partners at the time. This was partly due to internal upheavals and external conflicts associated with wars of independence and the consolidation of nations and borders.

The twentieth century, by contrast, is the period in which economic growth accelerated and became widespread across all regions of the world, with the sole exception of Eastern Europe, whose cumulative growth in the twentieth century was lower than in the previous century. In this century, East Asia stands out as the fastest-growing region in the world, managing to narrow the per capita income gaps accumulated in previous periods. Some countries, such as Japan, South Korea, and Taiwan, have even joined the group of countries with the highest per capita incomes globally.

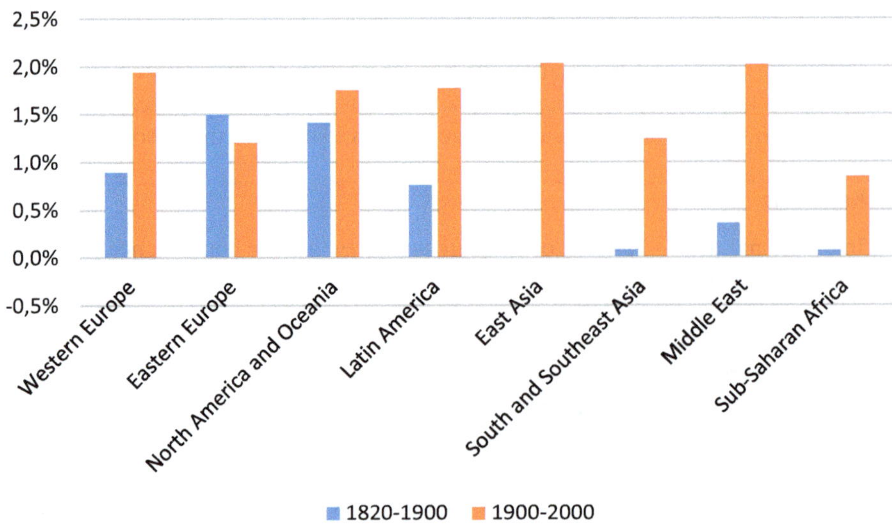

Fig. 2.1 Annual growth rates of real per capita income in geographic areas. (Source: Maddison Project Database (MPD) 2020)

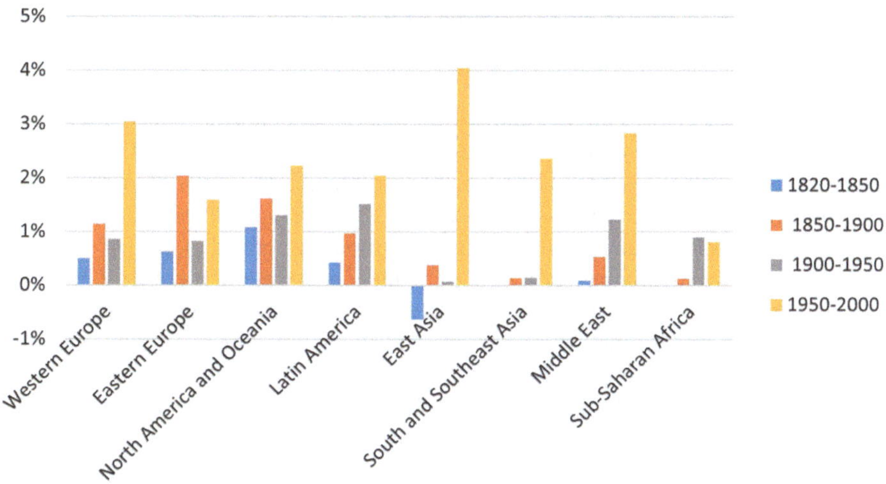

Fig. 2.2 Annual growth rates of real per capita income by geographic area: subperiods. (Source: Maddison Project Database (MPD) 2020)

Latin America posted high growth rates, like those of North America, but these were not sufficient to close income gaps. Africa, while growing, did so at the slowest pace in the world. In this case, decolonization and the formation of independent countries are very recent, occurring mainly after World War II. Moreover, this process was even more chaotic than in Latin America, with acute conflicts both within and between countries, because of territorial divisions imposed by European colonial powers that disregarded ethnic and cultural differences as well as geographic barriers. Added to this were the devastating impacts of the AIDS epidemic, which severely affected the region at the end of the twentieth century.

Figure 2.2 allows for a clearer identification of the periods of change in the trends shown in Fig. 2.1. When growth rates are examined by subperiod, it is evident that, although Latin America's economic growth has gradually accelerated, it has still not reached rates sufficient to compensate for its historical lag. In the first half of the twentieth century, not only did growth accelerate, but the region also began to close gaps with the developed world—though this trend did not continue in the second half of the last century.

In the case of Asia, the slow pace of its economic takeoff is notable, becoming widespread only from the second half of the twentieth century, but at an extraordinarily rapid rate, especially in East Asia, where Japan, South Korea, Taiwan, and more recently, China, stand out.

The Middle East grew primarily at the pace of increasing oil wealth in a group of countries, favored by greater autonomy in managing these resources after independence. Africa grew at a consistently low rate throughout the twentieth century.

2.2 Asia's Convergence and the Shift in the Axes of the Global Economy

Economic growth, both globally and within individual countries, has led to fundamental changes in their economic weight. One of the most notable phenomena of the twentieth century has been the rebalancing of the relative sizes of countries and regions. Those that successfully integrated into the development processes initiated by the Industrial Revolution came to have enormous weight in global income generation, which was accompanied by a significant concentration of political, military, and even cultural power. However, since the mid-twentieth century, rising internationalization of production, consumption, and finance has enabled new countries and regions to join these processes and begin to close the gaps. This has also triggered a significant shift in the global distribution of economic power. In the mid-twentieth century, when the emerging countries that benefited from globalization were neither numerous nor large, this did not have transformative consequences for the global economy. But by the end of the twentieth century, with the economic resurgence of China, the growing momentum of India, Brazil, Indonesia, and other large and medium-sized countries, the change was dramatic, as discussed in Chap. 1.

Figure 2.3, presented below, shows how much the absolute weight of each region has changed relative to the size of the global economy. Still, the economic weight of the developed world remains much greater than that of all other regions combined. The figure shows that the world economy grew 6.5 times over the past six decades. In other words, what would previously have represented millennia of progress has now been achieved in a single generation. Another notable point is that in the first 40 years, there were no significant changes in relative positions, except for Japan's recovery. In contrast, in just over 20 years so far this century, the relative weight of the richest countries has fallen from 77% of global GDP to 64%, reflecting the growing share of the middle-income countries in the global economy.

As discussed in Chap. 1, convergence in income levels between countries has been rather modest. However, it has been sufficient to provoke major tensions in the global economy and politics, due to strong pressures on natural resources and the environment, given the size of the countries that are accelerating to close the gaps. The most notable aspect of Fig. 2.3 is that it shows how the engine of global growth has shifted rapidly from the wealthiest countries to those with middle incomes, which in the last decade have contributed nearly two-thirds of global GDP growth, primarily driven by China, India, and other Asian countries. Looking ahead, and in contrast to the experience since the Industrial Revolution market growth in the twenty-first century will not be centered on wealthy countries and consumers, but rather on the emerging middle classes of the developing world. These new consumers will have lower incomes, but they are far more numerous. Moreover, they are located far from the North Atlantic and come from diverse cultures that are very different from those of the West.

An important dimension of this shift, explored in greater detail in the following sections, is that economic power has begun to move toward the most populous

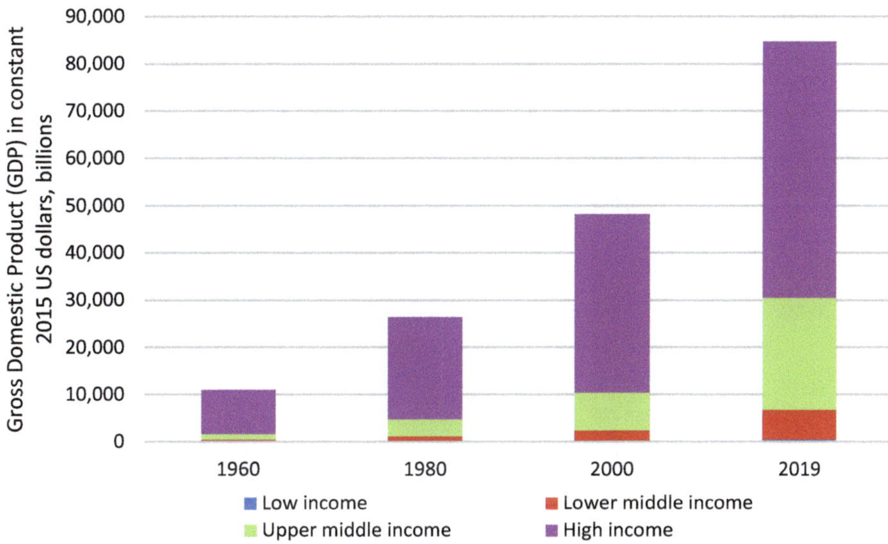

Fig. 2.3 Composition of world GDP by region according to income level. (Source: World Bank: World Development Indicators)

regions of the planet. This is illustrated in Figs. 2.4 and 2.5, which present the same information as Charts 14 and 15, but now refer to the world population instead of GDP.

Most of the growth in the world population in the latter part of the twentieth century was concentrated in middle-income countries, particularly lower-middle-income countries such as China and India, which also coincide with those experiencing the fastest economic growth. Toward the end of the period, their population growth slows, but remains the largest in absolute terms, while population growth in the lowest-income countries begins to gain importance[1].

Figure 2.5, in turn, shows that the relative share of the population living in the wealthiest countries has steadily declined, due to their lower population growth rates, especially in comparison with those of middle- and low-income countries.

In the last two decades, with the "rediscovery" of China, India, and other developing countries by major corporations, we have witnessed a preview of what will be seen in the years and decades to come. But this will not stop here: what is now evident in the economic sphere is also beginning to be felt in the political, military, and cultural arenas, generating tensions and shifting the dominant reference points that have prevailed since the onset of Western hegemony with the colonial expansion of European countries.

As shown in Chap. 1, this geopolitical shift coincides with the severe global environmental crisis resulting from the sharp increase in the human footprint and its pressure on the planet's resources, as well as on its capacity to absorb the impacts of that footprint. As this is a global issue that transcends national borders, it requires

[1] The determinants of these differences are examined in the next section.

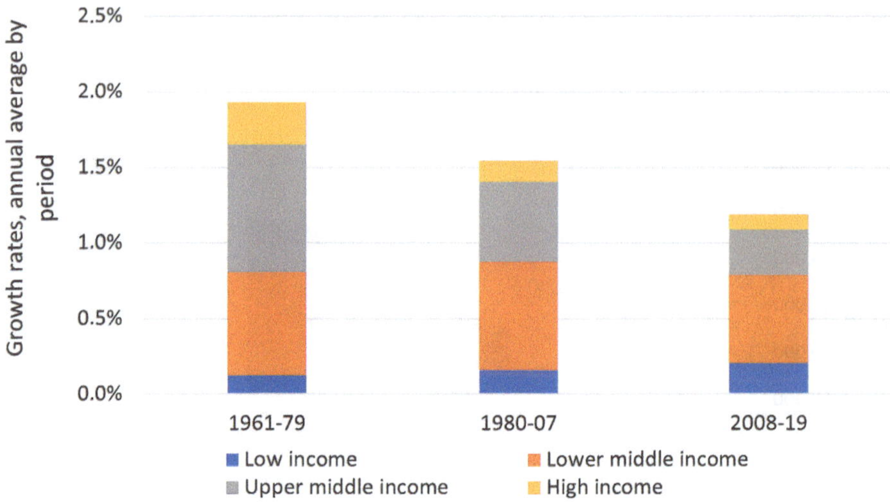

Fig. 2.4 Source of world population growth by region according to income level. (Source: World Bank: World Development Indicators)

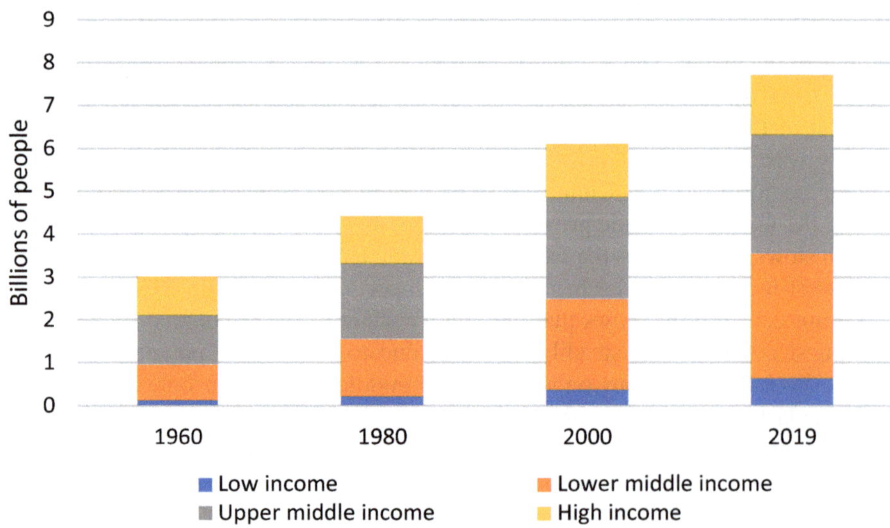

Fig. 2.5 Composition of the world population by income level. (Source: World Bank: World Development Indicators)

unprecedented supranational coordination efforts, precisely at a time when the prevailing power relations since World War II are undergoing radical change.

This is one of the most serious challenges that individuals, governments, businesses, and multilateral organizations will face soon. Adjusting to this new reality will not be easy: changes of this magnitude are never simple, especially in the political sphere. The closest parallel in recent history was the emergence of economic and political powers in Central and Eastern Europe between the late nineteenth and early twentieth centuries, which led to the bloodiest conflicts in history, as well as 45 years of "Cold War" that brought humanity to the brink of self-destruction (Ferguson 2006).

How does the emergence of these "new actors" affect the trajectory of the world economy? Or put another way, to what extent does this represent a trend that will change the environment in which medium-sized or small emerging countries will operate in the coming decades?

A first effect, which is already clearly visible, is the radical change in international trade patterns. The rapidly growing countries in Asia have achieved this based on a major shift in their pattern of international integration, which enabled rapid industrialization initially based on the advantages provided by an abundant supply of cheap labor. Given the importance of this process, it is worth pausing to examine in some detail the evolution of Asia's most dynamic economies in recent decades.

Faced with the threat of Western expansion in the mid-nineteenth century, Japan's leaders chose to abandon their previous isolationism and actively engage in the global economy. After centuries of turning its back on the world, Japanese society was shaken in 1853 by the arrival of American warships commanded by Commodore Perry, whose military technology gave them an overwhelming advantage in the event of conflict. This triggered a social and political revolution, replacing the medieval shogunate with a central authority backed by the emperor, thus ushering in the Meiji era. The new leaders committed to the massive importation of advanced Western technologies to achieve Japan's industrialization[2]. A vital component of this effort was the dispatch of official missions to study and acquire technologies, as well as sending large numbers of students to further their education at Western universities.

Being a poor country, with abundant labor but very scarce natural resources with export potential, it was natural to specialize in labor-intensive manufactured exports. On the other hand, the resources needed to finance the investments required for industrialization were obtained locally by limiting consumption opportunities, with strong protection of the domestic market. This made it possible to achieve high levels of domestic savings as workers' incomes grew. At the same time, there was significant investment in education and the strengthening of human capital.

Over time, as the local industry benefited from improvements in human capital, gaining technological capabilities and improving the quality of manufactured products, this enabled a shift toward specialization in the export of increasingly sophisticated goods, intensive in capital and knowledge. In parallel, higher wages priced out of the market traditional exports intensive in low-skilled labor—.

Today, Japan is an exporter of goods at the technological frontier, integrated into production chains spanning various countries, mainly in Asia, and where the majority of GDP is generated in the services sector. Moreover, Japan has also shifted from

[2] For further details, see Diamond (2019).

being an importer of natural resources for domestic consumption and manufacturing to an importer of parts and components for its industry, as well as sophisticated consumer goods and services. Changes in the institutional framework of international trade, first with the GATT agreements and later the WTO, have led Japan to gradually remove barriers to imports.

Other Asian countries that also enjoyed ready access to the North American market, such as South Korea and Taiwan, followed very similar paths. They also shared some characteristics with Japan, such as abundant labor and scarce natural resources, and followed the lead of the regional leader. Thus, they also focused on developing manufactured exports, suppressing consumption through protectionism, and making strong investments in human capital formation. These countries added an additional component to the Japanese integration model: the implementation of policies aimed at attracting foreign investment. This played a very important role in developing their manufacturing export capacities. Singapore followed a similar path, taking advantage of its privileged location on global maritime trade routes, as did Hong Kong, which, in addition to being the informal gateway to the Chinese market, was already a commercial and financial services center for all of Asia.

In the 1980s and 1990s, other medium- and large-sized Asian countries joined international integration processes with similar characteristics. Among them are Thailand, Indonesia, the Philippines, and Malaysia. Like the previous countries, they had abundant cheap labor and bet on attracting foreign investment to develop their manufacturing industries. Unlike the earlier group, however, these "new tigers" possessed significant natural resources and integrated into the global economy at a time when world trade rules limited the imposition of strong protectionist barriers to suppress domestic consumption.

Financial liberalization with relatively weak banking systems left these countries vulnerable to external financial shocks, which materialized in the late 1990s, triggering the so-called "Asian Crisis." At that very moment, China was beginning its integration into the international economy, and Japan and the original "tigers" were seeking to reorganize their production chains to take advantage of lower labor costs in neighboring countries.

With varying degrees of success, the new emerging Asian countries have managed to sustain high growth rates by integrating into the production chains of China, Japan, Korea, and Taiwan.

The case of China deserves special mention, both for the speed of its development and for the impact due to its size. The latter makes it a very attractive market for trade and investment but also allows to achieve large economies of scale, making it a global supplier of manufactured goods. This country was outside the circuits of the global economy for almost the entire twentieth century, except for the commercial enclaves imposed by Western powers on the Chinese Empire, taking advantage of its virtual disintegration at the end of the nineteenth century.

After unprecedented failures in economic and social engineering in the 1950s and 1960s, which resulted in the deaths of tens of millions of people from famine, the new leaders who succeeded Mao Zedong decided to liberalize the economy and integrate into the world economy. Although this proceeded gradually, opening special zones for the location of Foreign Direct Investment and gradually liberalizing internal migration, the impacts were soon felt as shown in Fig. 2.6, China went from nothing to accounting for more than 10% of global export capacity. A key step in this process was its accession to the WTO in December 2001, committing to adopting the rules of the global trading system, which allowed it to access markets that had previously been closed to it.

As in Japan and the "Asian tigers," China's initial export expansion was based on labor-intensive manufactured products, but quite remarkably, it shifted to sophisticated, capital-intensive manufactures in less than 20 years. The attraction of multinational companies to set up in China, taking advantage of abundant labor, good infrastructure in selected zones, and the great potential of the Chinese domestic market, were important factors in this process. Today, Chinese companies are expanding around the world, whether in search of food, natural resources, or cheaper labor, or to take advantage of technological or capital access advantages.

This industrial boom in China resulted in a shift of the global manufacturing capacity from developed countries to China and to companies, mainly Asian, that have integrated into the production chains of North American, European, and Japanese firms. Consequently, the demand for fuels and industrial raw materials has also shifted from developed countries toward China and other Asian countries. This can be seen in Fig. 2.7a–c.

These graphs, which cover just four decades, show the rise of China and other Asian countries in raw materials markets, thus indirectly reflecting the effect of their specialization in manufacturing activities that supply the rest of the world as well as their increasingly significant domestic markets.

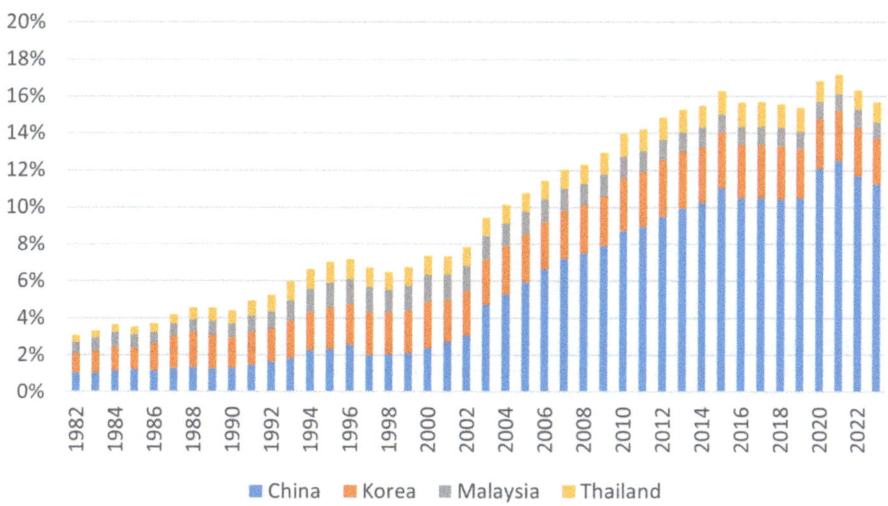

Fig. 2.6 Share of exports from selected Asian countries in world trade. (Source: World Bank: World Development Indicators)

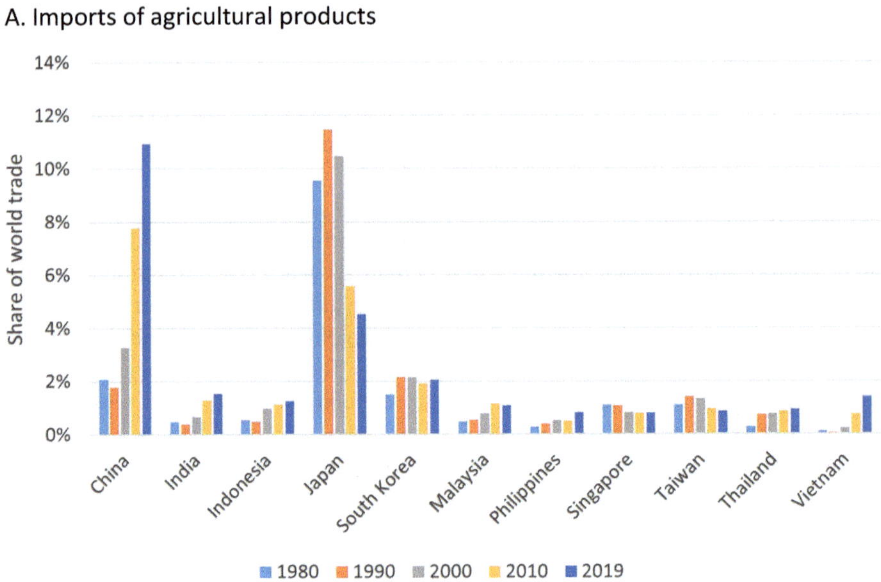

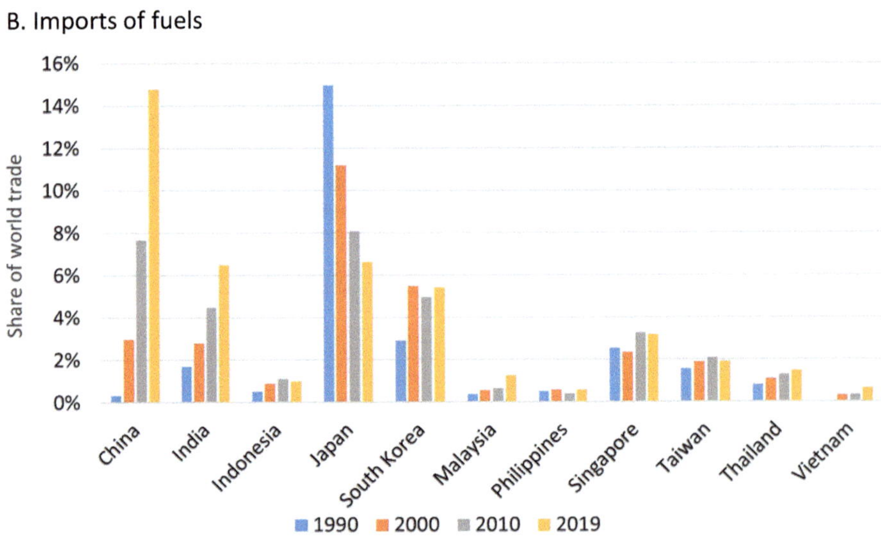

Fig. 2.7 Demand for raw materials by selected Asian countries. (**a**) Imports of agricultural products. (**b**) Imports of fuels. (**c**) Imports of minerals. (Source: WTO)

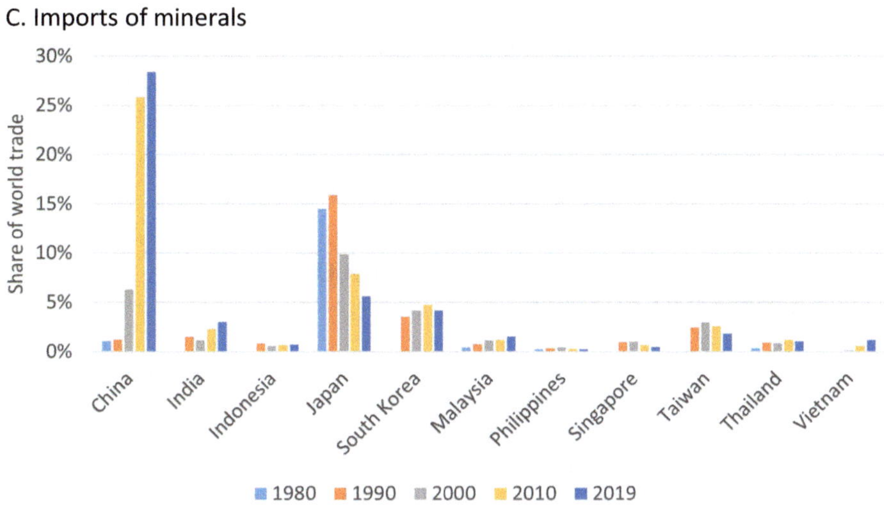

Fig. 2.7 (continued)

The new economic weight of emerging countries, which account for more than 70% of the world's population, represents a very radical shift in the economic trends prevailing since the Industrial Revolution. While it is difficult to imagine that per capita income gaps with developed countries will close soon, the sheer weight of large numbers (population, total GDP, among others) is already changing the economic environment in which emerging countries are operating, especially those specialized in the export of raw materials, food, and fuels, as well as those with the potential to integrate into manufacturing production chains.

A similar phenomenon is occurring with scientific and technological advances. There is no doubt that the United States, Europe, and Japan will continue to lead these advances in the near future, but this does not mean they will maintain a monopoly on innovation. In fact, we are already seeing how China, Israel, Korea, and Taiwan are beginning to exploit niches, challenging traditional leaders. This is especially true in practical applications of new technologies, where the sheer number of engineers and technicians dedicated to these tasks gives countries like China and India a competitive advantage.

An important conclusion is that the process of spreading economic progress to the world's poorest regions is closely linked to the development of economic ties between the most advanced and the poorest countries (Sachs 2015). In the initial stage, this was closely tied to physical proximity, which facilitated the development of trade, labor, and investment flows, but as transportation and communication technologies advanced, other variables gained importance. Among these are connections to maritime trade routes and the availability of globally scarce resources, whether energy sources, materials, food, or human capacities with specific advantages. These characteristics enabled the development of specialization patterns to compete successfully in international markets and to attract capital and technology.

Those countries that systematically sought to expand and deepen their links with the rest of the world have made the fastest progress in their economic development. In contrast, those that sought to develop by protecting their domestic markets eventually ran into the limits of such markets for growth, and their industries, isolated from external competition, failed to develop the capabilities needed to compete successfully in global markets.

2.3 Impacts of Asian Convergence on Other Emerging Countries

It is worth asking what difference this new trend makes for the progress of emerging countries, especially in Latin America and Africa, which specialize in the export of raw materials, fuels, and energy to the United States, Europe, and Asia, and have strong financial ties and dependence on foreign investment from these regions.

The first dimension relates to changes in the determinants of economic cycles, and there is no doubt that diversification of demand sources has reduced direct dependence on the economic cycle of developed countries.

The experience of the so-called Great Recession of 2008–09 provides a very interesting example. This event began with the deterioration of the U.S. financial sector in 2007, because of mortgage loan defaults in that country, which then led to bailouts of several large banks throughout 2008 and culminated in the collapse of Bear Stearns in September of that year. This quickly led to a global financial panic, with a contraction of interbank credit in the developed world. Most emerging countries did not experience problems in their own financial markets, although many were affected by their access to short-term external financing (such as trade credit lines) due to liquidity problems at international banks that usually provide such credit. In these cases, central banks provided short-term dollar liquidity until this market normalized.

The fear of a major crisis resulted in a collapse of global consumption and investment and, consequently, of international trade. In this context, some emerging countries that did not face financing constraints, particularly China, implemented strong public spending expansion programs and monetary stimulus to revive consumption and investment. This led to a rapid recovery in demand for raw materials and food, and as a result, many emerging economies experienced a rapid recovery starting in the second half of 2009, culminating in 2010. Developed countries had a much slower recovery, and by the middle of the decade, a significant number of them had not regained their pre-crisis per capita GDP levels.

Figure 2.8 shows the evolution of GDP growth rates and of imports for three major country groups (the 20 most advanced, developing Asia, and Latin America). This clearly shows the strong growth momentum prior to the crisis, the fact that the crisis triggered recessions in the developed world and in Latin America, but, contrary to what happened in previous decades, this recession was milder and shorter than in advanced countries, and in the subsequent recovery, the process of income convergence that preceded the crisis continued.

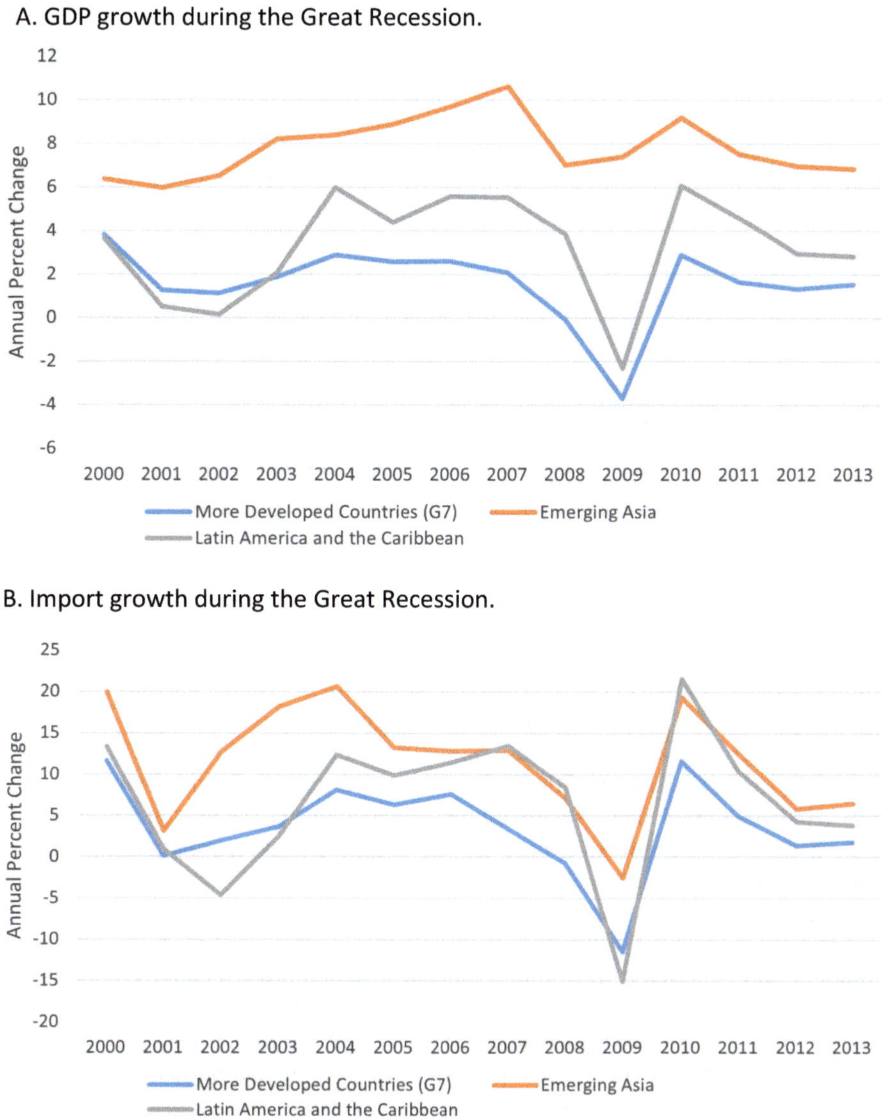

Fig. 2.8 Evolution of GDP and imports during the Great Recession (annual percentage change). (**a**) GDP growth during the Great Recession. (**b**) Import growth during the Great Recession. (Source: IMF, WEO Database)

When examining the evolution of China's economy, the impact of investment stimulus stands out, as it offset the sharp decline in export volumes, thereby cushioning the effect on imports, which quickly regained a high growth rate, as seen in Fig. 2.9. This, in turn, facilitated the recovery of demand for manufactured inputs as well as raw materials from supplier countries.

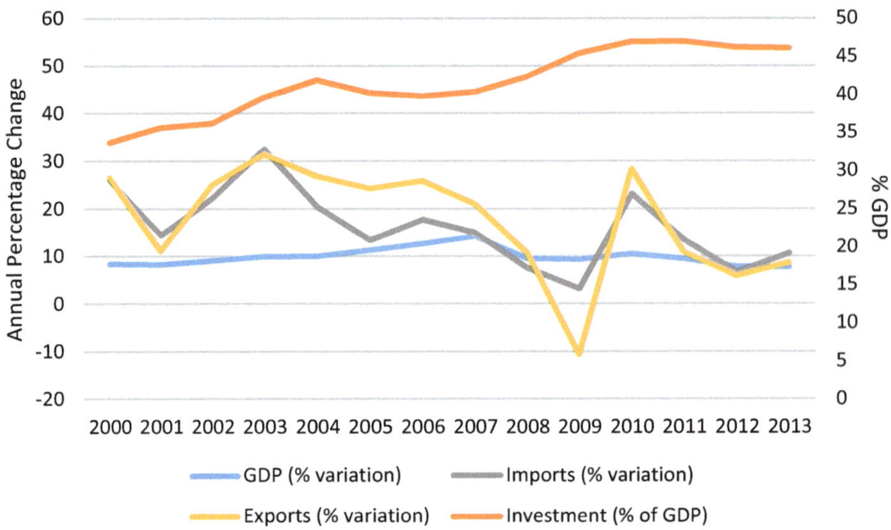

Fig. 2.9 The Chinese economy during the Great Recession (2011 international dollars). (Source: IMF, WEO Database)

The impact of these events on a selected group of emerging countries in Asia and Latin America is shown in Fig. 2.10 and the similarity in cyclical impacts is striking, especially as reflected in the evolution of GDP growth rates.

Although growth rates in Asia are higher than those in Latin America, the cycle around the crisis at the end of the 2000s is virtually identical, with a sharp drop in growth between 2008 and 2009—which in some countries resulted in a brief recession—and a very strong recovery in 2009–2010. Subsequently, growth rates returned to their previous range[3].

When examining the impact on foreign trade, and exports in particular, it is evident that the impact on volumes was much more pronounced in Asian countries, while in Latin America it was milder, especially in countries specialized in mineral resources (Chile, Colombia, and Peru). In contrast, these latter countries experienced a much greater impact on their terms of trade, because after the initial price drop, this type of product saw a rapid price recovery when markets became convinced that China would be able to sustain high growth.

In summary, the origin of external shocks faced by emerging countries has changed radically: from being linked almost exclusively to the business cycle of developed economies, particularly the United States, today the economic cycle of China and Asia in general has also become relevant. Of course, China is not immune to developments in the advanced economies of the West and Japan, but it is large

[3] It is noteworthy, by contrast, the difference with what happened in 2001, during the "dot com" crisis in the U.S. That event had a significant cyclical impact on developed countries and Asian manufacturing exporters, but was virtually imperceptible in Latin American countries, which are more specialized in resource-based products.

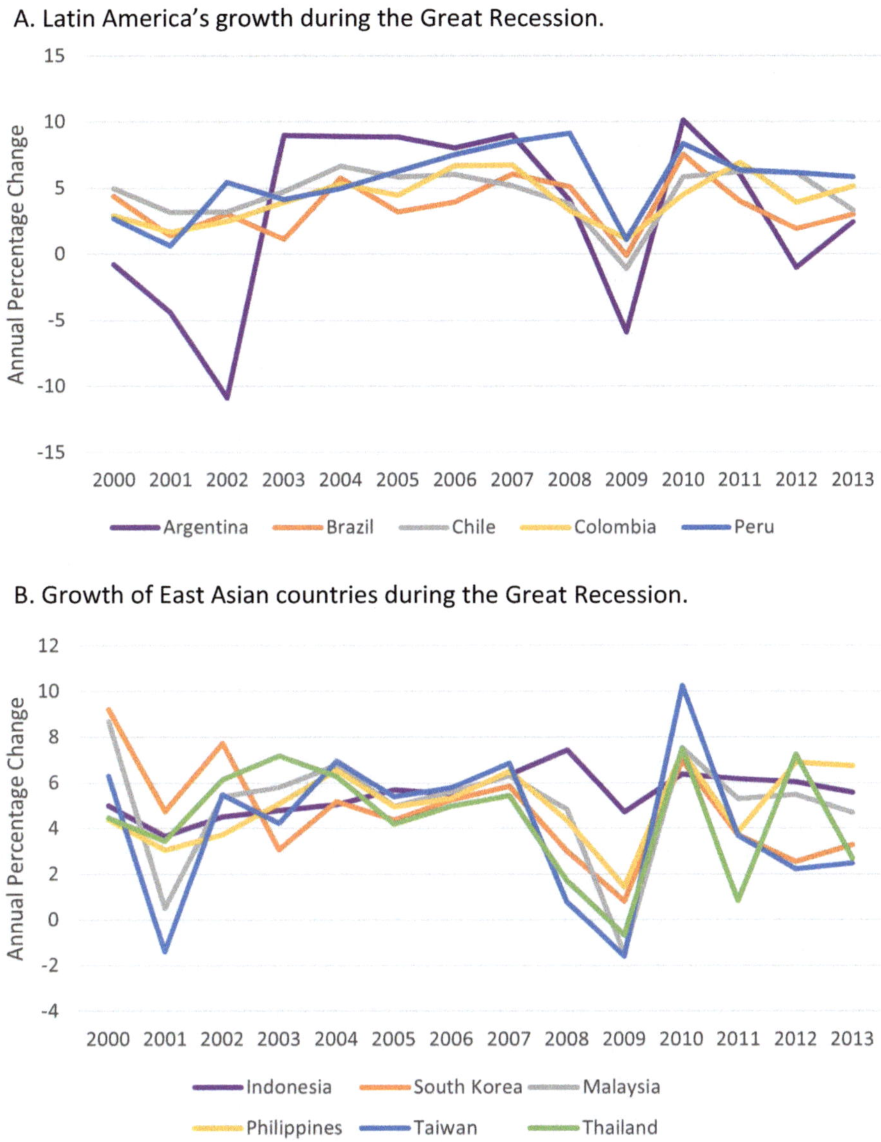

Fig. 2.10 Evolution of GDP in selected countries of Latin America and East Asia during the Great Recession (annual percentage change). (**a**) Latin America's growth during the Great Recession. (**b**) Growth of East Asian countries during the Great Recession. (Source: IMF, WEO Database)

enough to have the capacity to cushion these shocks, which allows for risk diversification.

Emerging countries, especially in Latin America, which have had a special economic relationship with the West and particularly with the United States, must pay close attention to developments in China, among other reasons because it has some vulnerabilities of its own and could also become a source of risk.

The global financial crisis provides further evidence of the interconnectedness of financial markets across countries. At the end of the twentieth century, this was a reality only for developed countries. Today, most of the developing world has become integrated into these markets. Greater political and economic stability in Asia and Latin America has been a very important factor. Another key element was the economic reforms undertaken by most of these countries in the final decades of the last century, in the spirit of the so-called "Washington Consensus" that emerged at that time as a reaction to the "debt crisis" of the early 1980s[4].

In the financial sphere, there has been a trend toward more widespread access by governments, banks, and companies to private financing intermediated by the major global financial centers (New York, London). On the other hand, capital flows from emerging countries to these financial centers have also increased significantly, whether for the accumulation of international reserves, sovereign wealth funds, or by private fund managers and companies (both financial and non-financial). There have also been significant changes in direct investment. One particularly interesting development is the emergence of multinationals in emerging countries, especially in Asia and Latin America, that invest beyond their borders, both in the production of goods and services.

A final aspect to highlight is the shift in perceptions about globalization and the resurgence of protectionist pressures, especially since these have emerged in countries such as the United States, which were at the forefront of trade and investment liberalization. It is difficult to predict whether these sentiments will be strong enough to reverse the previous trend, just as the influence of traditional powers is waning.

There are two strong reasons to be relatively optimistic about a potential reversal toward freer international trade. On the one hand, import tariffs are a tax paid by individuals and businesses in the country that raises tariffs, which affects the cost of living for families (who vote in elections) and the competitiveness of companies. On the other hand, it is worth recalling that large companies from developed countries, which are global leaders in their respective fields, have relied precisely on the globalization of their suppliers and customers, and should therefore be a force in favor of the continuation of this process.

Nevertheless, in the short and medium term, the threats of tariffs and other restrictions, as well as the implementation of some of them, have generated enough uncertainty to impact foreign trade and investment, even beyond the countries directly affected. Moreover, not all tariffs will be reversed, and the mere fact that the events seen at the beginning of 2025 have occurred has introduced a new source of

[4]The general principles of this "consensus" were set out in the book edited by John Williamson (1990). This was the result of an academic conference held in Washington, D.C., which, among other objectives, sought to assess how macroeconomic policy orientations had changed in multilateral economic organizations (IMF, World Bank, IDB, etc.) in response to the debt crisis.

uncertainty that will certainly lead to a greater perception of risk in activities related to international trade.

An additional factor to consider is the impact of these disruptions on the global financial system and the payment mechanisms that underpin international trade. While it is virtually impossible for this to lead to the replacement of the dollar in these areas in the short term, it is highly likely that its absolute dominance will be weakened and that other competitors will begin to emerge, adding an extra layer of uncertainty to cross-border economic and financial transactions. All of this has been accentuated in recent years by the reemergence of armed conflicts in Europe and the explicit threat of possible military conflicts between superpowers, particularly between China and the United States.

All this is happening very soon after the recent Covid-19 pandemic and its aftermath, which affected international trade flows. Increased human mobility was a key factor in the rapid spread of the disease. On the other hand, the vulnerabilities of global production and distribution chains became evident when some of these links were affected by quarantines or other restrictions on mobility and interactions. However, the effectiveness of international supply of medical equipment and supplies was also demonstrated. Furthermore, the development, production, and distribution of vaccines was made possible by instantaneous information flows, as well as the integration of laboratories with pharmaceutical companies operating production plants in widely diverse locations. The complex logistics required for the storage and distribution of vaccines at extremely low temperatures would not have been feasible without the pre-existing global transport and distribution chains with extensive experience in handling frozen goods.

The retreat of globalization presents new challenges on two fronts: on the one hand, emerging middle classes around the world, who will drive future consumption, want to expand their access to goods and services beyond what is available within their own countries. They, like those in wealthy countries, will not remain indifferent to the price increases that will follow a large-scale protectionist offensive. On the other hand, the magnitude of the problems facing humanity makes global cooperation and coordination of efforts indispensable. Common challenges such as global warming, ocean acidification, or the massive loss of biodiversity cannot be effectively addressed except through collaboration among the most diverse countries. Reduced commercial and financial interaction between countries undermines both objectives and will make it more difficult to establish mechanisms for cooperation to advance on both fronts.

While it is almost inevitable that groups harmed by the loss of competitiveness in advanced countries will try to resist change, they are already leaving the workforce due to age. Moreover, new activities and employment opportunities are emerging because of technological advances. In addition, the domestic markets of the wealthiest countries will gradually lose their preeminence as economic dynamism shifts toward other, much more populous countries where there are greater opportunities for growth. The dependence of Western countries on critical materials for the development of new technologies could be another factor that helps curb protectionist impulses. In fact, it is quite evident that a race is already underway between China and the West to secure access to critical resources.

What can Emerging Countries do to face the threats of trade conflicts and mutual restrictions imposed by larger countries? In the case of small countries far from major markets, there are no alternatives to joining globalization, in which case the relevant question is what can be done to minimize the damage from potential setbacks.

Several Latin American countries, particularly Chile, developed an extensive network of bilateral free trade agreements starting in the 1990s, which today serves as their first line of defense to circumvent barriers that arise because of conflicts between third parties. Other options include seeking regional agreements or partnerships with blocs of countries that have similar development strategies and growth potential, which can at least partially offset the disruptions to foreign trade caused by trade wars. A diversified foreign trade portfolio in terms of countries of origin and destination is possibly the best insurance against a slowdown in global trade.

Paradoxically, the fact that many emerging countries are major producers of industrial raw materials—such as copper, oil, timber, and pulp—as well as food, helps cushion the adverse effects of these conflicts. If, for example, the United States imposes restrictions on imports of manufactured goods from China, this will likely result in trade diversion to third countries or to the United States itself, but as long as the final goods consumed in the United States require raw materials such as those mentioned above, final demand for them will recover after an adjustment period.

In any case, it cannot be ignored that recent tensions between China and the West are especially complicated for countries that have developed by taking advantage of globalization. On the one hand, trade restrictions will result in global losses of efficiency and income, which will affect demand for all types of exports from emerging countries. On the other hand, it could eventually happen that they are forced to choose to align with one bloc or the other, risking high economic and social costs from losing access to the other bloc.

2.4 Natural Resources and International Integration of Emerging Countries

Up to this point, little has been said about the productive specialization of emerging countries that are rich in natural resources. There is extensive literature on the important role that natural resources play in the international integration of most emerging countries, and it is frequently cited as one of the causes of the slow pace of convergence in these countries, and even of the weakness of their institutions.

However, when looking at the big picture, there is no evidence that abundance of natural resources has been an impediment to achieving high levels of per capita income. If we take the group of countries that the United Nations classifies as "Developing" and disaggregate them according to the income categories used by the World Bank, we see that, in the group of High-Income Developing Countries, 14 out of the 49 countries listed there (28%) are generally considered to be rich in natural resources[5]. It should be noted, however, that 9 of the 14 are oil exporters. In the group of Upper-Middle-Income countries, 42% are rich in natural resources (a little more than half are oil exporters). In Lower-Middle-Income countries, a little more than a third are rich in natural resources, and among the poorest countries, the scarcity of natural resources appears to be part of the problem (only 25% are rich in natural resources).

In other words: a large proportion of developing countries are rich in natural resources, and within that group, those with highly valuable resources (mainly oil and gas) have achieved relatively high levels of per capita income. However, there are also many cases of low-income countries that are rich in natural resources, suggesting that abundance of these resources does not guarantee a rapid transition to development.

In times when productivity was low because capital (in all its forms) was scarce, it made sense to think that natural wealth was a straightforward path to high income levels. However, after the Industrial Revolution, when produced capital and human capital began to play a fundamental role in income generation, the abundance of natural resources started to lose relevance as a driver of economic progress, apart from energy, which underpins increases in physical productivity.

Raúl Prebisch (1959), one of the founders of ECLAC and a leader of the so-called structuralist school, pointed to a systematic deterioration in the terms of trade for countries specialized in raw materials, stemming from less dynamic demand for these goods in major consumer markets. According to him, goods such as food would lose value relative to manufactures and services as incomes rose. Over the years, it has become clear that this hypothesis is, at the very least, controversial. This is due, in part, to the fact that what matters for resource-rich countries is the evolution of unit rents, that is, the income surpluses remaining after deducting operational and capital costs, which, while possibly correlated with prices in the short term, may diverge over longer horizons as technological changes or other factors affect these costs.

Notwithstanding the above, earlier studies also showed a negative correlation between the growth rate of emerging countries and their dependence on natural resources (Sachs and Warner 2001), which has led to the proposition of a "natural resource curse." Among the possible explanations for this curse, the following are noteworthy (Lederman and Maloney 2007):

- Rents from natural resource exports lead to a more appreciated real exchange rate, making labor more expensive and discouraging the emergence of other exports, the so-called "Dutch Disease".
- Specialization in natural resources leaves little room for technological innovation (Prebisch 1959) and has few linkages to the rest of the economy, resulting in a low level of complexity in the productive structure and limited opportunities

[5] For example, because they are included in the IMF list in Gruss and Kebhaj (2019).

for the development of new export lines (Hausmann and Hidalgo 2011; Diodato et al. 2022).

- Competition to capture rents produces tensions and conflicts, fostering corruption and weakening institutions, leading to societies with greater concentration of economic and political power (Sachs and Warner 2001).
- A unique aspect of natural resources is that they are tied to specific physical locations, which vary in population density and in the characteristics of the ecosystems in which they are found, affecting the availability of complementary resources (water, energy, land use). In many cases, this can give rise to conflicts with local communities (Peru, Panama, and Papua New Guinea, for example, due to mining developments) that may eventually escalate to the national level (the partition of Congo, for example) or even internationally (the invasion of Kuwait by Iraq in 1990).

As mentioned above, the first econometric studies, using cross-sectional data cross-country studies supported the hypothesis that countries rich in natural resources had grown less than other emerging countries (Sachs and Warner 2001; Sachs and Vial 2002). However, subsequent studies using panel data and taking advantage of greater data availability to control for other variables have pointed in the opposite direction[6].

On the other hand, studies by Leamer and colleagues (Leamer et al. 1999) tend to suggest that achieving systematic economic progress depends heavily on the relationship with other factors of production. More than the abundance or scarcity of natural resources per se, what appears to be most decisive is how these natural resources are combined with other factors of production. In particular, the expansion of resource-based sectors whose exploitation is intensive in low-skilled labor helps lift many people out of poverty and accumulate human capital, which in turn facilitates the transition toward a more capital- and human-capital-intensive economy. Conversely, specialization based on natural resources that are highly capital-intensive (of all types) requires conditions that are often lacking in poor countries, a situation further complicated by the fact that it tends to exacerbate income inequality.

The following section aims to illustrate different types of specialization in selected emerging economies and examine their evolution over time, also including the case of some developed economies rich in natural resources. For this purpose, reference will also be made to the work of Leamer et al. (1999), who constructed a taxonomy to classify internationally traded goods. This classification focuses on the factors of production, which ultimately determine each country's comparative advantages. Although they define nine categories of goods, for the sake of simplity, these have been grouped into five: *raw materials*, which include minerals, fuels, and timber; *crops*, including grains, fruits, and others; *labor-intensive*, corresponding to manufactures such as textiles, clothing, and others; *capital-intensive*, which are manufactures such as machinery, equipment, and durable consumer goods; and finally, *other manufactures*.

[6] See Lederman and Maloney (2007) and Bravo-Ortega and De Gregorio (2005).

The second contribution from Leamer and colleagues being incorporated is an empirical definition of specialization, which is reflected in markets (including all their imperfections, regulations, subsidies, etc.) and is calculated based on the trade balance in each of these categories of goods. For example, a country that shows net exports of labor-intensive manufactured goods would be considered relatively specialized in that type of manufacture. Appendix 2 presents the precise definitions of the categories of goods, according to the United Nations SITC classification and the correspondence with Leamer's nine categories and the five used here.

First, Fig. 2.11 shows the case of four developed countries rich in natural resources (Australia, Canada, Norway, and New Zealand). In all of them the pattern of specialization did not change over a period of nearly 30 years. Nevertheless, these countries remained among those with the highest per capita income in the world.

Figure 2.12 shows the evolution of these variables for five Latin American economies, and although some changes are observed, they are not very significant, except in the case of Mexico.

In these charts, no significant changes are observed in the trade specialization of Chile, Colombia, and Peru. In Brazil, however, there is a change, mainly due to the discovery of oil fields that reversed the net balance in raw material exports, shifting from being a net importer to a net exporter of oil. Other changes relate to the declining importance of net exports of both labor- and capital-intensive manufactures. The latter remains positive, but the net balance is much smaller in relative terms. Mexico also underwent a change originating in the oil sector, but this was the opposite of Brazil: it ceased to be a net oil exporter due to the decline of this sector. Another important change in Mexico is the emergence of positive net exports in "other manufactures," a phenomenon associated with greater integration into manufacturing production chains, induced by free trade agreements with its North American neighbors.

Finally, it is interesting to observe what has been happening in Asia in recent years, with the most emblematic case being China, which, following its opening to international trade and accession to the WTO, has become a major global exporter of manufactured goods. Here, the abundance of cheap labor, the large size of the domestic market, and strong investment in infrastructure and human capital have all converged. All these factors have contributed to a profound transformation in the international specialization pattern of this economy, as shown in Fig. 2.13.

This figure shows that China has shifted from being an almost autarkic economy at the end of the twentieth century to one that is highly integrated into the global economy, with a high degree of specialization in all types of manufacturing and has become a net importer of raw materials and energy. It is very likely that in the coming decades, the net export balance grouped under "crops" will decline.

China's trajectory is not very different from those observed first in Japan and later in Korea and Taiwan during the second half of the twentieth century. All these countries faced a scarcity of natural resources and, at the onset of their industrialization, relied on abundant cheap labor. Over time, and thanks to massive efforts to

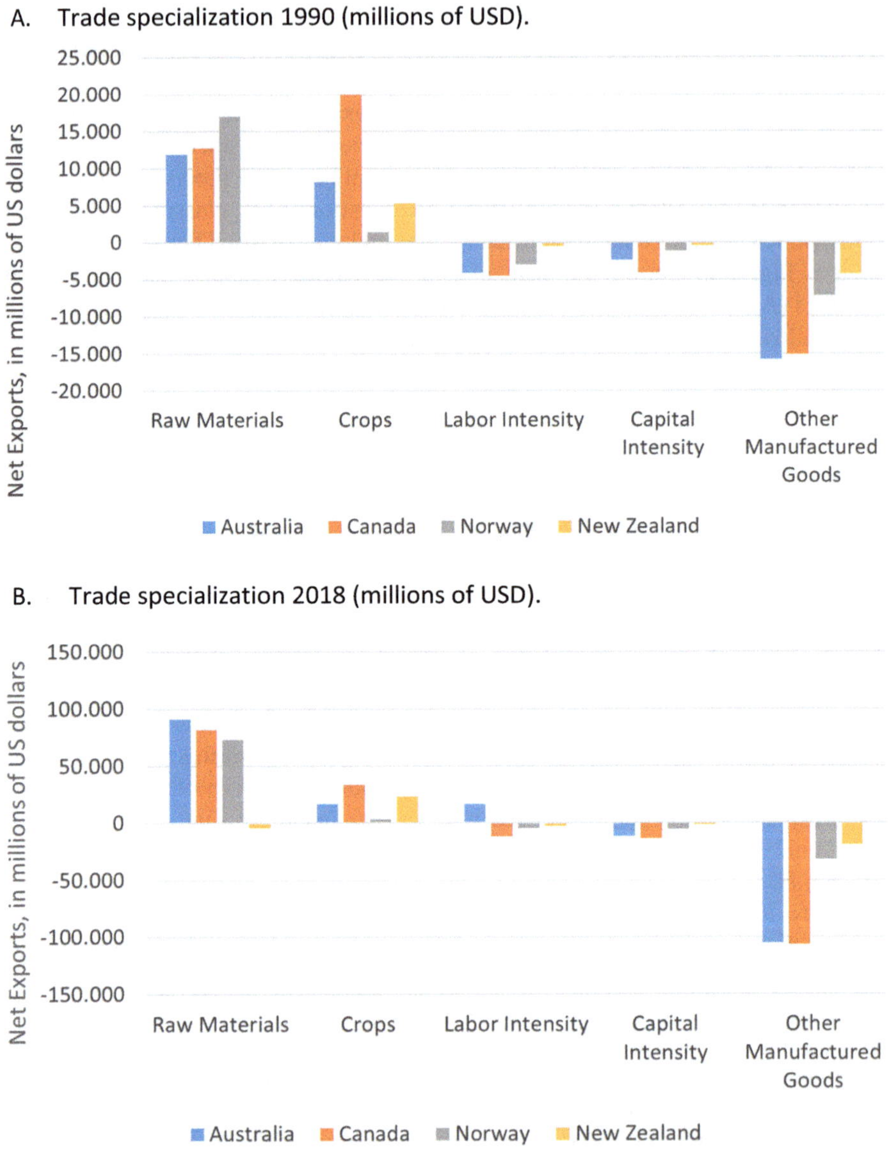

Fig. 2.11 Trade specialization of developed countries rich in natural resources. (**a**) Trade specialization 1990 (millions of USD). (**b**) Trade specialization 2018 (millions of USD). (Source: Own elaboration, WTO)

improve human capital, they were able to move beyond labor-intensive manufacturing to other types of manufacturing, including those that are capital-intensive.

The experience of Chile—a country that led trade liberalization in Latin America—and that of other resource-rich countries, suggests that it is difficult to change such specialization when it is driven by objective factors.

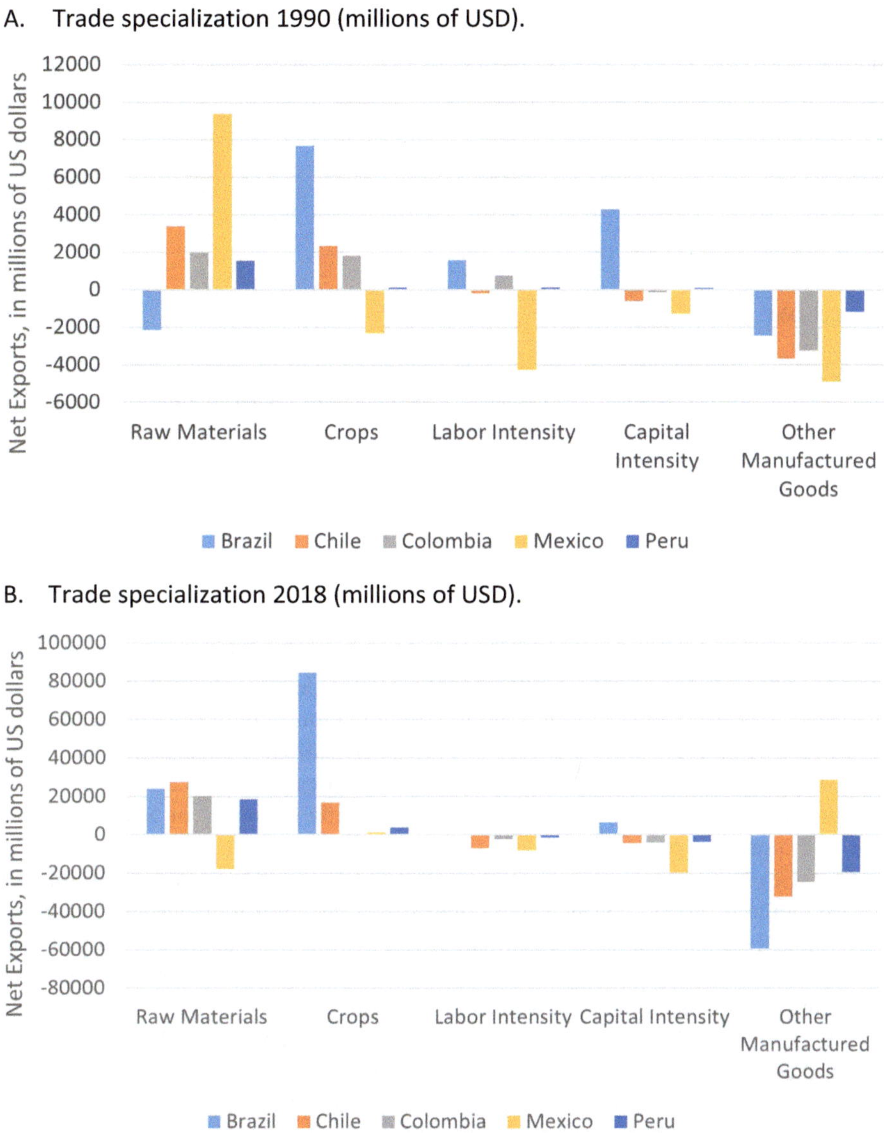

Fig. 2.12 Trade specialization of Latin American countries. (**a**) Trade specialization 1990 (millions of USD). (**b**) Trade specialization 2018 (millions of USD). (Source: Own elaboration, WTO)

Successful cases of changes in specialization have resulted from profound transformations, often linked to strong liberalization of international trade, enabling countries to better exploit their comparative advantages. When these advantages have been associated with an abundance of cheap labor, they have facilitated a process of industrialization that, when accompanied by significant investments in human capital, have led to fast and sustained development, as seen in Asia

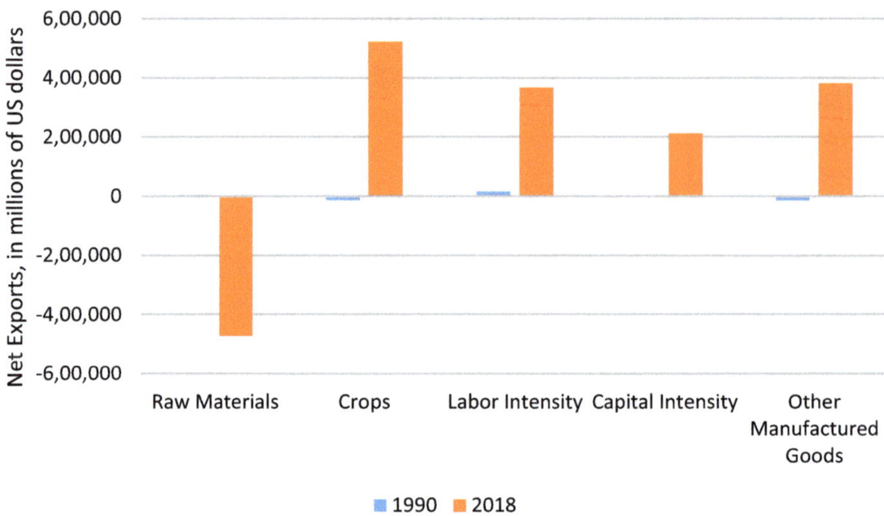

Fig. 2.13 Evolution of China's international specialization (millions of USD). (Source: Own elaboration, WTO)

throughout the twentieth century. However, when advantages have been more closely tied to an abundance of natural resources—especially those that are capital-intensive—trade liberalization has led to specialization in the exploitation of those resources, without a distinct path toward industrialization. This is particularly relevant for small countries that are distant from both destination markets and global manufacturing production chains, such as the Pacific countries of South America and central and southern Africa.

Historically, remoteness was an obstacle to trade, as it limited the possibilities for exchange to goods whose value could offset the high transportation costs by slow ships that, until the early twentieth century, were wind-powered and had very limited cargo capacity. This restriction was lifted by advances in maritime transport, including refrigerated cargo, as well as air transport.

The size of the domestic market also plays an important role. When markets are small, there are few opportunities to take advantage of the economies of scale that emerged with the industrial revolution, when the country is relatively isolated.

The combination of these factors hindered the development of specialized human capital, with the habits and skills required for manufacturing, so that even when advances in transportation significantly expanded trade opportunities for many goods, small and remote countries were unable to capitalize on them. In those cases where natural resources could support exports and generate foreign exchange income, that "natural" advantage proved unbeatable.

The Asian experience is remarkable because it demonstrated that there was an alternative path for poor countries without significant natural resources, by leveraging the abundance of low-cost labor, which in just over a generation was able to acquire the skills needed to produce competitively high-quality manufactures, allowing them to move up to increasingly sophisticated stages of production.

That experience is not replicable in resource-rich countries, as labor costs are comparatively high, precisely because of the exploitaition of their natural wealth. The major challenge for these countries is to sustain economic progress when the limits imposed by nature begin to constrain production expansion.

2.5 Environmental Impacts of Changes in the Global Economic Architecture

Since its origins, international trade has responded to the relative scarcity of resources among different countries. In the early days, and up to the Industrial Revolution, the main constraint on trade between cities and countries was the capacity of transportation, which greatly limited the range of tradable goods. Over long distances, such as those covered by mule or camel caravans, or sailing ships, small, lightweight, and high-value products like spices predominated. This constraint gradually eased, allowing for the exchange of higher-volume products.

With the advent of the Industrial Revolution and the use of new, high-power energy sources, trade expanded significantly, now including all types of manufactured goods and heavy, bulky raw materials such as minerals, timber, and grains. This, in turn, made it possible for manufacturing production to become physically separated from the availability of raw materials and natural resources, which had previously determined patterns of specialization.

This separation did not prevent some countries—such as Australia, Sweden, or Argentina—from achieving development levels like the richest nations at the beginning of the twentieth century, by meeting the raw material needs of these latter countries. In turn, they supplied their own domestic consumption of manufactured goods by importing them from the major industrial powers. In this context, it is not surprising that, by the mid-twentieth century, Marxist theories of class struggle within industrial societies were extrapolated to the relationship between rich countries, exporters of high-value manufactures, and poor countries that supplied them with (low-value) raw materials. Certainly, this paradigm began to falter when various Asian countries started to grow based on industrialization driven by cheap labor, and then rapidly moved up the value chain to more complex and valuable manufactures.

With the advance of industrialization and rising incomes, pressure on nature increased because of the growing demand for raw materials (animal products such as leather, grains, timber, minerals, and, later in the twentieth century, oil and natural gas to supply energy). As these resources became scarcer near the industrial centers of Europe, they began to be imported from elsewhere, giving rise to the specialization previously mentioned.

Until well into the twentieth century, there was little concern for the general state of nature, which was perceived as almost limitless. This perception began to change when the "population explosion" combined with the "economic miracles" of Europe and Asia. This concern has intensified as evidence of global environmental degradation has accumulated.

Although the destruction of emblematic animal species, such as large fauna in Africa and Asia, as well as the rapid disappearance of forests and woodlands, was already drawing attention, it was only at the end of the last century that strong concern emerged for the state of the remaining natural areas, located mainly in poor and sparsely populated regions of the world.

In more advanced countries, higher income levels, greater awareness of the problem, and the existence of stronger institutions have contributed to better containment of pressures on nature. This has also been possible in part because the expansion of the agricultural frontier ceased to be profitable and began to reverse as cheaper suppliers emerged elsewhere in the world.

Globalization has enabled more countries and people to participate in material improvements in living conditions, thereby extending and exacerbating environmental problems. However, globalization is the vehicle, not the original cause, of the environmental crisis.

Environmental degradation originates from the fact that markets often fail to signal the depletion of natural capital, either because there are no property rights over its services (air, the carbon and water cycles, for example), because property rights are acquired through exploitation (fish, water, forests in many cases), or because of negative externalities, such as air, water, and soil pollution, that exceed nature's absorptive capacity.

In the early days, when humanity was relatively dispersed and had limited mobility, these market failures were reflected locally, in declining crop yields, water pollution and the proliferation of infections resulting from the accumulation of all kinds of organic waste in soil and water. Increases in mortality served to keep population size within the limits of the local environment's capacity.

With the demographic, economic, and knowledge explosion following the Industrial Revolution, humanity acquired unprecedented abilities to control and modify its environment. International trade is one such manifestation. Others include advances in the development of sanitation infrastructure, the replacement of animal power with mechanical power in transport and production, advances in medicine, and many more. In practice, this meant that nature ceased to be the dominant constraint on the expansion of human activity.

Globalization is an integral part of this process and has played a key role in two respects: first, by enabling enormous productivity gains through new possibilities for division of labor, market expansion, and the explosive increase in real-time exchange of information and knowledge worldwide. Second, consequently, it has allowed for an extraordinary rise in per capita consumption levels in all regions of the world. Although nearly 15% of the world's population still lives in very precarious conditions, close to subsistence levels, it should not be forgotten that this was the situation for much of the world's population just over 50 years ago.

The impacts of this explosion in human consumption on nature are now felt everywhere on the planet, but they are most visible in the developing world.

In developed countries, nature changed radically in less than a century. The dense forests that once covered much of Europe and were a source of terror[7], had already been dramatically reduced by the mid-twentieth century, to the point that in less than a century they have become objects of veneration. The copper mines of Spain and Portugal remain only in the names of some multinational companies, and the coal mines of the United Kingdom, which fueled the industrial revolution, were exhausted by the end of the last century.

On the other hand, in many poorer countries that depend on raw material exports, there is still a strong dependence on nature because the subsistence and prospects for progress of their populations are closely tied to the jobs and income generated by the cultivation of grain and industrial fruit crops, industrial livestock production for human consumption, mineral extraction, and so on.

The most extreme case is that of countries that are no longer poor because they derive large revenues from oil extraction. Some of these, aware that they are rapidly consuming their natural capital, are attempting to replace it with produced capital and human capital that will allow them to maintain their standard of living once oil is depleted and the energy transition is realized.

Although these countries dependent on nature's services may differ greatly in their income levels and in their endowment of other forms of capital, their decisions are also distorted by the fact that market failures do not signal the depletion of their natural capital stock. This is leading them to continue consuming it at an unsustainable rate. Thus, we see fisheries being overexploited and forests in South America being "cleared" to expand cropland, just as the rainforests of the Asia-Pacific islands are destroyed for palm cultivation or, more recently, to open spaces for mining, and so it is across the entire planet.

On another front, the contribution of emerging countries to greenhouse gas (GHG) emissions has also increased enormously, to the point that China is now by far the largest emitter of these gases, and India is moving in the same direction, both driven by their heavy reliance on coal for energy generation. Amazonian countries, as well as those in Central Africa and the Asia-Pacific, also contribute to this pollution due to the release of carbon stored in their forests because of deforestation and land-use change.

It is easy to blame this environmental degradation on globalization. Indeed, without it, there would be no pressure on the remaining natural areas in these regions, nor would there be such GHG emissions, but the vast majority of the world's population would be poorer, and middle- and high-income consumers (currently the majority of the global population) would not have access to the food, vehicles, and smartphones produced with these raw materials.

International trade is the vehicle that enables the flow of goods and services that ultimately meet the needs of the growing number of increasingly affluent consumers around the world. Market failures that prevent "penalizing" these consumers for the excessive pressure they exert on nature's capacities are at the root of its

[7] This was reflected in children's stories that warned children of the dangers of venturing into unknown and threatening spaces (wolves, witches, and all kinds of mythological beings).

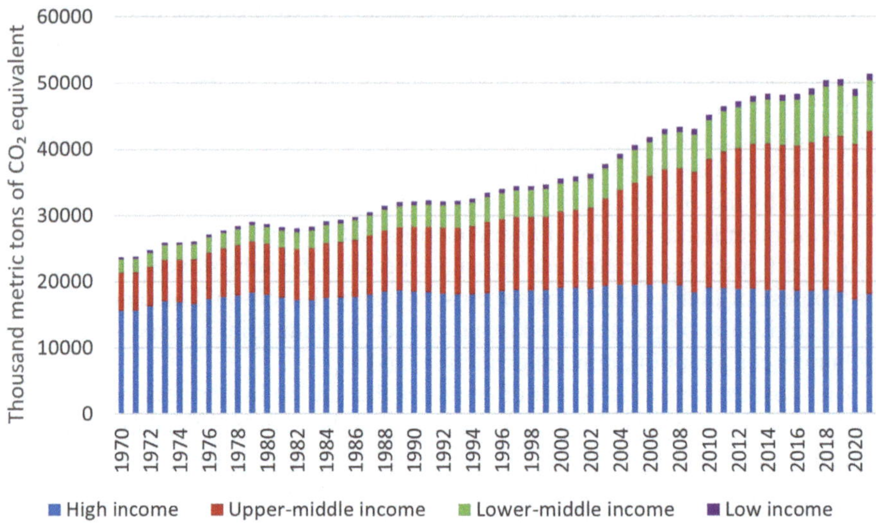

Fig. 2.14 GHG emissions by region, defined by income levels. (Source: World Bank: World Development Indicators)

degradation. When markets fail, it is up to countries' political systems and institutions to compensate for these shortcomings.

In most cases, the means to make these corrections exist, but so far, the problems arising from environmental degradation have not been severe enough to drive the behavioral changes needed to address them.

A reversal of globalization would result in lower global incomes, with significant setbacks in economic well-being in emerging countries. It should not be forgotten that this material progress has supported the increase in the world's population from just under 1 billion people a century and a half ago to 8 billion today. Without a global exchange network, it would be unthinkable to feed, clothe, and provide acceptable living conditions for so many people.

One of the consequences of generalized economic progress resulting from globalization is that the adverse environmental impacts that were once concentrated in the most advanced countries are now also a global responsibility.

Various indicators reflect this phenomenon. For example, the Human Footprint Project (Wildlife Conservation Society, n.d.) shows that most of the human population lives in areas where environmental damage exceeds the regenerative capacity of their ecosystems. The Living Planet Report 2024 (WWF 2024) informs a 73% decline in the populations of land species monitored in its Living Planet Index since 1970, although the latest measurements show stabilization.

The most informative global indicator of regional contributions to global environmental degradation is the evolution of GHG emissions, shown below.

Figure 2.14 shows that, despite all the agreements from Kyoto in the 1990s to the present, global GHG emissions have continued to rise rapidly. Moreover, it reveals a major shift in their composition: middle-income countries (especially upper-middle-income) have increased both the volume and their share of total emissions, surpassing those of the highest-income countries. In fact, the latter have slightly reduced their annual emission flows. This suggests that, looking toward a sustainable future, it is no longer enough for developed countries to make a rapid energy transition. Developing countries must also make a similar effort to achieve the common goal.

2.6 Why Is Global Economic Growth Needed Today?

This is a question that would have been unthinkable just a few decades ago. However, it is something that many young people, even in developing countries, are asking today.

If humanity's impact on the systems that make life on the planet possible is as great as described in the previous sections, this is a legitimate question and must be answered with more than just referring to the natural curiosity and ambition of the human species.

The main reason is that our species, in addition to being ambitious, is also compassionate. Evolution led us to act in groups and support each other to survive, sharing both efforts and rewards[8]. In this context, it is difficult to remain indifferent to the daily reality that many people and families live in misery, with survival as their most immediate concern. Poverty is distressing and painful because it goes against our deepest instincts, so, even without appealing to ethical values and principles, it is perceived as detrimental to society. Unfortunately, despite the enormous material progress of the past couple of centuries, there are still many people who are far from living under the conditions indicated by the averages.

To gauge the scale of this problem, one can begin by examining differences in average per capita income between countries, as discussed in previous sections. This yields a first relevant number: Fig. 2.5 in Sect. 2.2 showed that about 800 million people (10% of the world's population) live in countries where average annual per capita income is less than $1145, that is, just over $3 a day. It is virtually impossible for countries with this level of average income to meet the needs that are now considered minimally acceptable, ranging from nutrition to access to health care, housing, or education.

Figure 2.15a shows the map of countries worldwide classified by the average income level used by the World Bank. In most of Africa and much of South Asia, there are still many people living in conditions of extreme poverty, and their governments are unable to finance actions to help them overcome this situation.

While these differences are difficult to accept, it is worth comparing them to the situation just 32 years ago. In 1991, the situation was much worse, but thanks to economic growth, enormous progress has been made in the world's poorest regions.

[8] See Harari (2015).

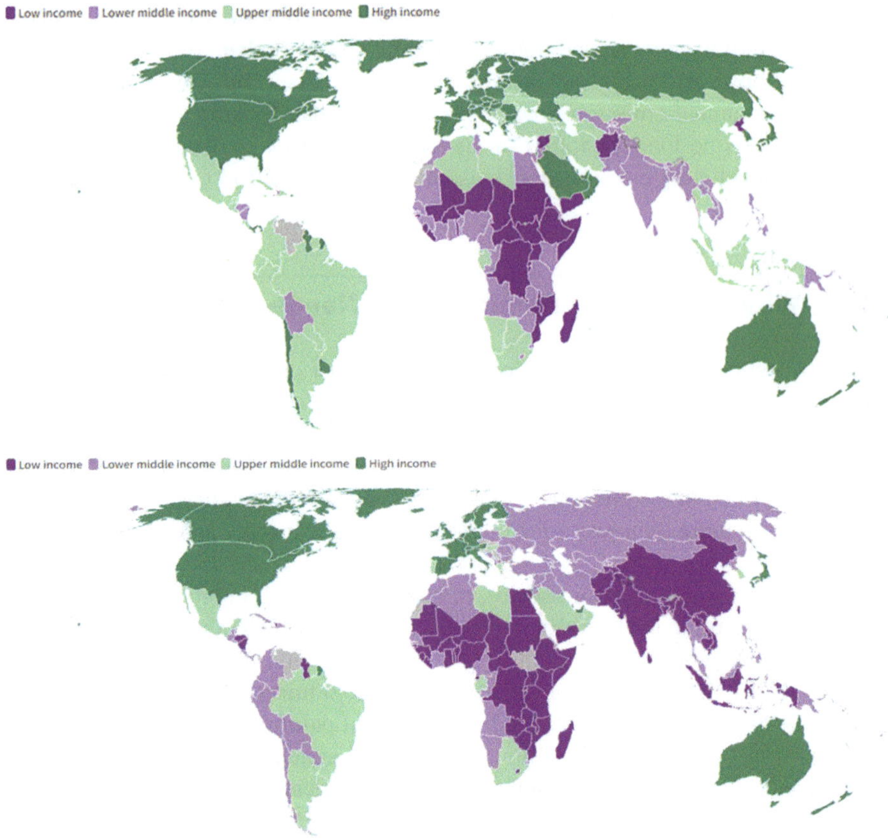

Fig. 2.15 (**a**) Map of countries by income group per capita in the world in 2023. (**b**) Map of countries by income group per capita in the world in 1991. (Source: World Bank. https://datatopics.worldbank.org/world-development-indicators/the-world-by-income-and-region.html)

Although growth in Latin America and Africa has not been sufficient to close the gap with the developed world, it has nonetheless made a huge difference in reducing poverty levels in these regions. For example, most of the population in South America has moved from living in lower-middle-income countries to upper-middle-income countries, and in a few cases, even to high average income levels.

However, the story of poverty is more complex than what is reflected by average incomes per country, as these do not account for the great diversity of living standards within countries. One way to go beyond average incomes is to analyze the extent of poverty in different countries. It is important to note that poverty measurement is a relatively recent development worldwide. In most of the developing world, this only began at the end of the last century. The basis for these studies are household surveys, which are expensive to conduct and require qualified human resources for their design and implementation, something scarce in poorer countries.

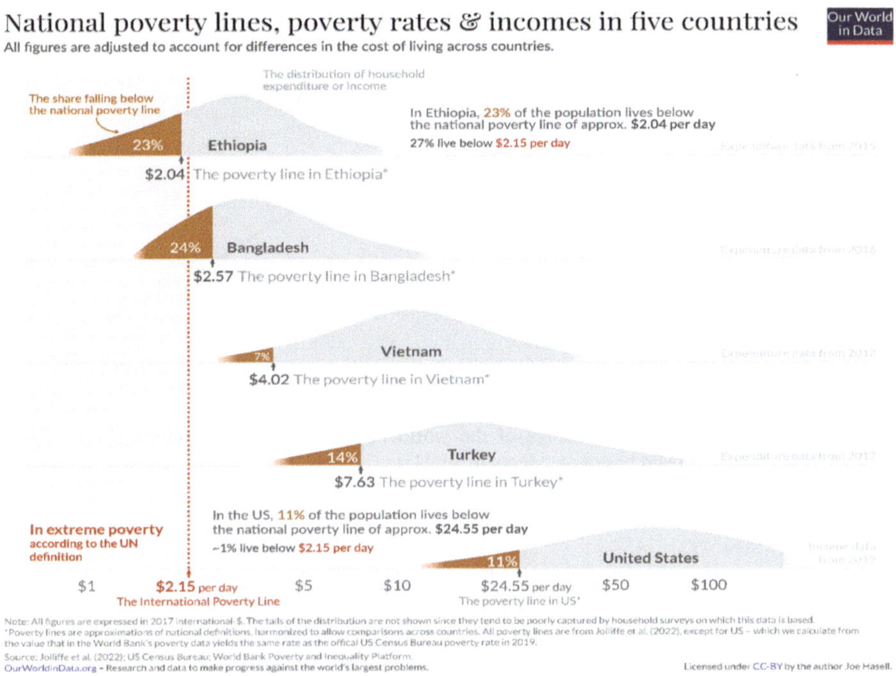

Fig. 2.16 Poverty lines in five very different countries. (Source: Our World in Data based on Jolliffe et al. (2022); US Census Bureau; World Bank Poverty and Inequality Platform. Own translation)

A commonly used measure is the percentage of the population living below the poverty line. Typically, each country defines its own poverty line, considering the cost of a basic consumption basket and household incomes. This is essential for the definition of social policies in each country, but it is not as useful for making cross-country comparisons: countries differ in household size and composition, as well as in consumption patterns and prices of the various items included. This becomes evident when comparing poverty lines across different countries, and even for the same country at different times (Hasell et al. 2022) as shown in Fig. 2.16.

An alternative way to represent this issue for comparisons among countries is to arbitrarily define a threshold in international dollars for a given year and then identify the percentage of the population falling below that threshold. For example, the World Bank uses an international poverty line of USD 2.15 per day for its international comparisons. According to this metric, among the five countries included in Fig. 2.16, only Ethiopia and Bangladesh have significant percentages of their populations living in extreme poverty.

A third option is to measure the percentages of the world population below various poverty lines, which provides an idea of global progress in this area, regardless of country. The most notable aspect of Fig. 2.17 is that, under any definition of the poverty line, the progress achieved since the second half of the twentieth century has been extraordinary.

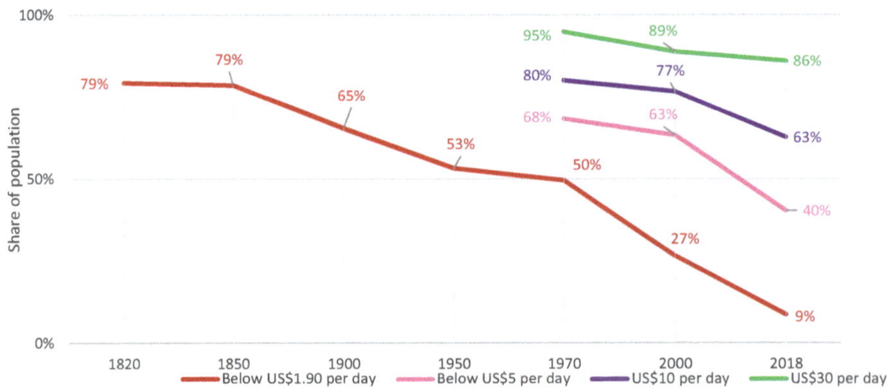

Fig. 2.17 Evolution of the percentage of the world population below different poverty lines. (Source: Our World in Data based on Moatsos (2021))

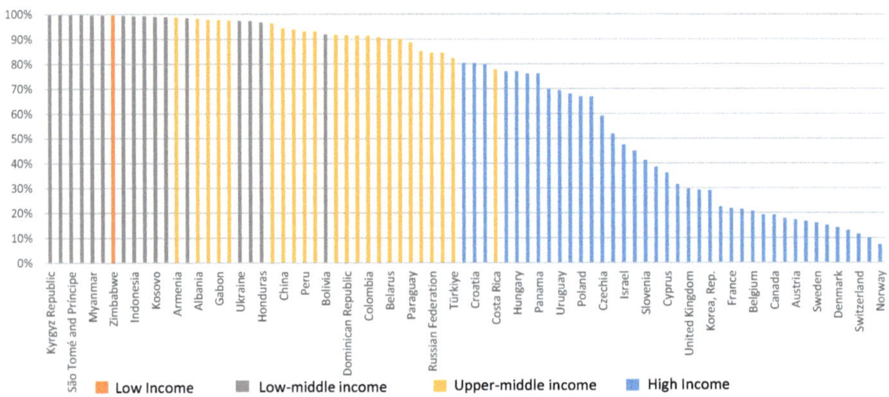

Fig. 2.18 Proportion of the population living below USD 30 per capita in 2017. (Source: Own elaboration based on information from Our World in Data, in turn based on World Bank data (PovcalNet))

The last figure presented here shows the distribution by country, ordered according to their average per capita income, of the population living below a poverty line somewhat higher than that defined for the United States.

Figure 2.18 shows, first, that the relationship between average income and lower incidence of poverty is unequivocal: there are no peaks or valleys in the proportion of the population living in poverty. Second, there is a clear discontinuity, and once a certain per capita income level is reached—around 45,000 PPP dollars in 2021—the percentage of the population living below the 30 dollars per capita line falls below 50%[9].

[9] This threshold should not be interpreted mechanically. It may well be that, if income distribution is highly concentrated, it becomes very difficult to advance beyond that average income level. Many countries that have reached middle-income ranges have fallen into the so-called "middle-income trap."

When income levels are translated into well-being indicators such as infant mortality, malnutrition, educational attainment, and many others, the conclusion is always the same: it is very difficult to achieve acceptable levels of well-being if a large proportion of the population lives in poverty. Without economic growth, the vast number of people still living in unacceptable conditions will not see improvements in their quality of life.

While it is true that the planet's capacity to sustain life is severely threatened by the enormous impact of consumption of material goods and energy, it is essential to maintain positive growth rates in the incomes of families concentrated in the world's poorest regions, so that all people can aspire to minimum levels of security and material well-being.

Of course, economic growth in the world's poorest regions must follow a different path than that of past successful cases, and for this it is crucial that they rapidly incorporate technological advances occurring in the most developed nations.

A key area for a sustainable convergence process is for the poorest countries to leapfrog the stage of dependence on fossil fuels by promptly adopting new renewable energy sources, especially when these become economically competitive. An interesting case to watch is the rapid penetration of electric vehicles in China, which are already entering international markets at prices competitive with conventional gasoline models. If this trend continues, there is a very real possibility that the poorest developing countries could bypass the stage of mass mobility based on fossil fuels. Moreover, many of these countries are in tropical regions with high solar radiation, which should facilitate electricity generation based on these resources.

Other key areas for action include improvements in water use systems to rapidly increase agricultural productivity, an essential requirement for reducing pressure on forests and wild areas. In this regard, it is very important that countries with large areas of wild forests focus their climate change mitigation efforts on preserving these territories and develop the capacity to attract financial resources as compensation for emissions reductions through this means. This could be complemented by biodiversity conservation in these areas if financial and certification mechanisms are developed for these purposes.

From the consumption perspective, progress could also be made if governments prioritize investments in infrastructure to develop efficient public transportation and large-scale freight movement, rather than focusing on building roads and highways that facilitate private vehicle use, for example.

In summary, growth in countries where a significant portion of the population lives in poverty, or is at risk of falling into it, is a necessity that cannot be ignored in civilized societies. However, this also requires making every effort to ensure that the environmental footprint of such growth is as small as possible, which in turn demands investment and cutting-edge technology, something that will not be possible without strong commitment from local authorities and a fundamental reassessment of development aid priorities.

References

Bravo-Ortega C, De Gregorio J (2005) The relative richness of the poor? Natural resources, human capital, and economic growth, Policy Research Working Paper Series 3489. World Bank
Diamond J (2019) Upheaval: How nations cope with crisis and change. Penguin UK
Diodato D, Hausmann R, Schetter U (2022) A simple theory of economic development at the extensive industry margin (CID Faculty Working Paper No. 416). Harvard Kennedy School. https://growthlab.hks.harvard.edu/files/growthlab/files/2022-09-cid-wp-416-extensive-industry-margin.pdf
Ferguson N (2006) The War of the World: Twentieth century conflict and the decline of the West. Penguin Press
Gruss B, Kebhaj S (2019) Commodity terms of trade: a new database. International Monetary Fund
Harari YN (2015) Homo Deus: A brief history of the future. Penguin Random House
Hasell J, Roser M, Ortiz-Ospina E, Arriagada P (2022) Poverty. Our world in data. https://ourworldindata.org/poverty
Hausmann R, Hidalgo CA (2011) The network structure of economic output. J Econ Growth 16:309–342
Jolliffe D, Mahler D, Lakner Ch, Atamanov A, Tetteh-Book S (2022) Assessing the impact of the 2017 PPPs on the international poverty and global poverty. Policy Research Paper 9941. World Bank
Leamer EE, Maul H, Rodríguez S, Schott PK (1999) Does natural resource abundance increase Latin American income inequality? J Dev Econ 59:3–42
Lederman D, Maloney WF (2007) Neither curse nor destiny: Introduction to natural resources and development. In: Lederman D, Maloney WF (eds) Natural resources: Neither curse nor destiny. World Bank, pp 1–14
Maddison A (2000) The world economy: A millennial perspective. OECD Development Centre Studies
Moatsos M (2021) Global extreme poverty: Present and past since 1820. OECD (2021), How Was Life? Volume II: New Perspectives on Well-being and Global Inequality since 1820, OECD Publishing, Paris
Pomeranz K (2000) The great divergence: China, Europe, and the making of the modern world economy. Princeton University Press
Prebisch R (1959) Commercial policy in underdeveloped countries. Am Econ Rev 49(2):251–273
Pritchett L (1997) Divergence, big time. J Econ Perspect 11(3):3–17
Sachs JD (2015) The age of sustainable development. Columbia University Press
Sachs J, Vial J (2002) Can Latin America compete? In: Vial J, Cornelius P (eds) The Latin American Competitiveness Report 2001–2002. Oxford University Press
Sachs J, Warner AM (2001) The curse of natural resources. Eur Econ Rev 45(4–6):827–838
Williamson J (1990) Latin American adjustment: How much has happened? Institute of International Economics
World Bank (2024) The changing wealth of nations: Revisiting the measurement of comprehensive wealth. World Bank
WWF (2024) Living Planet Report 2024: a system in danger. WWF

Sustainability and Natural Capital

3

Abstract

The concept of Natural Capital facilitates the re-introduction of natural constraints in the analysis of economic development. This allows us to bring the notion that nature is a limited asset that provides essential services for human well-being and survival. This is important to reincorporate Nature's limitations into the standard framework of economics, the discipline studying and providing advice on how to address scarcity. One step in this process is to put values on the services provided by Nature, something relatively straightforward for those materials, from water to minerals, providing marketable inputs for production processes, but far more challenging for those that are non-marketable but play a critical role in maintaining the integrity of ecosystems. When thinking about natural resources, Economic Theory should incorporate their heterogeneity and limited availability, so many of the standard assumptions must be revised and models must be adapted to deal with issues such as rents and declining productivity. Related issues such as the substitutability of natural and produced capital, and the related concept of weak sustainability are also addressed in this chapter. Methodological challenges to measure and set values to natural capital are also discussed here.

Keywords

Natural capital · Comprehensive wealth · Economic rents · Sustainable development · Valuation of ecosystem services

3.1 Sustainability and Comprehensive Wealth

The traditional measure of economic performance is GDP. As discussed in the previous chapter, this measure leaves much to be desired when analyzing whether such performance is sustainable over time, since it is a "gross" measurement that does not include the depreciation of physical capital. Strictly speaking, then, one should

J. Vial, *Sustainable Development and Natural Resources*, Integrated Science 39,
https://doi.org/10.1007/978-3-032-12263-6_3

start by considering Net National Product (NNP) even in the most standard version of national accounts. But even this concept is insufficient because it does not include the depreciation of human capital or natural capital, which is not surprising since neither is measured in these versions of national accounts.

One way to address this problem is to calculate what is known as Genuine Savings, that is, Net National Savings accounting for the depreciation of traditional capital, as well as human capital and natural capital.

For produced capital, there are standard estimation techniques, and its calculation is routinely included in national accounts. With respect to human capital, aggregate depreciation estimates are not usually made, since it is assumed that new generations are acquiring more knowledge and skills than those retiring, so it would not be a particularly relevant issue. On the contrary, it is usually assumed that human capital per capita tends to grow precisely for those reasons, and this customarily included as corrections to the flow of human labor in the traditional estimation of aggregate production functions. However, the combination of population aging and migratory movements suggests it would be advisable to incorporate these phenomena into the measurements.

Finally, regarding natural capital, measuring its depreciation is more complex and requires distinguishing at least three types of services provided by nature that contribute to human well-being:

- Provisioning services, which are natural resources such as agricultural land, water, minerals, various energy sources, fish, etc., that are used directly by humans to meet their material needs. Most of these services are appropriable and traded in markets, and their economic value can be determined based on the net income they generate over their useful life. As these resources are depleted or overexploited, this should be considered a form of depreciation (or consumption) of natural capital.
- Supporting and environmental regulation services, which allow nature to regenerate its capacity to process waste, maintain ecosystem integrity, and help sustain the functioning of systems that make life on the planet possible. Many of these services are not appropriable and fall under the category of "externalities," making them very difficult to value. On the other hand, the level of scientific knowledge available to identify the interrelationships between species and ecosystems is insufficient for the purpose of valuing them. Here, too, a kind of depreciation occurs, either through the destruction of ecosystems or through the deterioration of their capacity to provide ecosystem services because of pollution or the destruction of key components. Starting with the pioneering work of Daily and co-authors (Daily 1997) there has been a significant effort to design and apply novel methodologies to value these services.
- Cultural services, which comprise the contributions to human well-being derived from the existence of nature, whether for aesthetic, religious, or psychological reasons. These are particularly difficult to value because they are inherently subjective. Normally, these services are not depleted by this type of use, but they can certainly be devalued to the extent that human activities alter their essential characteristics.

Setting aside for the moment the problem of measuring the economic value of natural capital depreciation—which will be addressed later—the calculation of Genuine Savings is a first step toward assessing the sustainability of the economic performance of a country or group of countries.

If such savings are positive, this means that what businesses, households, and the government of a country are saving is sufficient to offset the depletion of all forms of capital, including the depreciation of natural capital associated with the extraction of non-renewable resources, among other factors. This is what is commonly known as Weak Sustainability. Why this qualifier? Simply because it assumes that if a resource is being depleted, its scarcity can be compensated in other ways, either by importing it from other countries or by increasing contributions from other forms of capital. In general, this concept applies reasonably well to the analysis of the sustainability of provisioning resource use at the national or local level, but it is much more debatable to extend it to other ecosystem services, where the possibilities for substitution are far more limited[1].

In any case, this approach to the problem suffers from the flaw of remaining focused on the evaluation of flow variables: NDP, savings, and depreciation. As Dasgupta (2021) has pointed out, the problem of sustainability is associated with the preservation and increases in the stocks of the various forms of capital. This is why Dasgupta has proposed shifting the focus of sustainability analysis to the evolution of Comprehensive Net Wealth, in the sense that it incorporates all forms of capital. This is what the World Bank and various international agencies involved in the definition of statistical standards have been putting into practice in recent decades, materialized in the Changing Wealth of Nations (CWON) project, which presents the best existing international estimates of these variables.[2]

Does this mean that traditional national accounts, and in particular the use of GDP to assess economic performance, should be discarded? The answer is a categorical no: the analysis of flows is essential for evaluating the performance of the economy in the short and medium term, which was the original motivation of those who designed and built the statistical foundations of national accounts. The focus of macroeconomic policy remains the smoothing of economic cycles and the prevention of crises over a short- and medium-term horizon, which is most relevant for the decisions of households, businesses, and government. Traditional national accounts are irreplaceable for these purposes.

However, smoothing the cycle does not ensure the sustainability of economic well-being over longer timeframes, since these depend on the evolution of net assets, that is, comprehensive wealth. Just as distance is measured in meters and kilometers, and weight in kilograms and tons, GDP and national accounts are used to assess economic performance over the cycle, while Comprehensive Wealth should be used to assess the sustainability of economic well-being over longer periods. Both are necessary and complementary.

---

[1] For a discussion of this topic, see World Bank (2024).

[2] For further details on the methodology and results, see: https://documentos.bancomundial.org/es/publication/documents-reports/documentdetail/099101024111541936/p17844618018390a41b7e61b28be7f722d1

As will be seen later in the analysis of Genuine Savings and Wealth estimates developed in recent decades, there has been more progress in valuing the provisioning services of natural capital. While there have been advances in valuing some supporting and regulating services, there is still considerable work to be done in this area, and regarding cultural services, there is little to show so far.

This means that there is still a long way to go before we have reliable measures of Comprehensive Wealth[3] at the country or global level, but the growing interest in seeking information and developing measurement methodologies, as well as the resources of all kinds devoted to these purposes, suggest that significant progress can be expected in the coming years.

Nevertheless, what is already being done is useful for addressing one of the original concerns in the study of these issues: to what extent do natural resources (that is, the provisioning services of natural capital) allow for the maintenance or increase of the material well-being of future generations?

Of course, if the systems that support human life on the planet collapse, it matters little that natural resources are sufficient, but answering this question at least addresses the main concern of Malthus, as well as the authors of "Limits to Growth" (Meadows et al. 1972).

3.2 What Is Special About Natural Resources?

Natural resources such as minerals, hydrocarbons, forests, water, arable land, fish, and others are very different, but they are provided by nature, meaning that they are heterogeneous in their characteristics, their availability is conditioned by the environment, and they are subject to appropriation, so there are markets where they are traded. These services that humanity obtains from the environment are called provisioning services, precisely because they are used or transformed into materials to meet the material needs of the specie. In contrast, there are other ecosystem services that enable their own subsistence, which often also benefit humanity, but which are not appropriable and therefore are not traded in markets and do not have observable prices[4].

In many cases, the homogeneity of some inputs based on natural resources is confused to the point of calling them "commodities," overlooking the fact that they are obtained from highly heterogeneous materials.

To illustrate this with an example, consider the case of metal markets such as copper, where tons of refined copper with very specific characteristics of purity and homogeneity are traded. This is essential for their efficient use in industrial processes to produce various types of electrical cables and other intermediate goods.

[3] Dasgupta, in his report on the Economics of Biodiversity, referred to Inclusive Wealth. Subsequently, the World Bank has replaced this with Comprehensive Wealth to avoid possible confusions arising from the use of the term "Inclusion," which in current language has a connotation very much focused on human diversity.

[4] As will be discussed later, there are sometimes tensions arising from the fact that some natural resources are critical components of ecosystems. Forests are among the clearest examples.

However, at their origin are copper deposits distributed around the world with different grades of purity, different combinations with other minerals—some very valuable, such as molybdenum, and others detrimental, such as arsenic. Each location has its own environmental characteristics: in Zambia, large underground copper mines must deal with excess water that threatens to flood the shafts. In the deserts of northern Chile, water scarcity is a problem for metallurgical processes to process the material extracted from the mines. Furthermore, location plays a key role in transportation costs, as these are heavy and bulky loads. Returning to the same example, Zambia has no direct access to the sea and lacks good railway networks to the ports of neighboring countries, so most transport is by truck on poor-quality roads. In Chile, the major copper deposits are near the sea, and the volume of cargo allows for the financing of railway operations or even the use of "mineral pipelines" to send the load to ports.

This heterogeneity means that the costs faced by producing companies differ, and the technologies used to extract the resource and convert it into homogeneous goods for international markets are also different. Moreover, in all these cases, costs increase as the resource is extracted. Graphically speaking, in any mine, each additional ton of material extracted is farther from the surface and tends to have a lower mineral content than the previous ones.

Since all producers sell at the same price, companies with lower costs obtain economic rents, that is, revenues that go beyond what is necessary to finance the inputs and factors (including capital) used to generate those revenues. These rents are due less to innovation or effort by the companies and more to the characteristics of nature itself. This is very different from what is observed in normal industrial processes, where location can be adjusted to the needs of inputs or workers, or to proximity to consumer centers. Production can be expanded or reduced by adjusting the quantities of capital, labor, and raw materials (homogenized in previous stages) so that costs remain approximately constant. In these cases, rents are eliminated by competition, and factors such as innovation in production or in the attributes of final goods are decisive for generating profits (quasi-rents) that eventually dissipate as competitors adopt them.

In any case, rents are not immutable. As natural capital is consumed or degraded through use, operational costs and the need for investment in productive capital to compensate for this deterioration increase, and rents begin to decline. Discovery of additional resources through exploration and innovation in technologies or in new ways of organizing production and the use of inputs also plays an essential role in maintaining productive capacity and "revaluing" natural capital, by allowing more provisioning services to be extracted from the same physical stock of natural capital.

The existence of rents linked to the characteristics of nature underlies the legitimacy problems often faced by companies focused on the use of natural resources. In most countries in Latin America, Africa, and Asia, the ownership of these resources, or at least the rights to exploit and use them, originated in privatization and allocation during colonization, or are based on rules established by colonizers and their descendants.

Although today the exploitation of natural resources requires large initial investments—whether in land preparation for crops, exploration for mineral resources, or increasingly sophisticated capital goods and technologies—this does not prevent a significant portion of the population in many of these countries from perceiving that, at the very least, that natural wealth rightfully belongs to the countries where the resources are located. In practice, this means that the issue of the distribution of natural resource rents has an important place in public debate, becoming a source of instability in the rules of the game and social conflict.

Certainly, this same phenomenon is reproduced within companies, where the struggle between unions and employers for the capture of rents is often a matter of great importance.

The fact that the location of natural resources is specific to a particular place defined by nature also has implications for their environmental consequences. Large-scale agricultural, forestry, or mining operations, for example, can have major impacts on environmental quality and the living conditions of local populations. Displacement of families, pollution, truck and machinery traffic, competition for water or other natural services (such as grazing land), can trigger conflicts that can end in high compensation or damage mitigation costs. They vary greatly between different activities and tend to resurface from time to time.

Another characteristic of natural resources is that they are limited, both at the level of a country or other jurisdiction, as well as globally. This limitation at the national or global level can be relaxed by the incorporation of new resource sources because of technological advances. The extraction of natural gas and oil from shale, or the eventual exploitation of mineral deposits on the ocean floor, are examples of this.

However, at the level of individual operations, such as a mine or a plot of farmland, the natural limitation plays a critical role. Investments in exploration may lead to the discovery of new reserves and thus extend the useful life of a mine, similar to the acquisition of adjacent land to expand a crop or plantation, but ultimately there are natural limits, which tend to be more pressing in the case of resources that are depleted through exploitation, such as minerals and hydrocarbons[5].

This, in turn, leads to an analysis of how to address the closure of operations, both from the perspective of society as a whole and of the operation itself. When it comes to large-scale operations with significant economic and social impacts at the local and national levels, the issue of closure often involves the relevant political authorities, adding an additional layer of complexity. In any case, this also occurs with large industrial or commercial operations not linked to natural resources. What is specific to natural resources is that, in many cases, these operations are the sole reason for the population to be in their vicinity, as is the case with a mining facility

[5] In practice, depletion or extinction of the resource rarely occurs; rather, the exploitation of the remaining stock ceases to be profitable or lies beyond the capabilities of available technologies.

in the desert or mountains. Therefore, their depletion leads to the relocation of the population, unless other independent sources of employment have emerged in the meantime.

3.3 Measuring Natural Capital: An Unfinished Agenda[6]

To assess the long-term sustainability of individual countries or the world, it is necessary to measure Net Wealth, which implies valuing produced capital, human capital, and natural capital, at a minimum[7].

In an initial stage, efforts focused on measuring large aggregates based on the calculation of Genuine Savings for countries, inspired by the pioneering work of Richard Peskin (1976) and Salah El Serafy (1989). These calculations centered on valuing the stock of natural resources that provide provisioning services, adjusting for their "depreciation" resulting from exploitation, as well as for increases in reserves or revaluation of existing assets. Currently, the World Bank regularly publishes measures of Adjusted Net Savings, which include deductions for depreciation of produced capital, depreciation of natural resources, environmental damage costs, and add education expenditure as a share of GDP as an approximate indicator of human capital growth. Even so, these figures have significant gaps, as, for example, they do not include depreciation of mineral capital (other than the extraction of coal and hydrocarbons).

Figure 3.1 shows the levels of gross and net savings (adjusted for depreciation of certain components of natural capital) for countries grouped by income level in 2019. It is worth noting the greater relative importance of natural capital consumption in low- and lower-middle-income countries, which are heavily influenced by deforestation and pollution.

Previous estimates that included calculations of mineral and fisheries capital, whose methodologies and data are under review, also show the significant weight of natural capital and its depreciation in many middle- and low-income countries (Vial 2023a).

The methodology for calculating the value of different types of assets is based on the classical principle that this corresponds to the present value of future net income flows. In the case of produced capital, this requires estimating the flow of future profits, which is normally associated with a rate of return representative of the opportunity cost of capital. The flows of future labor income are estimated based on

[6] For a discussion of these topics, see Daily, G. (1997), particularly Chap. 3 of that volume by Goulder and Kennedy (1997).

[7] Dasgupta himself (2021) analyzes the importance of institutions and the way societies operate, what some refer to as "social" capital, as determinants of economic development per se. D. Acemoglu, Johnson, and J. Robinson were awarded the Nobel Prize in Economics in 2024 precisely for their contributions in this field, whose importance is undeniable. Its omission from Net Wealth is due to the difficulty of measurement.

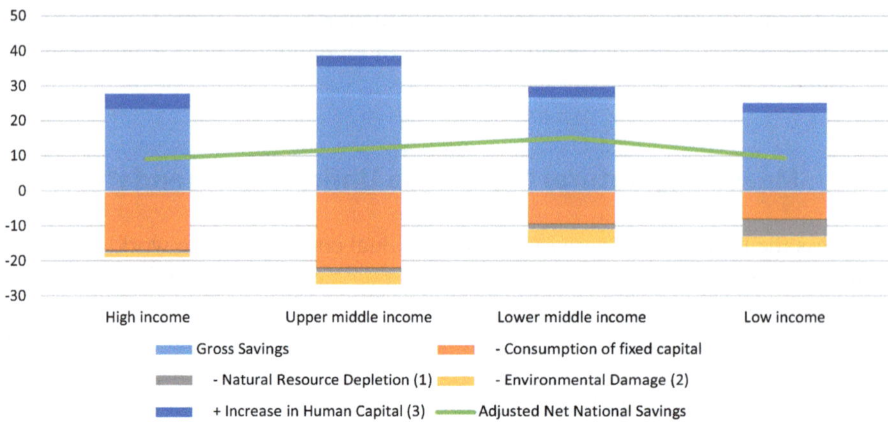

Fig. 3.1 Composition of Gross National Savings and Adjusted Net Savings by countries grouped by income level in 2019 (% of GDP). (Source: World Bank: World Development Indicators)
Note:
1. Includes change in forest mass and fossil energy
2. Includes costs of CO_2 emissions and particulate emissions
3. Includes education expenditure as a percentage of GDP

assumptions about wage trends, drawing on historical trends in labor productivity growth.

The case of natural capital is more complex. For provisioning services, it would suffice to project the rate of change of unit rents into the future, which requires starting from an estimate of current rents, where operating costs and the opportunity cost of capital must be subtracted from revenues from the sale of these product. The estimation of the cost of capital is not trivial, due to the reasons explained in the previous section. The future projection of these unit rents is also a complex exercise: if an approach based on the Hotelling rule is adopted, where the consumption of various forms of natural capital reduces their future availability, this should be reflected in greater scarcity and, therefore, higher unit rents. On the other hand, technological advances may allow the exploitation of resources that were previously unprofitable or extend the operational lifespan by lowering costs, which in turn would generate new rents, even in older operations.

But this is the easy part of the problem, because there are market prices for all these variables. The problem becomes truly complicated when it comes to estimating the value of ecosystem services that do not have market prices. This is a complex issue with profound ethical dimensions (Does the value of an ecosystem or a species depend exclusively on its contribution to human well-being? How should differences in utility among different people be weighed?), as well as technical difficulties arising from the lack of knowledge about the consequences of interventions in ecosystems (What happens if a wetland is drained to build housing? What happens to people's quality of life if potentially harmful medicines are applied to prevent diseases in livestock?)[8].

Over the past decades, general principles and practical approaches have been developed to value ecosystem services, and these have gradually permeated the estimates regularly produced by the World Bank in the context of the CWON project. However, there are still too many gaps in scientific information to obtain comprehensive measurements of these values.

These valuations, like those carried out by the World Bank, follow the usual practice of considering the value of marginal changes in resources and ecosystems. This implies that as an ecosystem degrades and part of the flow of services it provides is lost, the unit value of these services increases, either because it is more expensive to replace them by artificial means, or simply because the costs of losing them rise. For example, if a mangrove deteriorates, the biodiversity it harbors will decrease, and the risk of damage from storm surges will increase. Consequently, the income of fishers who depend on catching fish and shellfish will decrease, and the economic losses from the destruction of buildings, or even human lives, caused by more damaging storm surges will increase. Therefore, it cannot be inferred that the values thus calculated represent the cost of the disappearance of ecosystems: if one disappears and is indispensable for supporting life, its value would be infinite.

Taking all these limitations into account, it is nevertheless useful to briefly review the main results emerging from the latest update of CWON (World Bank 2024).

A first result to highlight is that, although per capita wealth worldwide has systematically increased over the past 30 years, this is mainly due to the increase in produced capital and human capital. Surprisingly, per capita exhaustible natural capital has remained approximately constant, while there has been a sharp decline in renewable natural capital, as can be seen in Fig. 3.2.

What explains this apparently anomalous evolution of natural capital? In the case of non-renewable resources, it is mainly the incorporation of new reserves of mineral and energy resources as a result of technological advances (Vial 2023b), while the decline in renewable resources is due to market failures arising from the absence of property rights over free-ranging fish in the case of marine resources, and competition with more profitable uses (at market prices) of land in the case of forests, mangroves, and grasslands. Underlying this latter problem are externalities and the pressure for present consumption, especially among those living in poverty and dependent on nature to meet their basic needs (Vial 2024).

---

[8] Regarding this last question, an interesting New York Times report on a particular case in India is recommended reading: https://nam10.safelinks.protection.outlook.com/?url=https%3A%2F%2Fwww.nytimes.com%2F2024%2F07%2F29%2Fclimate%2Fvultures-india-poisoning.html%3FsearchResultPosition%3D1&data=05%7C02%7Cjvialu%40uc.cl%7C3577e7b7d72e4723dce808ddca2aee6c%7C5ff5d9faf83f4ac1a4d2eb48ea0a00d2%7C0%7C0%7C638889007920600645%7CUnknown%7CTWFpbGZsb3d8eyJFbXB0eU1hcGkiOnRydWUsIlYiOiIwLjAuMDAwMCIsIlAiOiJXaW4zMiIsIkFOIjoiTWFpbCIsIldUIjoyfQ%3D%3D%7C0%7C%7C%7C&sdata=vAIJ9m%2FNQfZV1d5MXqSVwKN4XLSIVKwC47fntMfk%2BWY%3D&reserved=0

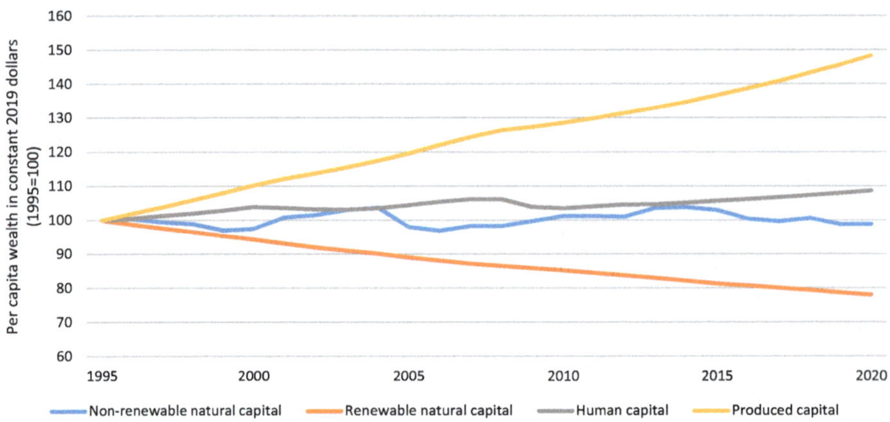

Fig. 3.2 Evolution of Comprehensive Wealth per Capita between 1995 and 2020. (Source: World Bank (2024). Reproducibility package for The Changing Wealth of Nations 2024—Revisiting the Measurement of Comprehensive Wealth. World Bank)

When examining the evolution of the different components of comprehensive wealth per capita, according to groups of countries classified by income level, several interesting phenomena can be observed in Fig. 3.3:

- Produced capital is the component of wealth that increases the most for all country groups, being relatively less important only in the case of high-income countries.
- Human capital increases in all country groups, being the second most important factor in wealth growth in all groups except the high-income group, where it is the fastest-growing component.
- Renewable natural capital per capita declines in all country groups, but this decline is more pronounced in low- and lower-middle-income countries.
- In the case of per capita exhaustible natural capital, there is a marked disparity between income groups: in low- and lower-middle-income groups, it falls significantly, indicating that they are financing part of their (low) per capita consumption by using up their stocks of non-renewable resources. On the other hand, high-income countries and, to a lesser extent, upper-middle-income countries experience an increase in their wealth in non-renewable resources.
- Finally, when focusing on total wealth, it is evident that only the upper-middle-income country group is closing the gap with higher-income countries, while the poorest countries essentially maintain the same initial differences,

In parallel with these "bottom up" natural capital measurements, which start from the different types of capital in individual countries, Costanza and collaborators (Costanza et al. 1997) have also been developing efforts to measure the value of ecosystem services at the global level. To do this, they identified the main types of ecosystems within the major biome categories and, using studies on ecosystem services from various locations around the world, calculated service flows and their

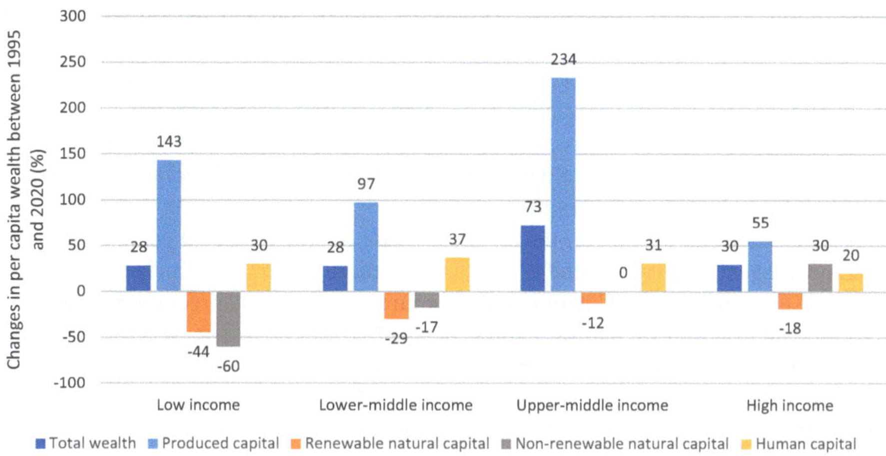

Fig. 3.3 Evolution of wealth components per capita, by country groups according to income levels. (Source: World Bank (2024). Reproducibility package for The Changing Wealth of Nations 2024—Revisiting the Measurement of Comprehensive Wealth. World Bank)

Table 3.1 Valuation of global ecosystem service flows

Biomass	Area billion ha. $(10)^9$	Annual flow value trillion USD 2007/ year $(10)^{12}$
Marine	**36.3**	**49.7**
Open Ocean	*33.2*	*21.9*
Coastal	*3.1*	*27.8*
Terrestrial	**15.3**	**75.1**
Forests	*4.3*	*16.2*
Grasslands	*4.4*	*18.4*
Wetlands	*0.1*	*26.4*
Other	*6.5*	*14.1*
TOTAL	**51.6**	**124.8**

Source: Costanza et al. (2014)

unit values, which were then extrapolated to global biomes. Later, in 2014, they updated the results, which changed significantly thanks to a much larger base of studies and, at the same time, a significant increase in the estimated unit values for these services. Table 3.1, taken from Costanza et al. (2014), summarizes the main quantitative results of the update carried out that year.

To provide some sense of scale, 124.8 trillion 2007 US dollars are equivalent to approximately 4.5 times the estimated value of global GDP for that year and are much higher than the World Bank estimates, which are conducted at the country level and do not have full coverage of countries or biomes.

Building on this approach, the World Bank, in collaboration with teams from several universities, has advanced the development of global models to assess the economic impacts of environmental degradation under different development pathways and mitigation actions (Johnson et al. 2021). These studies have highlighted the importance of actions that developing countries can take to prevent further degradation, especially regarding biodiversity protection and the preservation of ecosystem services.

The measurement of natural capital is an ongoing project, in which methodological challenges still need to be addressed, even to obtain measurements of capital associated with natural resources that provide provisioning services. However, the greatest challenge lies in the valuation of ecosystem services, especially those associated with biodiversity preservation.

Notwithstanding the above, the progress made over the past 30 years is remarkable and suggests that by the end of this decade, there should be a fairly comprehensive measurement of the value of natural resources and their depreciation, as well as a significant fraction of natural capital derived from ecosystem services that are directly linked to human well-being in the short term, such as soil protection, pollination, water regulation and purification, waste absorption and degradation, among others. While it will be very difficult to value services that are still "invisible" to humans, it seems reasonable to expect that, in the not-too-distant future, sufficiently robust environmental sustainability indicators will be available to support judgments and recommendations for the design of public policies at the country level.

References

Barry JM (2021) The great influenza: the deadliest pandemic in history. Capitán Swing Libros

Costanza R, d'Arge R, de Groot R et al (1997) The value of the world's ecosystem services and natural capital. Nature 387:253–260. https://doi.org/10.1038/387253a0

Costanza R, de Groot R, Sutton P, var den Ploeg S, Anderson ShJ, Kubiszewski I, Farber S, Turner T (2014) Changes in the global value of ecosystem services. Global Environ Change. 26:152–158

Daily G (1997) Nature's services: Societal dependence on natural ecosystems. Island Press

Dasgupta P (2021) The economics of biodiversity: the Dasgupta review. HM Treasury

El Serafy S (1989) The proper calculation of income from depletable natural resources. In: Yusuf A, El Serafy S, Lutz E (eds) Environmental accounting for sustainable development. World Bank

Goulder LH, Kennedy D (1997) Ch. 3: Valuing ecosystem services: Philosophical bases and empirical methods. In: Daily GH (ed) Nature's services: Societal dependence on natural ecosystems. Island Press

Johnson JA, Baldos UL, Cervigni R, Chonabayashi S, Corong E, Gavryliuk O, Hertel TW, Nootenboom C, Gerber JF, Ruta G, Polasky S (2021) The economic case for nature: a global earth-economy model to assess development policy pathways. World Bank. https://hdl.handle.net/10986/35882

Meadows D, Meadows D, Randers J (1972) The limits to growth. Potomac Associates

Peskin HM (1976) A national accounting framework for environmental assets. J Environ Econ Manag 2:255–262

Vial J (2023a) Desafíos para un Desarrollo sostenible de Chile en el Siglo XXI. Ediciones UC

Vial J (2023b) Desarrollo sostenible y capital natural. https://scioteca.caf.com/handle/123456789/2030

Vial J (2024) Capital forestal y sus desafíos en Chile y el mundo. (Working Paper No. 133). Clapes UC

World Bank (2024) The changing wealth of nations: Revisiting the measurement of comprehensive wealth. World Bank

Part II

The Demographic Lever: The Transition Toward a Sustainable Path for the World Population

Part I described the unprecedented increase in the world population, which in just under two centuries rose from around 800 million people to 8 billion. This, combined with the enormous increase in per capita consumption over the same period, has led to significant deterioration in the systems that govern the planet's capacity to support life in general, and humanity in particular. These impacts continue to intensify, and several safe operating limits of these systems have already been surpassed, as described earlier.

As mentioned in Part I there are three levers that must be activated to put global economic development on a sustainable path over time: one is the size and growth rate of the world population, another is the efficiency with which nature's services are used, and a third, directly linked to the previous one, is the rate and composition of the growth in material consumption by the population.

This Part examines in greater depth the determinants of demographic trends, as well as the most likely trajectories of the world population, both in terms of its size and its age composition. Regarding size, a gradual reversal of previous trends is anticipated, with an absolute decline in the size of the world population. As for its composition, the increase in the proportion of the elderly is the other major trend. Both phenomena are already occurring in countries that are furthest along in the demographic transition, highlighting the challenges it poses, which are addressed in the final sections of this part.

General Trends in the Evolution of the World Population

4

Abstract

After the "demographic explosion" that took place in the last two centuries, the pressure of population growth became a driving force of environmental degradation at the global level. This is the first of two chapters dealing with this issue. In this one the general trends of the global demographic transition are described and explained, with a special focus on the interactions between population dynamics and economic development. The well stablished relationship between economic and social progress and the reduction of fertility rates is a critical one to explain the current evolution of population across the world. The positive feedback from human progress into lower fertility is one key reason to promote investment in human capital—specially among young women—in the poorest regions of the world as a pre-condition for sustainable development at the global level. The other key demographic issue addressed here is the challenge posed by aging populations to those countries that are farther ahead in their demographic transition, including now several large and small emerging countries.

Keywords

Demographic transition · Economic development · Fertility rates · Infant mortality · Population aging · Dependency ratios

To project GDP over a 20-year horizon, for example, it is necessary to make assumptions about the evolution of investment, employment, and productivity, which in turn requires assumptions about technological change. For all these variables, forecasting capacity is very limited beyond the current cycle (4 or 5 years).

In contrast, projecting population is a much more reliable exercise: the number of women of childbearing age in the next couples of decades are known, and the

J. Vial, *Sustainable Development and Natural Resources*, Integrated Science 39,
https://doi.org/10.1007/978-3-032-12263-6_4

trends in fertility rates (the average number of children born to women of childbearing age) are quite stable, which allows for fairly accurate projections of birth rates.

To complete the calculation, mortality rates must also be projected, which is somewhat more challenging. On the one hand, infant mortality rates have fallen significantly worldwide, and in most middle- and high-income countries, it is unlikely they will fall much further. This variable is key because it allows for reasonably accurate estimates of how many women born today will reach childbearing age. The other key variable to estimate is the average age at which those who reach adulthood will die. Contrary to common belief, longevity is quite stable and is heavily influenced by the health and nutritional conditions of the mother and baby, from gestation through the early years of life (Fogel 2004). All these variables at the aggregate level tend to change gradually in line with long-term trends. With well-informed assumptions regarding their determinants, it is possible to make much more accurate projections than for economic variables over a 20- or 30-year horizon.

Of course, when variables such as migration are introduced, the exercise becomes more complex. In ancient times, migration occurred at the pace at which people could walk from one place to another, carrying their belongings and having to survive along the way. The current situation is very different. Who would have imagined that more than 100,000 people from a Caribbean island (Haiti) would arrive at the southernmost corner of South America (Chile) in just a few years? The efficiency and low cost of transportation have made possible what was unthinkable just a few decades ago, and this, of course, introduces significant complexity into these projections and can be a source of changes in demographic trends at the country or regional level.

Another source of uncertainty can be new diseases, or medical innovations that reduce the impact of old ones. As experienced in recent years, a pandemic caused by a new virus has significantly increased mortality in many countries around the world. In the past, major pandemics decimated the populations of entire continents as seen in Europe with the Black Death, and the Americas with the various pathogens brought by European explorers and conquerors.

This is much more difficult to predict, but at least since the twentieth century their demographic impacts at the global level are relatively limited and transitory. For example, the so-called Spanish Flu,[1] which caused between 40 and 80 million deaths (3.3% of the world population at that time, taking the midpoint of the range), had a limited and transitory impact on the trend of global population growth.

Of course, their economic impacts are another matter: with a highly interconnected and complex global economic system, where the continuity and timeliness of the flow of goods and services play a critical role, even a much milder epidemic like Covid-19 caused major economic disruptions worldwide.

Coinciding with the Industrial Revolution, the so-called "demographic explosion" also began. A global phenomenon, but one that has occurred at different

[1] The evidence suggests that it originated in the western United States and from there quickly reached Europe, taking advantage of the movement of large contingents of soldiers to the theaters of war, from where it spread throughout the world (Barry 2021).

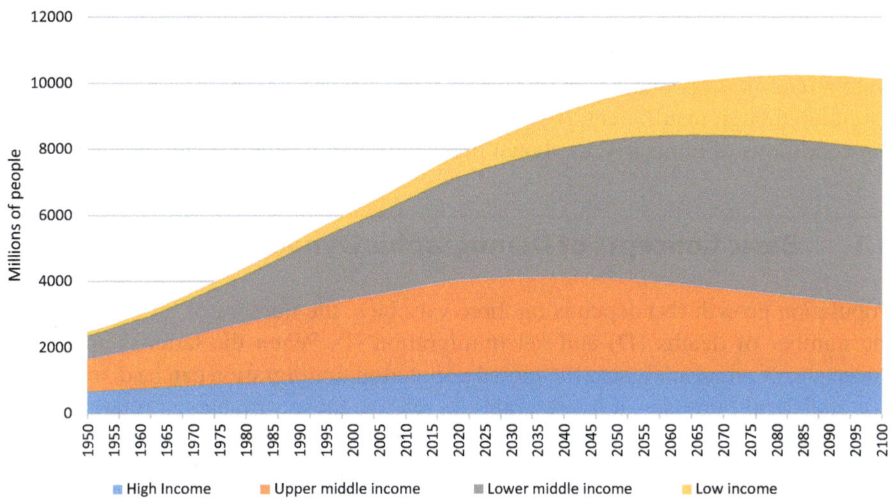

Fig. 4.1 Actual and projected evolution of the world population by country income level (medium projection variant). (Source: United Nations 2024)

speeds and times around the world, as explained in Part I. In the mid-twentieth century, it was anticipated that this could cause a resource crisis, which has not occurred so far for reasons that will be discussed in the next part of this book. What has occurred is what has been termed the demographic transition, from a global population equilibrium at around 800 million inhabitants to another equilibrium, the level of which is still unknown and about which there is a wide disparity of opinions, but which will certainly be lower than the nearly 10.3 billion currently projected by the United Nations as the maximum population level to be reached in the last quarter of this century, after which the world population will begin to decline, as shown in Fig. 4.1.

The great surprise of the last century and a half has been the dramatic decline in infant mortality, followed—at varying intervals—by an equally rapid decrease in the number of births, first in the most advanced countries, to then spread even more rapidly to middle-income countries and is now starting in many low-income countries.

The decline in infant mortality is the direct cause of the so-called "demographic explosion," while the decrease in birth rates is first reflected in the slowdown of population growth and later in an absolute reduction in its size. In just a few decades, the world has shifted from concern over a demographic explosion to fear of a virtual implosion, described as a demographic crisis due to its potential economic effects, caused both by the expected decrease in the number of workers and the sharp increase in the proportion of elderly people in the population.

On the other hand, if this evolution is examined from the perspective of its impact on the sustainability of the economic progress achieved, this adjustment in population comes at a time when widespread evidence shows that the systems governing the functioning of the biosphere have deteriorated to the point of endangering the persistence of the environmental conditions that enabled the success of mammals, and humans in particular. From this perspective, the demographic crisis is simply an essential, though insufficient, adjustment to achieve a sustainable balance in the living conditions of humanity, as shown in the simulation exercises in Sect. 1.5.

4.1 Basic Concepts of Demographic Dynamics[2]

Population growth (N) depends on three variables: the number of births (B) minus the number of deaths (D) and net immigration (I). When the latter is small and uncertain, or when analyzing the global population, immigration can be disregarded. For analytical purposes, it is preferable to refer to these variables in relation to the size of the population, noting that demographers tend to use rates per thousand inhabitants, while economists prefer to work with percentages. As the audience of this book is almost certainly more familiar with percentages, with apologies to the demographers, that is the metric that will be used here.

Starting from the initial definition, the absolute growth of the population in period t can be represented as:

$$\mathrm{N_t = N_{t-1} + B_t - D_t + I_t} \tag{4.1}$$

Dividing by the population size in the previous period, this can be expressed in rates using the nomenclature defined above:

$$N_t / N_{t-1} = \left(N_{t-1} + B_t - D_t + I_t\right) / N_{t-1} \tag{4.2}$$

Which in turn corresponds to:

$$\mathrm{n_t = b_t - d_t + i_t} \tag{4.3}$$

Where n is the population growth rate, b is the birth rate, d is the death rate, and i is the immigration rate, all expressed as a proportion of the population.

To understand the population dynamics reflected in n, it is necessary to examine how and why the birth, death, and immigration rates are changing.

The birth rate, since times immemorial, was something beyond individual control and—as Malthus said—subject to the "passions between the sexes." Although there was some degree of influence over the number of children per woman (fertility rates), due to cultural reasons such as tolerance of out-of-wedlock pregnancies, age at marriage, etc., as well as natural methods (such as sexual abstinence), the number of births was mainly regulated by nature. This changed radically in the twentieth century when the availability of reliable birth control methods allowed women, for the first time in human history, to control the number of children they would bear. From that moment on, and as cultural and social changes have given women greater

[2] For a more in-depth discussion of these topics, supported by abundant statistical evidence, see Livi-Bacci (2012).

decision-making power in various societies, birth rates have evolved according to conscious decisions by individuals, particularly women.

The second key factor in population dynamics is the death rate. The major change in this dimension has been the sharp decline in infant mortality, especially in the early years of life. This occurred gradually, beginning with improvements in maternal and child nutrition supported by increases in agricultural productivity, accompanied by better sanitary conditions, starting with greater access to clean water and the development of sanitation systems. This decline meant that, within a few generations, the number of babies surviving to reproductive age increased dramatically, and with it, the pressure to feed, educate, and care for them until they became self-sufficient.

In more backward rural societies, children contributed significantly to domestic and productive tasks from a very early age, in a world where all activities—from farming to warfare—were highly "labor-intensive," and thus children were highly valued. The Industrial Revolution and urbanization changed this: everything began to be mechanized and mass-produced, so many of the contributions children made to the material well-being of families lost relevance, while the time and resources devoted to raising and educating them increased. It is not surprising, then, that declines in fertility followed decreases in infant mortality, a phenomenon observed worldwide throughout the history of the nineteenth and twentieth centuries.

The shift from an initial demographic equilibrium with a low population growth rate characterized by high birth rates and high infant mortality, to another equilibrium, also with low population growth but now characterized by low birth and infant mortality rates, is known as the demographic transition. A simple way to represent this phenomenon is shown in Fig. 4.2.

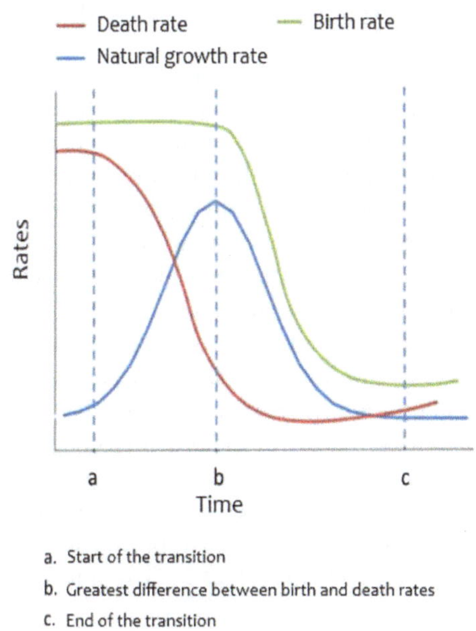

Fig. 4.2 Schematic representation of the demographic transition. (Source: Based on Livi-Bacci 2012)

The demographic transition was a relatively gradual process in the nations that advanced most rapidly after the Industrial Revolution, but it accelerated dramatically in the developing world from the mid-twentieth century onward. Poor countries such as China, India, and Korea in Asia, and Brazil or Chile in Latin America, still exhibited infant mortality rates above 10% and fertility rates of 5 or 6 live births per woman in the mid-twentieth century. Within a few decades, they reduced infant mortality rates to about 1 or 2%, like those of developed countries, and within a few more decades, fertility rates fell below 2 children per woman, that is, below the so-called "replacement rate" required to maintain a stable population size in the long term. Africa and many parts of Asia and the Middle East have progressed more slowly in this process, mainly because economic development and urbanization have advanced at a slower pace, delaying the decline in infant mortality rates.

Finally, it is important to highlight that these changes in infant mortality and fertility rates also have a significant impact on the age composition of the population. In the initial stage, dominated by declining infant mortality, the proportion of young people increases. Later, as infant mortality stabilizes and fertility begins to fall, coinciding with the aging of the first survivors, the overall population begins to age. In recent decades, this process has intensified as advances in nutrition and health have enabled a greater proportion of the population to live longer.

4.2 Demographic Transition and Economic Development

This section will address the factors underlying the demographic transition described in the previous section and its relationship with economic development. These demographic changes are radically altering the natural and social environment, as well as key aspects of the functioning of economic activities, starting with human labor.

The demographic explosion is clearly visible in the data, as shown in Fig. 4.1. After centuries of very slow growth in the world population, it began to accelerate at an exponential rate in the seventeenth century, rising in just over two centuries from around 800 million people to the current 8 billion.

This phenomenon has coincided with monumental economic progress following the Industrial Revolution, which is no coincidence since both are related and mutually reinforced each other in the twentieth century. The combination of improved nutrition, urbanization, advances in sanitation, access to safe drinking water, and later higher levels of education among women were decisive factors in the sharp declines in infant mortality rates achieved throughout the twentieth century in different parts of the world (Deaton 2013). The most recent part of these changes is shown in Fig. 4.1 of the previous chapter, which depicts the evolution of the world population relative to the level of development of countries.

In the initial stage, population growth enabled the expansion of markets and provided labor to increase production, which helped sustain economic progress. However, this alone does not explain why the previous trend was broken.

The explanation lies precisely in the Industrial Revolution: the combination of increased global food production capacity with the incorporation of the Americas, the development of trade between the colonies and European countries (Pomeranz 2000), as well as technological advances, the use of more powerful energy sources, and greater capital intensity, made it possible to break the so-called "Malthusian Equilibrium" that had prevailed until then.

Malthus (1798) posited the existence of a stationary equilibrium in which the capacity to produce food limited the population, and in an environment of low capital accumulation and little innovation, this resulted in almost zero GDP growth. This was undoubtedly an accurate description of trends prior to the Industrial Revolution, but it became obsolete precisely when the theory was formulated, as the Industrial Revolution broke this equilibrium through sharp increases in the production capacity of all types of goods, including food.

The world population stood between 800 million and 1 billion people at the beginning of the nineteenth century, when declining infant (and maternal) mortality led to an increase at the base of the population pyramid (children and youth). This process subsequently continued at exponential rates, as a growing number of girls survived childbirth and diseases during their most vulnerable years, reached childbearing age, as was typical at the time, had a large number of children. By the mid-twentieth century, the world population had more than doubled, and then the process became "explosive" as declines in infant mortality spread to the developing world, where women still had an average of 5 or more children in their lifetimes.

The data from Maddison and his collaborators[3] show that the world population took a little over 1500 years to double from the beginning of the Common Era, when it reached approximately 230 million inhabitants. The next doubling required about 300 years, and in the nineteenth century the third doubling occurred, bringing the world population to nearly 1.8 billion people. From the beginning of the twentieth century to the present, it has multiplied by more than four times.

Figure 4.3 shows that there has been a continuous decline in infant mortality rates worldwide. However, the pace of these declines has varied greatly depending on the income level of countries[4] and the initial levels of these rates. In upper-middle- and high-income countries, these rates have tended to stabilize at levels that are unlikely to decrease much further in the future. In contrast, in lower-middle- and low-income countries, there is still significant room for progress, although rates have already fallen to around 5%.

The main cause of this slowdown in population growth is that, alongside the process of economic development, very significant social changes occur, particularly affecting the fertility rate (the number of live births per woman over her reproductive lifetime). Among these factors are the increasing difficulty of supporting

[3] Maddison (2006) and Maddison Project. University of Groningen. https://www.rug.nl/ggdc/historicaldevelopment/maddison/releases/maddison-project-database-2020?lang=en

[4] The anomaly observed in infant mortality rates between 1959 and 1961, affecting upper-middle-income countries and even the world as a whole, is due to the famine that took place in China during those years.

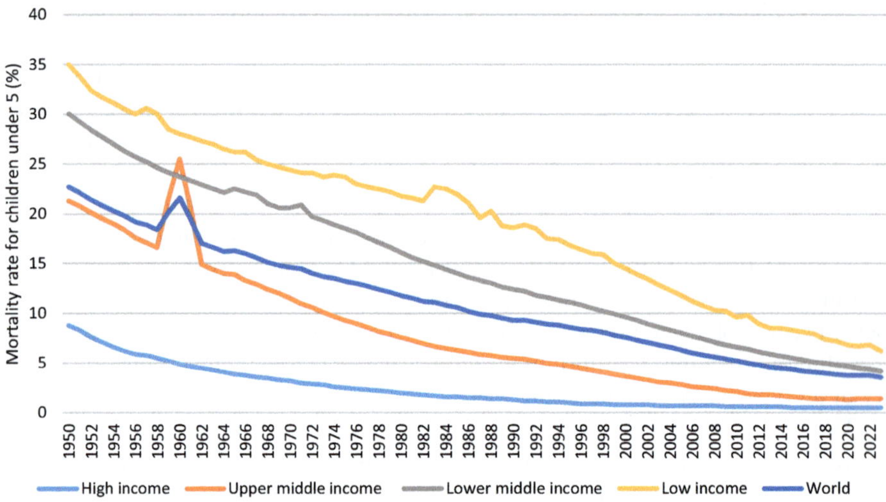

Fig. 4.3 Evolution of infant mortality rates by country income level. (Source: United Nations 2024)

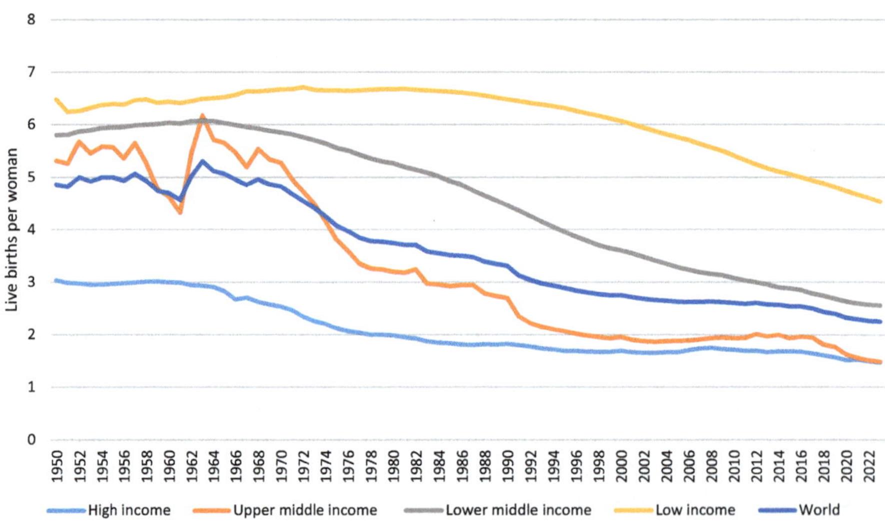

Fig. 4.4 Evolution of fertility rates by country income level. (Source: United Nations 2024)

and educating a larger number of children who survive early childhood, higher levels of education and autonomy among women, supported by better employment opportunities outside the home and access to contraceptive methods, among others. This is illustrated in Fig. 4.4.

As with infant mortality rates, fertility rates have declined worldwide, but with a different timing depending on income. By 1950, mortality rates were already falling and continued to do so thereafter. In contrast, fertility rates remained stable, albeit at different levels among various groups of countries. Only in the mid-1960s did they begin to decrease—and very rapidly—in high- and upper-middle-income countries, with both regions falling below replacement rates by the end of the century. The adjustment in lower-middle-income countries began around the same time but proceeded much more slowly, while in the poorest countries it only started at the turn of the century. This also explains why future world population growth in the coming decades will be concentrated precisely in the lowest-income countries.

When considering the timing of the acceleration in fertility decline, it is impossible to ignore the coincidence with the period when the use of artificial birth control methods became widespread. However, it should be remembered that these are merely tools to align the number of pregnancies with the desires of parents, and particularly of women. There may be debate about how widely available these methods are in poorer countries, but it is likely that the slow decline in their birth rates reflects a lower demand for contraceptives compared to higher-income countries.

In more economically advanced societies, there has been a sharp decline in fertility rates, which are now below the replacement level (approximately 2.1 children per woman) required to maintain a stable population in the long term.

Possibly the greatest driver of this change in reproductive behavior is the radical shift in the range of personal development opportunities experienced by women since the twentieth century, which now compete with motherhood for their time. Higher education and better employment opportunities have led women to significantly postpone their first pregnancy to pursue university studies and begin a professional life outside the home.

If the fertility rate falls below the threshold of 2.1 children per woman, as has already occurred in most developed countries and many middle-income countries, the population first stabilizes and then begins to decline. This is an empirical regularity observed in societies of all cultures and regions of the world, as can be seen in Fig. 4.3. This process was more gradual in currently developed countries and more abrupt in those with more recent development.

Thus, as countries progress in their economic and social development, fertility rates tend to decline, which after some time translates into a slowdown in population growth rates. Figure 4.5 shows that an indicator associated with the level of development, such as the level of formal education among women, is inversely associated with the birth rate in a selected group of countries.

It should be noted that this association should not be confused with a univocal causal relationship, as there are effects operating in both directions: lower birth rates, associated with a more permanent decline in fertility rates—due, for example, to postponement of the first pregnancy—help women to extend their education. In any case, more comprehensive studies tend to show that there is a causal relationship from more years of schooling to lower fertility rates (Kim 2016) (Fig. 4.6).

As can be seen in Fig. 4.1, the demographic transition has intensified in the second half of the twentieth century, although it has not occurred simultaneously and synchronously worldwide, but rather has accompanied advances in the level of development.

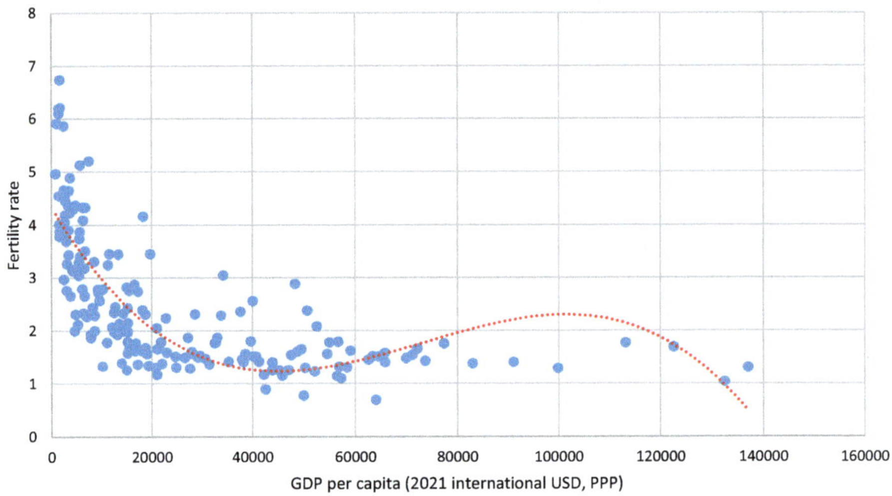

Fig. 4.5 Relationship between fertility rate and per capita income for a selected group of countries in 2022. (Source: World Bank: World Development Indicators)

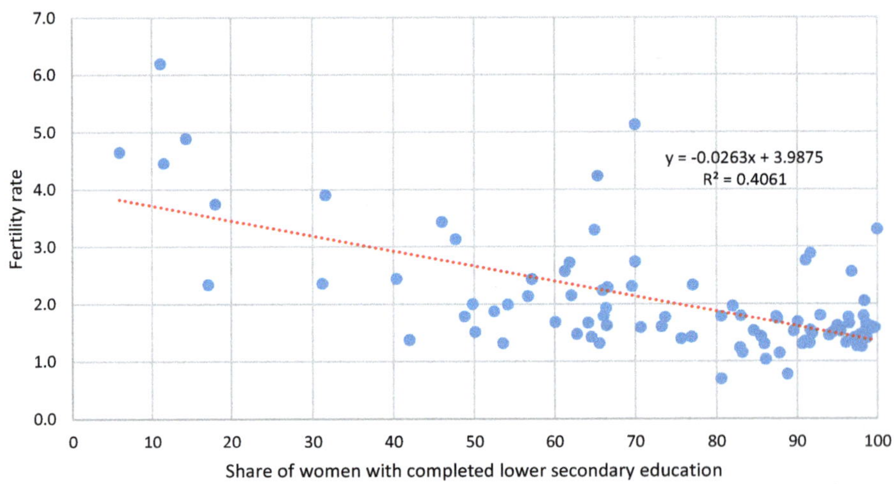

Fig. 4.6 Relationship between secondary education among women and birth rates in selected countries, 2022. (Source: World Bank: World Development Indicators)

Because of declining birth rates, population growth is slowing down, then decreasing, and even turning negative, as is already happening in some Asian countries (Japan is the clearest case, closely followed by Korea and China) and in Europe, where Germany and the Mediterranean countries stand out.

Figure 4.7 shows the actual evolution of the world's demographic transition, along with projections for its future development for the remainder of this century.

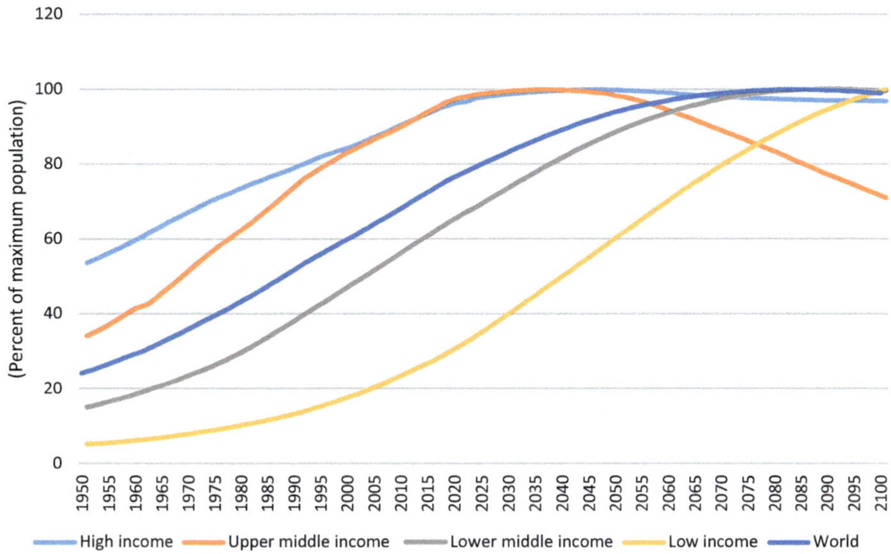

Fig. 4.7 Evolution of the demographic transition by country income group. Medium variant of the 2024 projection. (Source: United Nations 2024)

Here, too, the main country groupings are used according to their per capita income levels at the beginning of the present century, following the World Bank classification. The comparison covers a century and a half, starting in 1950 and extending to 2100, based on United Nations estimates up to 2024 and their projections for the rest of the century under the medium scenario, updated that year.[5] As with any projection, this one is based on assumptions that may or may not prove accurate. The further to the right the peak occurs, the later the respective country groups are in the demographic transition.

As shown in this last chart, the level at which the world population stabilizes and the timing of this event depend critically on the speed of the demographic transition in middle- and low-income countries, which account for most of the population. In high- and upper-middle-income countries, population dynamics suggest that the demographic transition is likely already concluding, as shown in Fig. 4.7, since fertility rates have fallen below 2 for more than a decade and have continued to decline recently. In contrast, there is still some uncertainty regarding the poorest regions, because although fertility rates have begun to fall, they remain above 2 in many of these countries. Geographically, this means that the population in Africa and South Asia will continue to grow, albeit at a slower pace. These regions, in particular, face the greatest challenges in feeding, caring for, and educating children, as well as providing well-paid jobs, which is already driving migration and political

[5] The figures are not presented in absolute numbers, as is customary, but rather as a proportion of the maximum estimated population size for each country during the period. These figures, including projections, incorporate migratory flows.

and social tensions that are likely to intensify in the coming decades (Ferguson 2006), as occurred in Europe in the first half of the twentieth century and in Latin America more recently.

A particular feature of the United Nations demographic projections is that they assume that the (natural) population growth rates in the least developed countries follow a trajectory like that of other developing countries, but with a lag of 30–40 years from now. Looking at the recent trajectory, with stable economic growth rates above 2.5% per year, this may seem optimistic, but some argue the opposite, assuming that in a more connected world, cultural changes leading to greater empowerment of women, as well as higher levels of schooling among the generations of girls now reaching reproductive age, will be reflected in faster declines in birth rates in the future.[6] If so, the world population could evolve at a pace more in line with the low variant of the United Nations projections, which are presented in Figs. 4.8a, b.

The main factor behind the differences in these two projections is the assumption regarding the future evolution of fertility rates in different countries and regions.

4.3 Demographic Transition and Population Aging

The demographic transition affects not only the overall size of the population, but also its age composition. The age structure of the population is usually represented by a pyramid, which is true only if the population growth rate is constant. But during the demographic transition the size of the population and the age structure change as the transition progresses, so the supposed pyramid changes significantly. If, in addition, for various reasons, longevity has increased in parallel with the demographic transition, the distortion of the pyramid becomes even more pronounced.

Typically, in an initial stage, when birth rates remain high and infant mortality rates decline, the base of the pyramid widens, and this bulge moves upward over time as the young population continues to grow. Once the birth rate begins to fall, the base of the pyramid starts to shrink, although with a considerable lag because the number of women of childbearing age is still increasing. Over the decades, the initially tapered pyramid tends to resemble more of a rectangle, but with a bulge that gradually shifts upward and, after many decades, eventually disappears.

This can be seen in the images presented in Fig. 4.9. They show that, at the global level, there is an aging process underway, with a reduction in the base of the population pyramid and a "thickening" at the upper levels over the past 70 years. Europe is an extreme case, with significant fluctuations in the proportion of young people, both male and female, in the cohorts born during wartime periods and with a high proportion of elderly people already in 1950. The demographic transition in that continent was completed in the twentieth century. Latin America and Africa present

[6] See "Thanks to education, global fertility could fall faster than expected." The Economist. March 26, 2019.

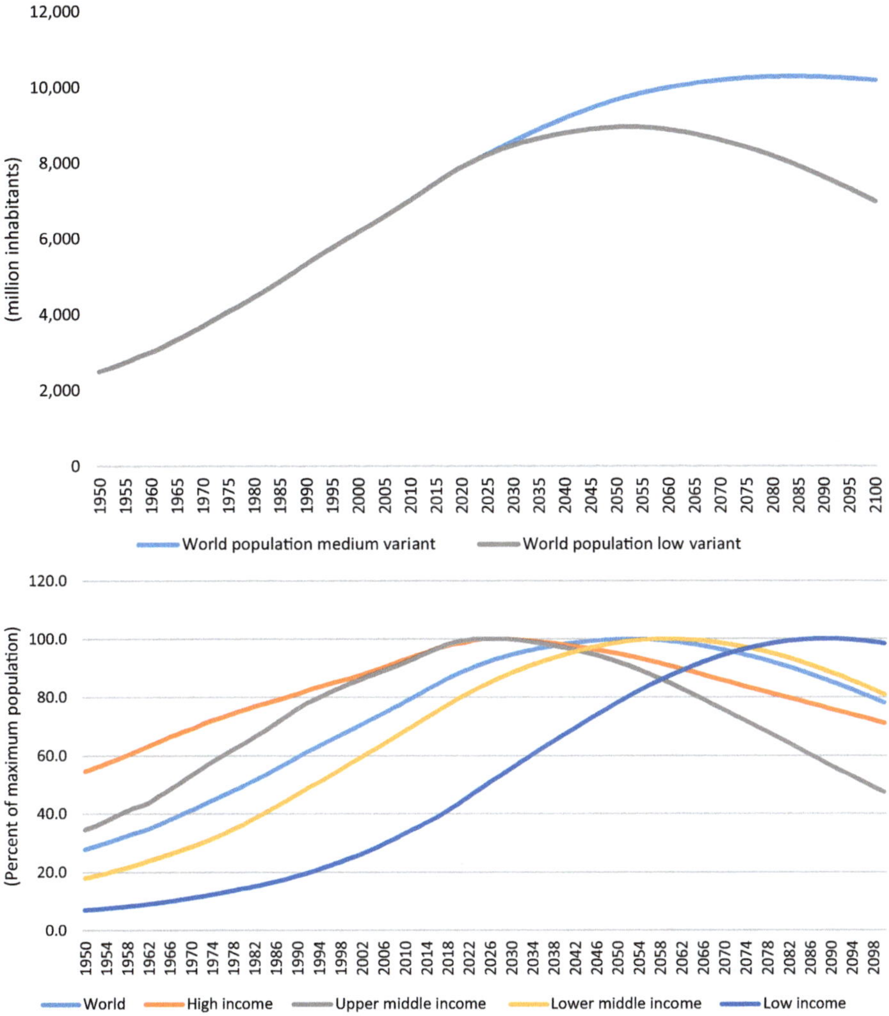

Fig. 4.8 (**a**) World population projections: low and medium variants. (**b**) Demographic transition dynamics: low variant projection. (Source: United Nations 2024)

an interesting contrast: both continents had a very similar age structure in 1950, with a high proportion of young people and very small share of the elderly. By 2020, they look completely different, with Africa showing virtually no change, while Latin America and the Caribbean display advanced population aging.

Asia exhibits sharp internal contrasts, and with two giants like India and China, the continent's demographic representation would be highly distorted. For this reason, two individual cases are shown in Fig. 4.10: two countries with very different income levels and economic and political systems, but which, from a demographic perspective, look very similar: China and Japan.

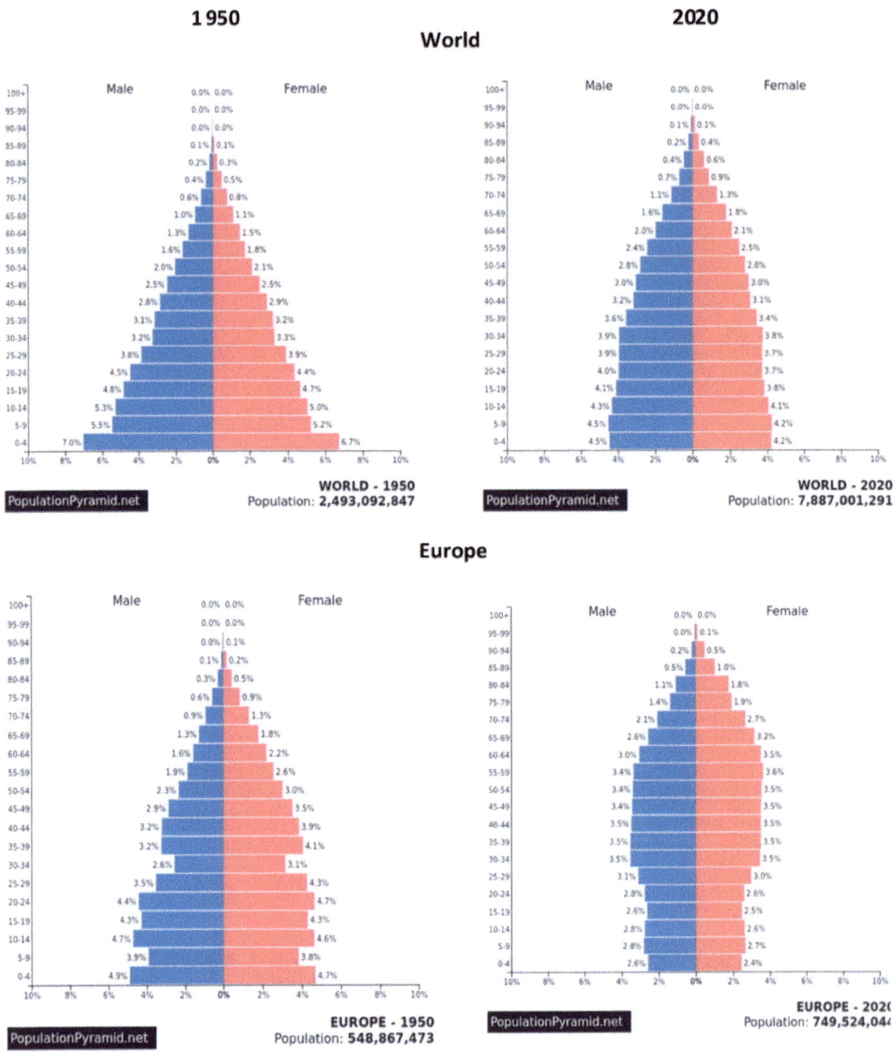

Fig. 4.9 Population pyramids for selected continents in 1950 and 2020 (proportions of each country or region's population by age and gender cohorts). (Source: Population Pyramid.net https://www.populationpyramid.net/)

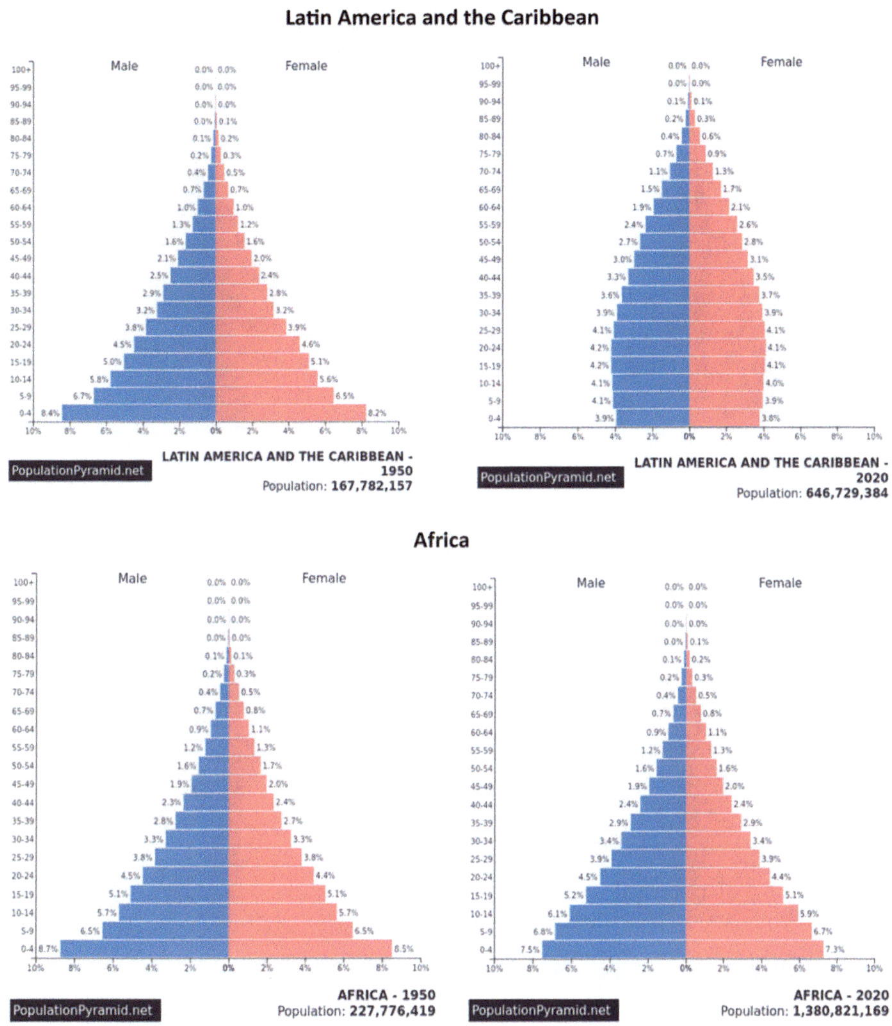

Fig. 4.9 (continued)

Both countries started with a profile like other developing countries, with predominantly young populations in the mid-twentieth century, but both quickly entered a phase of accelerated demographic transition, which in China was partly driven by deliberate birth reduction policies. Today, both countries have entered a phase of absolute population decline and significant aging, with an increasingly narrow base of young people.[7]

[7] These are not the only peculiarities of these countries. Note, for example, that in China all female cohorts are smaller than the male cohorts, which cannot be explained biologically, but coincides with what occurs in other countries whose cultures have a strong preference for male children.

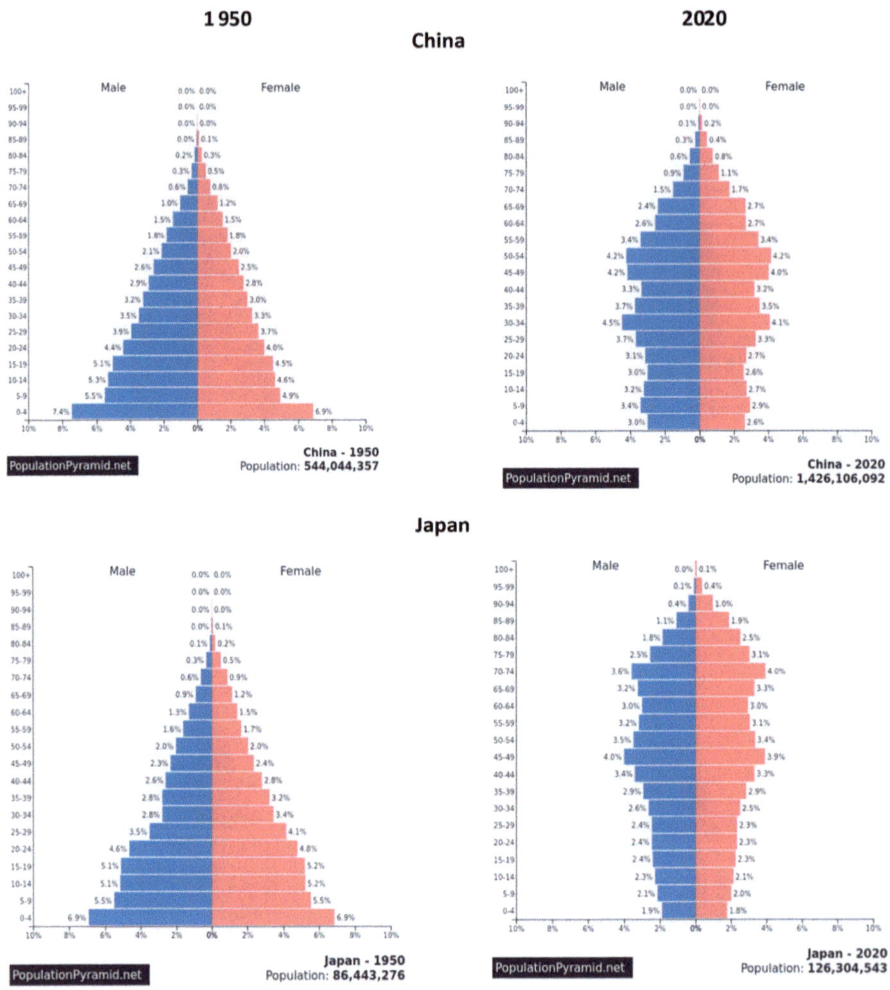

Fig. 4.10 Population pyramids for two selected Asian countries in 1950 and 2020 (proportions of each country or region's population by age and gender cohorts). (Source: Population Pyramid.net https://www.populationpyramid.net/)

This change in the age composition of the population has significant economic and social effects. In the current stage of the demographic transition, there is a rapid increase in the proportion of elderly people in the total population, to the detriment of the younger cohorts. This represents a major shift from the situation experienced in the twentieth century, especially in developing countries, where changes occurred more rapidly than in those that preceded them.

It is highly unlikely that this trend will reverse soon, in which case societies will need to adapt to this new age composition. In addition, there is a parallel increase in longevity, and it is expected that in the future the elderly will live even longer (at

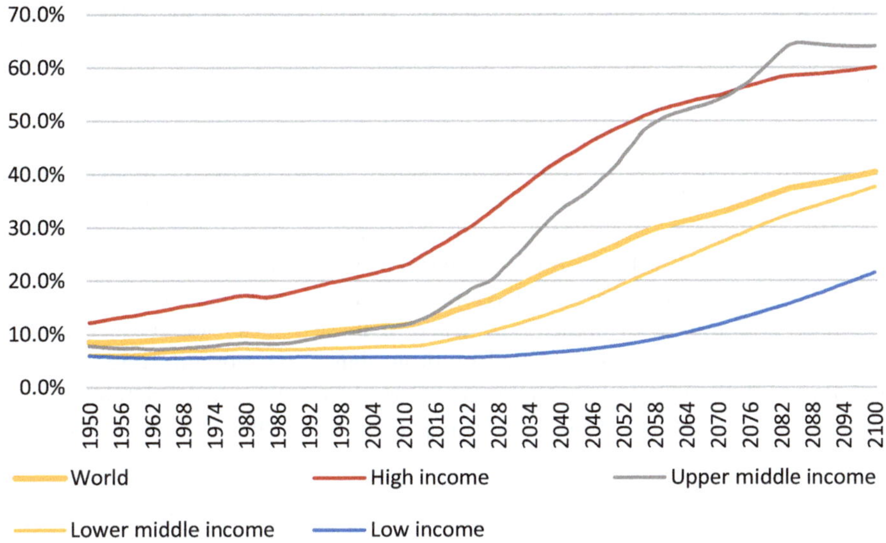

Fig. 4.11 Evolution of old-age dependency ratios by country income group (population over 65/ population aged 15–64). (Source: United Nations 2024)

least on average), thanks to improved living conditions (Fogel 2004) and advances in science and their applications in medicine. This will translate into growing pressure for increased spending on health, pensions, and elderly care, which in turn will be reflected in greater tensions for institutions responsible for providing economic protection in these areas, with a very strong component of public spending that will need to be financed.

One way to represent this is through the calculation of the so-called Old-Age Dependency Ratio (OADR), which shows the relationship between the number of elderly people and the working-age population according to the United Nations' 2024 medium projection.[8] Figure 4.11 shows that globally, the OADR has risen from slightly below 10% (that is, one person over 65 for every 10 people aged 15–64) to a level that could reach 40% by the end of the century. It is also evident that, for the past decade, high- and upper-middle-income countries have passed an inflection point in this ratio, which is now increasing much faster than in previous decades.

One consequence of progress in the demographic transition is that population aging is much more advanced in more developed countries, where the demographic transition occurred earlier.

In contrast, in lower-income countries, birth rates remain very high, and the young population continues to grow rapidly. These countries are only just beginning

[8] There is no single way to measure this relationship, as any general segmentation of the population is arbitrary. Here, the ratio between the population over 65 years of age and those in the 15–64 age group will be used.

to experience a slight aging in the population, if at all. However, they face the dual challenge of educating their younger generations and creating new employment opportunities to enable them to enter the workforce productively.

The coronavirus pandemic disproportionately affected older adults, but with nearly 20 million deaths globally, it is unlikely to alter the trend toward population aging. However, if Covid-19 becomes an endemic disease, resulting in higher mortality rates among the elderly, it is quite possible that this will ultimately lead to a lower life expectancy at age 65, partially reversing the longevity gains of recent decades. On the other hand, lifestyle changes in higher-income countries, with increased obesity and sedentary behavior, are beginning to affect some of the most developed countries with higher prevalence of these issues. For instance, in the United States, which is also affected by drug abuse, the mortality rates stopped declining in 2010, began to rise slightly in subsequent years, and accelerated during the pandemic, increasing from 7.9 deaths per thousand inhabitants that year to a peak of 10.4 during the pandemic, subsequently falling to 9.2 in 2023.

The counterpart to rising old-age dependency ratios is the decline in youth dependency ratios, that is, the proportion of young people (under 15 years old, by our definition) relative to those of working age (15–64 years). This will represent some relief in education spending, for example, but will place significant pressure on governments, which will need to make necessary adjustments in infrastructure, personnel expenditures, and other areas.

In this final stage of the transition, the elderly will gain relative importance in political and economic terms, which will almost certainly bias public policies in their favor, on the one hand, and induce companies to tailor their offerings to the preferences of this age group, which, unlike in previous centuries, now reaches old age in better physical and economic condition.

This is already causing profound changes in countries that are further along in this process, such as Japan, for example, where the working-age population peaked in the 1990s and is already declining.[9] A reduced availability of workers could be mitigated, at least initially, by extending working lives, which could be accompanied by shorter working hours for older adults, or by a combination of increased remote work and reduced physical presence in offices. Immigration is other way to compensate some of the problems of an ageing society, and Japan is now extending visas for people of working age. In more advanced countries there may be a mismatch between the skills required by the labor force and those brought by immigrants, not to mention potential challenges in adapting to different cultures and languages.

Other mitigating factors could include higher human capital per person, as people attain higher levels of formal education, which is then complemented by successive opportunities for education and training throughout their working lives.

Leading research countries, such as Japan, are attempting to address this potential shortage of workers through research and development of new forms of

[9] See, for example, Financial Times "How Japan's ageing population is shrinking GDP." May 16, 2018.

automation for tasks traditionally performed by humans, even in areas related to personal services.

In any case, many countries, especially those with middle and low incomes, still have significant room to increase female labor force participation, where, due to a lack of childcare options, cultural or even economic barriers, many women do not participate in the labor market.

One consequence of increased longevity, which works in the opposite direction, is a gradual reduction in average labor force participation rates, as well as a decrease in average hours worked, since older adults tend to have lower participation rates than average, and when they do participate in the labor force, they tend to work shorter hours or have higher absenteeism due to health reasons.

Greater longevity accentuates these challenges and has already began to take a prominent place on the political agenda in Japan and other countries.[10] Slightly behind Japan in the aging process are the main European countries, the most developed Asian nations such as South Korea and Taiwan, and China. In the latter case, the combination of the now-abandoned "one child per family" policy and rapid economic development have accelerated the demographic transition to such an extent that the working-age population is already past its peak.

All these problems are already emerging in middle-income countries. These nations have undergone a faster demographic transition, so the changes are more abrupt, and the resulting economic and social tensions are beginning to arise before these countries have reached income levels comparable to those of more developed nations, which have greater resources to address such challenges. Population aging will require a high degree of adaptability from institutions, practices, and customs, with little time to achieve it.

The challenges of the Demographic Transition are well illustrated by the evolution of the so-called "baby boom" generation, born in the postwar period (approximately between 1945 and 1960), which has experienced the final stage of the demographic transition, especially in the developed world. This generation grew up hearing that the world belonged to the young, played a leading role in the social movements of the late 1960s and 1970s, and both led and bore the costs of globalization in the 1990s and early twenty-first century. Now, it is aging in a world increasingly populated by the elderly.

The retirement of this generation in advanced economies and many middle-income countries, is having a significant impact on labor markets and Social Security systems. In the latter case, the problem is particularly acute, as the financial designs of pension and health systems did not account for the increase in both the proportion and life expectancy of the elderly, resulting in increasingly severe deficits, putting pressure on public and private finances, and straining political and social systems.

The truth is that the characteristics of the contemporary world and its short-term trajectory have been shaped by the "Baby Boom" generation, which has dominated politics, the economy, science, and technology since the mid-1960s. Their children will have to manage the adaptation to the new economic reality, navigating the

[10] See: Financial Times: "Japan begins to embrace the 100-year life." August 8, 2018.

political, social, and environmental "bottleneck" they have inherited. The younger generation, now in the driving seat of politics and business will also be responsible for driving and channeling the tremendous advances in science and technology that will shape the future evolution of humanity.

References

Deaton A (2013) The great escape: health, wealth, and the origins of inequality. Princeton University Press

Ferguson N (2006) The War of the World: Twentieth century conflict and the decline of the West. Penguin Press

Fogel RW (2004) The escape from hunger and premature death, 1700–2100: Europe, America, and the Third World. Cambridge University Press

Kim J (2016) Female education and its impact on fertility. IZA World of Labor

Livi Bacci M (2012) A concise history of world population, 5th edn. John Wiley & Sons

Maddison A (2006) The world economy: Volume 1: A millennial perspective and volume 2: Historical statistics. OECD Development Centre Studies. https://doi.org/10.1787/9789264022621-en

Malthus TR (1798) An essay on the principle of population

Pomeranz K (2000) The great divergence: China, Europe, and the making of the modern world economy. Princeton University Press

Ritchie H (2021) The world has lost one-third of its forest, but an end of deforestation is possible. Our world in data. https://ourworldindata.org/world-lost-one-third-forests

United Nations (2024) World population prospects 2024: Online edition

Challenges of the Demographic Transition in the Twenty-First Century

5

Abstract

Three main challenges are addressed here, with a special focus on emerging countries: one is the environmental impact of still growing populations in low-income countries on the use of the territory and its resources. In many of them, especially the poorest ones, this is a driving force behind deforestation and degradation of ecosystems, since a large fraction of their people depend on nature for their livelihood. So, there is positive feedback from social and economic progress to the protection of nature and reduction of greenhouse gases emissions. However, lower population growth also means a slower expansion of the labor force, and this could translate into slower economic progress if not compensated by higher productivity growth, something hard to achieve in less advanced countries. Finally, the financial impacts of aging on the social security systems are also addressed. This issue is reaching critical proportions in many advanced countries, where this process started earlier, and also in several large and medium size emerging countries where the demographic transition has proceeded very fast in the last few decades.

Keywords

Aging · Environment · Labor markets · Economic growth · Social security · Fiscal pressures

5.1 Environmental Impacts of the Demographic Transition

The population increase in the nineteenth and twentieth centuries had large environmental impacts: the occupation of habitable land has expanded, increasing the area used for raising livestock, agricultural production, and physical infrastructure for urban areas, transportation, and more. Moreover, this great change in land use has been accompanied by soil, water, and air pollution, along with the alteration of

J. Vial, *Sustainable Development and Natural Resources*, Integrated Science 39,
https://doi.org/10.1007/978-3-032-12263-6_5

essential habitats for the survival of multiple species. To illustrate the magnitude of this impact, it is worth noting that at the onset of the Industrial Revolution, about 10% of the total potentially habitable land surface was used to support human activity—a figure that remained stable for the previous 500 years. Today, it is estimated that human land use has increased fivefold, reaching 50% of that surface, with most of this expansion occurring in the last century.[1]

Most of the land occupied by the expansion of the human population and the biomass required to sustain it—whether for food, housing, or infrastructure—has come at the expense of replacing forests, grasslands, and wetlands, as shown in Fig. 5.1, with consequential losses of habitats for numerous species.

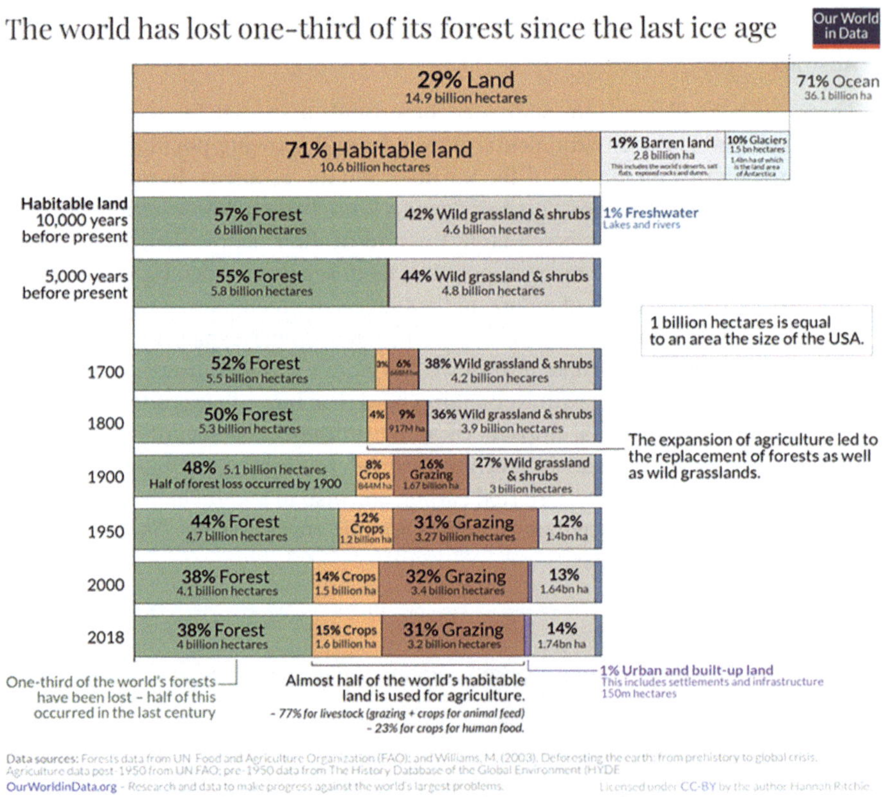

Fig. 5.1 Global changes in land use. (Source: Ritchie (2021) Our World in Data (author's translation))

[1] Figures based on Ritchie (2021). It should be noted that the concept of potentially habitable excludes only glaciers and high mountain peaks.

Another significant consequence of land use changes is the release of greenhouse gases stored in vegetation and soil, which is currently the main source of emissions in the world's poorest regions.

Another way to illustrate humanity's growing impact on the environment is to compare the physical mass of human-made structures with natural biomass, as Bar-On et al. (2018) have recently done. They conclude that humans account for a mass of about 0.06 GT, to which should be added 0.1 GT of livestock biomass and 0.05 GT of poultry. To put these figures in perspective, the biomass of wild animals is estimated at 0.07 GT and that of wild birds at 0.002 GT—that is, only 4% of the mass of domesticated birds.

Yet, as striking as these numbers may seem, they pale in comparison to the natural biomass of the planet, which amounts to 1100 GT, versus the mass of human-made structures and their waste, which now reaches the staggering figure of 1200 GT—surpassing the planet's natural biomass.[2] The landscape of today's world is very different from that in which our great-grandparents lived, and not only because of technological progress, but due to the enormous physical influence of human activity on the environment.

A larger human population also means higher demand for energy, food, and all kinds of materials, with a major impact on ecosystems and landscapes, with a significant effect on greenhouse gases emissions and a large accumulation of residuals of artificial materials, which are very hard to degrade by natural processes.

In wealthier countries, residential space is highly valued, which explains the low density of urban populations, despite the prevalence of high-rise buildings in major metropolitan areas. In poorer countries, densification means urban sprawl and greater crowding, which is typically not accompanied—until much later—by the provision of basic services such as potable water, primary healthcare, waste collection, schools, and so forth.

On the other hand, given the economic and social advantages of urban life, there has also been a significant concentration of human populations in large cities, as illustrated in Fig. 5.2.

This greater occupation of territory, combined with high mobility since the twentieth century, the coexistence with domestic animals and increased contact with species whose habitats are shrinking in poor rural areas, create favorable conditions for the emergence of new diseases and, eventually, epidemics. In fact, the twentieth century saw at least four pandemics, each causing more than one million deaths. The largest was the so-called Spanish Flu, which began in 1918 in the United States and caused between 50 and 80 million deaths; two influenza pandemics (1957 and 1968) with about one million deaths each; and HIV, which emerged strongly in the early 1980s and continues to this day, with more than 30 million deaths and a similar number of current infections. Covid-19 is the first major pandemic of the twenty-first century, which in two and a half years claimed nearly 20 million lives and caused much economic damage.

---

[2] A graphical representation of these magnitudes developed by a team at Johns Hopkins can be found at: A new visualization of everything on Earth | Hub . For all these measurements, the carbon content of species is used, which is standard practice in this type of comparison.

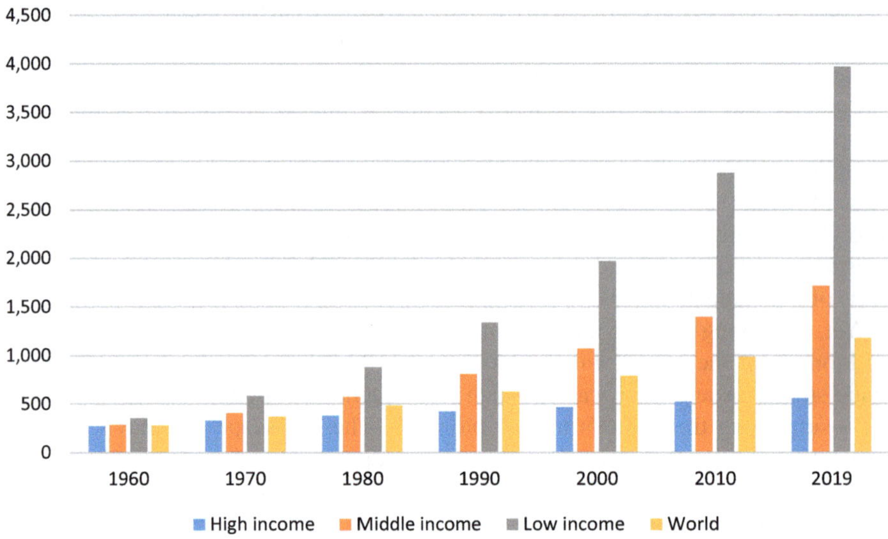

Fig. 5.2 Evolution of urban population density by development level (people per square kilometer). (Source: World Bank, World Development Indicators)

In parallel, these same factors are facilitating the movement of animal and plant species, with the proliferation of invasive species that adapt well to new territories where, at least initially, they face no predators, affecting native populations and associated food chains. The major driver of the introduction of new species was precisely the expansion of humanity and the economic opportunities that arose from access to new foods, from spices to potatoes, maize, and sugar, which played a very important role in food security in Europe, to name just one example. However, along with these came rats, rabbits, pigs, and numerous pathogens that decimated populations lacking natural defenses against them, which had a tremendous impact on the population of the Americas (Diamond 1997).

Progress in the demographic transition, marked by the rapid decline in fertility rates in middle-income countries, has a positive component because it anticipates a reduced human impact on the environment, by easing pressures to convert natural areas for food production, as well as to provide energy and material goods associated with the simple fact that there are more people in the world. This is already evident in the increasing forested areas in Europe and North America, where the agricultural frontier is receding.

The most remarkable aspect of this process is that it is occurring because of individual and voluntary decisions by young women and men across most of the planet, even in the face of very active public policies aimed at reversing this trend. This adjustment in the size of the world population is an essential step toward returning to a sustainable trajectory of progress in the material living conditions of humanity.

The great challenge is to successfully adapt to these new conditions, as population decline and aging have significant economic consequences that must be addressed to avoid crises that could make this process very costly and conflict-ridden. This is especially urgent in emerging countries (or those that have recently been so) as diverse as Korea, China, and Thailand in Asia; Turkey and Russia near Europe; or Brazil, Chile, and Costa Rica in the Americas, which have experienced abrupt and sustained declines in their fertility rates, now closer to 1.5 than to 2.1, the replacement rate that keeps the population size constant in the long term.[3]

5.2 Impacts of the Demographic Transition on Economic Growth

In recent years, demographic issues have also gained attention in the media due to the economic impact of the rapid decline in birth rates observed in recent decades could have.

The issue has become more visible in emerging countries because of the sharp decline in the number of births that many countries have been experiencing for some time now, which is already reflected in shifts in demand for related health services, as well as in the size of the cohorts of young people entering preschool and primary education. In response, alarm has been raised about the potential decline in working-age population and the economic effects it could have. Until a few years ago, the prevailing perception was that this was a problem for wealthy countries, or for poor countries that had taken drastic measures to address the population explosion of the mid-twentieth century.[4]

Economic concerns about the costs of a potential "demographic implosion" stem from three sources. One is the effect on the future growth of the labor force and its direct impact on productive capacity. Another source of concern are the adjustments that will be required to accommodate the change in the composition of the aggregate demand resulting from population aging, and the third is the threat to the solvency of public finances due to the rising cost of Social Security. This section will address the first two, while the next section will discuss the impact on public finances.

Concern about the evolution of the working-age population (conventionally defined as those between 15 and 64 years old) is highly significant because, in the absence of substantial changes in variables that are difficult to control or highly unpopular, as will be discussed below, it directly affects the productive capacity of the country. In the traditional Solow model, where GDP growth originates from increases in produced capital, labor, and total factor productivity, the contributions of the first two are directly proportional to the growth in both factors. Therefore, slower growth of the working-age population translates directly into lower projected

[3] See Appendix 3.

[4] China is frequently cited as the prime example of this, overlooking the fact that the fertility rate had already declined significantly before the policy was implemented, and that countries such as Taiwan and Korea are undergoing similar demographic adjustments without having adopted such policies.

GDP growth, unless labor force participation rates, hours worked per person, or worker productivity increase.

To illustrate these challenges 10 emerging countries with the lowest fertility rates listed in Appendix 3, were chosen, using the low variant of the latest UN's population projection for the 15 to 64 age group. The results are striking, as shown in Fig. 5.3a, b, since in all cases, by the end of the century, this age group would be much smaller than its current size.

Why use the low variant of the United Nations projection? Simply because actual fertility rate data from recent years are well below the projections of the medium variant, at least for this group of countries, and are more in line with the low variant.

One surprising aspect of these projections is the diversity of emerging countries in the group, which merely reflects that declines in fertility rates are due to universal causes, rather than those specific to a country or group, and are possibly difficult to trace beyond their correlation with the economic and social progress previously discussed.

A second point to highlight is that long-term demographic projections are actuarial exercises based on assumptions about key variables (such as the fertility rate) that typically do not incorporate endogenous changes in their evolution over time. This does not mean that demographers are unaware of the possibility of such changes, but rather that it is very difficult to predict them, both in terms of timing and intensity. This is why United Nations projections describe scenarios with many possible variants, precisely to convey the uncertainty surrounding these projections.

Nevertheless, as mentioned earlier, over short periods for demography (10–20 years) but very long for economic models, inertia dominates population projections: the number of women of childbearing age will not deviate much from what is projected in the various scenarios, and changes in reproductive behavior tend to manifest over longer periods.

What can be inferred from these figures?

First, it is not appropriate to take long-term projections literally: changes of the magnitudes cited will induce changes in reproductive behavior, either as a reaction to public policies or, more likely, as an endogenous response to the social and economic changes associated with this population evolution.

Second, there is a significant demographic challenge to sustain economic growth: the contribution of labor force expansion, which played a very important role in the growth of these countries in the last century, will no longer be available, and these economies will have to adapt to this new situation. There are several ways to do so:

First, it is necessary to assess labor force participation rates, that is, the proportion of the working-age population willing to work, as well as their evolution over time. Generally, data published by the ILO show divergent trends by gender. On the one hand, male participation rates are higher but tend to decrease, mainly due to increased years of schooling. Female participation rates are initially lower but increase as a greater proportion of women enter the workforce outside the home. At the same time, however, women also tend to delay their entry into the labor market

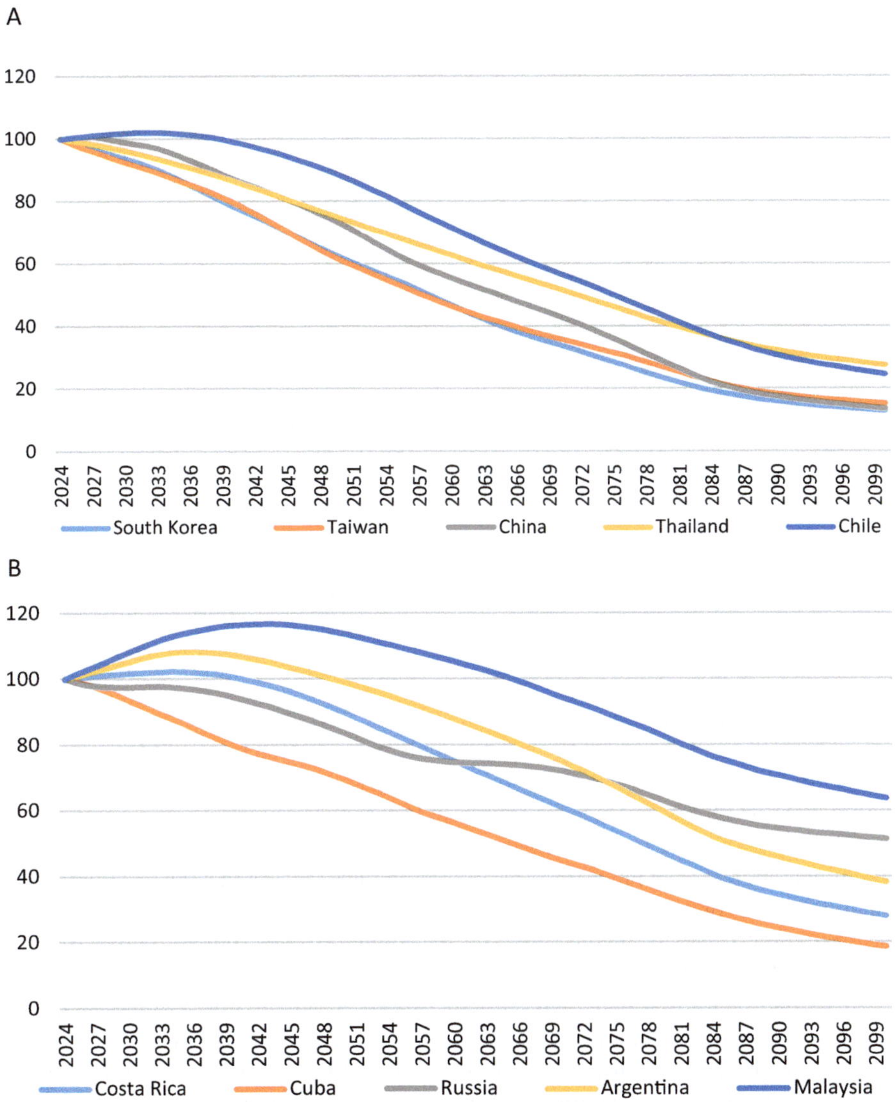

Fig. 5.3 Projected evolution of the working-age population in the 10 emerging countries with the lowest fertility rates (Index base 2024 = 100). (**a**) Working-age population in 5 emerging countries with the lowest fertility rates in 2023 (low variant). (**b**) Working-age population in the next 5 emerging countries with the lowest fertility rates in 2023 (low variant). (Source: United Nations 2024)

for educational reasons. In the selected group of countries, it is difficult to identify a common trend. Generally, the Asian countries in this group show higher aggregate participation rates, but with a slight downward trend, while the opposite occurs in Latin American countries. Compared to the OECD average, except for Cuba, labor

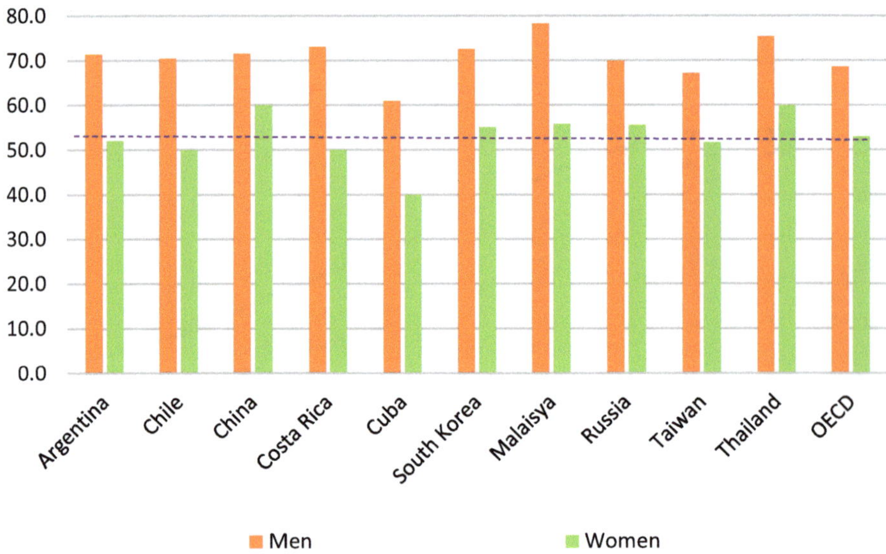

Fig. 5.4 Labor force participation rates by gender in countries with low fertility rates and the OECD average (percentage of the respective population by gender, ages 15 and older). (Source: ILO and Taiwan Statistics Office. https://statdb.mol.gov.tw/html/nat/natehidx01.htm)

force participation rates are higher than those in developed countries. This difference is mainly explained by a large gap in male participation rates, while for women, in 5 of the 10 countries (4 of them in Latin America), female participation rates are lower than the OECD average, as shown in Fig. 5.4.

What does all this imply? In general terms, there appears to be little room for significant increases in labor force participation rates, except in a few isolated cases, such as Cuba, for example. In the rest of the Latin American countries, there may also be some scope to increase female participation rates, but its impact will be limited and will not offset the effect of declining birth rates.

A second factor that may help mitigate or exacerbate the effect of the decline in the working-age population is the number of hours worked per week. In general, these have shown a downward trend over time, which is consistent with a greater preference for leisure as incomes rise. However, other factors are also at play, such as the increase in part-time workers as the population ages in physical and mental conditions that allow them to work longer. In addition, new work modalities, whether part-time, in-person, or remote, with the possibility of communication anywhere and at any time, make much more difficult to determine actual working hours. In fact, remote work may effectively result in longer working hours if it replaces commuting time that is shared between work and leisure. In any case, the data show that the group of countries with the lowest fertility rates work more hours per week than the OECD average, as can be seen in Fig. 5.5, and in some of these countries, formal working hours have been gradually reduced.

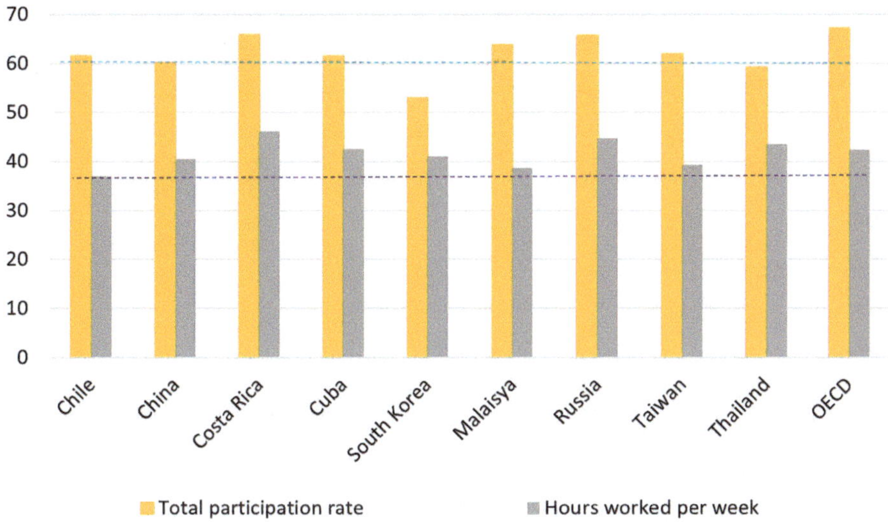

Fig. 5.5 Labor force participation rates and weekly working hours in low-fertility countries and the OECD in 2022 (percentages of the working-age population and average hours per worker per week). (Source: ILO)

Taking these factors into account, it seems unlikely that they will provide significant relief to the potential shortage of workers that could occur in the coming decades. On the one hand, labor force participation rates have tended to stabilize. On the other hand, hours worked show a tendency to decrease as income rises, while the time savings associated with remote work leave some room to share commuting time savings between work and leisure.

One factor that can help mitigate the decline in the number of births in a country is immigration. In fact, this has played an important role in alleviating labor shortages in the United States and Europe. However, it is precisely in these countries that social tensions have arisen, and the political climate has become strongly adverse toward immigration.

Among the group of countries with the lowest fertility rates, the migration phenomenon has been highly heterogeneous, as shown in Fig. 5.6. On the one hand, taking as reference the proportion of the population that reports being born abroad, most show low but significant percentages of migrant population. The exceptions are China, Cuba, and Korea. On the other hand, the net annual flow of migrants over the past 35 years relative to the working-age population, is not enough to compensate the fall in birth rates and several countries are losing population due to net emigration, notably Cuba, China, and, to a lesser extent, Argentina. In contrast, Malaysia, Russia, and Chile received significant inflows. In Russia, this flow dropped sharply in 2023, possibly due to war, while in Thailand, the net flow fell significantly in 2018 and has not recovered since. Only in Malaysia and Chile has there been a relatively high and continuous inflow in recent years, although in the latter case, net immigration is a third of what it was during the peak years of the Venezuelan diaspora. The experience of countries that received large migratory

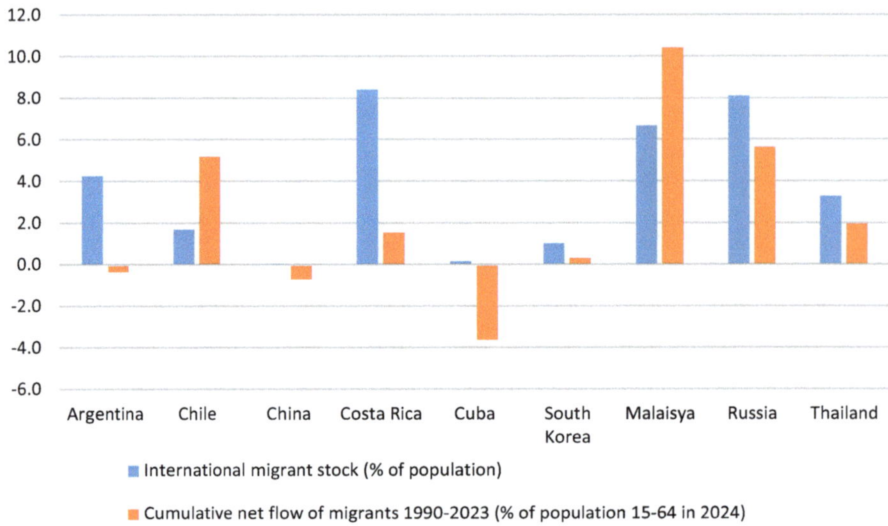

Fig. 5.6 Relative importance of migration in countries with the lowest fertility rates. (Source: World Bank, World Development Indicators, and United Nations (2024))

inflows in a short period, composed by people who do not speak the local language and come from very different cultures show it can be a source of economic and social tensions that tend to result in greater restrictions. This has been the case in Europe, some South American countries, and more recently the United States. Undoubtedly, immigration could be a palliative to declining birth rates in more prosperous countries, but not without social and political challenges, making it unlikely that it can be used on a scale sufficient to offset falling birth rates, unless the economic costs of "demographic implosion" make it unavoidable.

Finally, there remains one major unknown: the evolution of productivity per worker. In principle, a strong and sustained increase in productivity has the potential to support growth in per capita income, even if the number of workers is declining. However, projecting the future evolution of this variable is extremely complex. On the one hand, it will depend on the evolution of the stock of produced capital, since capital and labor are, first and foremost, complementary factors. Second, it depends on the endowment of human capital, that is, the skills of workers, their qualifications, education, and work habits, all of which are usually approximated by educational attainment. A third important factor is the efficiency of operation of the labor market: a doctor working as a taxi driver does not have the same productivity as in a hospital. One possible way to reflect this is the percentage of the workforce employed in the informal sector of the economy. In addition, natural capital should also be included, which, as previously discussed, is not easy to measure but is known to be rapidly deteriorating. Table 5.1 presents a set of indicators on the historical evolution of average productivity per worker in countries with low fertility rates, which is the relevant variable for assessing the potential to offset slower growth in the working-age population.

Table 5.1 Evolution of average labor productivity in low-fertility countries and related indicators[a]

	Average labor productivity				Annual growth of average labor productivity			Labor force composition	
	1990	2000	2010	2023	1990–23	2000–23	2010–23	Highly skilled	Informal
Argentina	41,416	49,769	62,723	55,928	0.9%	0.5%	−0.8%	80%	50.4%
Chile	30,081	42,887	51,135	58,023	2.0%	1.3%	0.8%	85%	27.4%
China	2,919	7,126	17,749	40,317	8.3%	7.8%	5.6%		
Costa Rica	29,839	34,586	41,952	56,692	2.0%	2.2%	2.0%	75%	37.1%
Cuba									
South Korea	31,163	51,872	73,938	87,381	3.2%	2.3%	1.1%	74%	
Malaysia	30,572	44,830	52,280	64,306	2.3%	1.6%	1.4%	78%	
Russia	58,012	40,401	62,816	79,469	1.0%	3.0%	1.6%	80%	17.2%
Thailand	16,430	20,896	29,235	37,221	2.5%	2.5%	1.6%	81%	65.0%
High Income Countries	72,746	83,881	95,000	111,330	1.3%	1.2%	1.1%	77%	13.0%

Source: World Bank, World Development Indicators

[a]Average labor productivity is measured as GDP (USD 2021 at PPP) divided by the number of people in the labor force

A comparison of the historical performance of productivity shows significant heterogeneity within the group, and only China performed in a way that would allow it to offset the effects of declining birth rates. The rest would find it difficult to achieve this, since virtually all of them exhibit a trend of slowing productivity growth over recent decades, a particularly challenging issue in the two South American countries. Apparently, the problem does not lie so much in workers' qualifications as in deficiencies in other areas, such as low investment (Argentina) or declining productivity in natural resource sectors (Chile). Another potentially relevant factor is informality, which is particularly high in Latin American countries but also affects countries in Asia and Africa.

One aspect not reflected in this table is the impact of new technologies, as their adoption depends on countries' capacities to do so. They could either accelerate productivity or lead to regression. Among these, robotization and the rapid adoption of artificial intelligence stand out for their potential to free up human resources across a wide range of activities and sectors. The usual focus of these processes is the threat to existing jobs, but they also represent a highly relevant option for increasing overall productivity, thereby at least partially offsetting the future decline in the workforce due to demographic reasons. The combination of these two trends offers great opportunities, but also very concrete threats to the ability to generate income in activities where workers are replaced by robots or AI. What is clear is that, at least in this area, current demographic trends will facilitate adjustment in the face of the emergence of technologies that substitute human capital with produced capital.

A final word on the future evolution of labor productivity and natural capital. One consequence already being observed because of climate change and actions to control emissions and to mitigate their impacts is that a large amount of investment is being directed toward obtaining the same amount of energy, but at higher monetary costs than those provided by fossil fuels. From the perspective of the monetary value of the services rendered by these investments, compared to energy generation from coal and hydrocarbons, this will be recorded as lower productivity. However, it should not be forgotten that the underlying problem is that markets have not valued the damage and costs associated with the use of fossil fuels for energy generation. If costs were properly measured, China's productivity gains, which relied heavily on the massive use of coal and hydrocarbons for energy generation, would certainly be lower than those reported in national accounts. But the problem is not merely an accounting one: governments, companies, and households are already investing large sums to prevent damage resulting from climate change. These investments are not aimed at providing more goods and services for families, but simply at protecting capital (both produced and human) from future climate-related disasters. An example is the current efforts to recover what was destroyed by wildfires in California or by floods in Spain. To this the direct effects of climate change on economic activities must be added, reflected in lower crop yields, port shutdowns due to storm surges, and so on. All of this results in lower labor productivity and makes it more difficult to offset the economic impacts of declining birth rates. It is not an impossible goal, but it will be highly demanding and will likely require a reassessment of future expectations for economic well-being.

5.3 The Challenge of Financing Old-Age Expenditures: The Dilemmas of Pensions

The aging of the population poses serious challenges for both personal and public finances. This is evident in various areas, but especially in all matters related to Social Security, where old-age expenditures are financed in part by the wealth accumulated by individuals during their working lives, contributions from current generations, and fiscal transfers.

Population aging, resulting from a sharp decline in fertility rates accompanied by increased longevity, is already leading to a significant rise in old-age dependency ratios, which poses serious challenges for the sustainability of these systems. As is well known, personal spending on health and long-term care increases significantly upon entering the so-called "fourth age," at a stage of life when individuals' incomes come, to an ever-greater extent, from pension systems and accumulated savings.

To illustrate the economic magnitudes involved in this phenomenon, Table 5.2 presents public pension expenditure levels for a group of countries, both for the current period and for long-term projections. These data are presented as a proportion of the GDP and come from two main sources: OECD studies on pensions and a regular publication produced by the European Commission to project the fiscal effects of aging.

Table 5.2 Actual and projected public expenditure on social spending components affected by population aging (percentages of GDP)

	Pensions		Health		Long-term care		Education	
	2020	2045/2050	2020	2070	2020	2070	2020	2070
Germany	10.3%	12.1%	7.4%	8.1%	1.6%	2.3%	4.0%	4.5%
Spain	12.3%	13.2%	5.7%	8.6%	0.7%	1.5%	3.6%	3.2%
France	14.8%	14.6%	8.4%	9.7%	1.9%	2.8%	4.4%	3.9%
Italy	15.4%	17.3%	5.9%	7.3%	1.7%	2.7%	5.0%	3.1%
European Union	11.6%	12.7%	6.6%	7.7%	1.7%	2.9%	4.1%	4.0%
United States	5.2%	6.1%						
Japan	9.0%	9.1%						
United Kingdom	7.2%	8.1%						
South Korea	1.3%	5.9%						
China[a]	4.1%	9.5%						
India[a]	1.0%	1.0%						
Indonesia[a]	0.8%	1.2%						
Russia[a]	9.1%	12.4%						
Brazil[a]	9.1%	16.8%						
Mexico[a]	1.8%	3.0%						
South Africa[a]	2.2%	3.3%						

Sources: Pensions: OECD (2023); Other: European Commission (2021)
[a]Source: Rouzet et al. (2019)

The table shows high initial levels of public spending in almost all countries and projects significant increases, especially in some emerging economies such as China and Brazil. The increase in pensions is the item with the greatest fiscal impact resulting from population aging.

This is a particularly relevant issue because, in practice, it has become a key component of the "social contract" in societies that have moved beyond extreme poverty. This originates from the design of the original pension systems, which assigned a central role to the State in collecting contributions from active workers to finance the pensions of retirees. These systems, known as "pay-as-you-go" schemes, effectively made the State the guarantor of income levels in old age sufficient to maintain living standards established during the working years. As will be seen, these systems are particularly vulnerable to increases in old-age dependency ratios, and has already caused parametric adjustments or even structural changes to preserve the solvency of the systems. For this reason, this chapter focuses exclusively on the issue of pensions, while recognizing that it is only one part of the challenges arising from the ongoing demographic adjustment.

In a world where people rationally plan for their future and where preferences are consistent over time, individuals would save enough during their working years to finance their consumption once they stop working. Since life is full of surprises, a prudent planner would also invest in insurance to protect consumption against the most likely adverse contingencies.

However, the real world does not work this way, and "impatience" combined with time inconsistency results in insufficient personal savings to ensure continuity of living standards during old age. Therefore, public policies are needed to secure savings from labor income to be allocated to finance old age.

The way in which people's expenses are financed during old age, when they are no longer able to participate actively in the labor market, varies greatly between countries depending on how their pension systems are organized.

Although these systems may differ in many respects (Arellano 1980), there are two aspects that are crucial in determining their vulnerability to population aging: one concerns whether the (mandatory) savings of workers are used to pay the pensions of the current generation of retirees, or whether they are accumulated to pay the future pensions of those saving today. The former are called pay-as-you-go systems, and the latter, funded systems.[5]

The other dimension concerns the distribution of risks regarding the value and certainty of pensions, which is fundamental when making decisions whose impact will be felt many years later.

In very general terms, two systems for determining pension value are usually distinguished: *defined benefit* systems, where the value is established by calculation rules for pensions linked to past labor incomes, that are known in advance by the generation of workers who save a fixed fraction of their income (generally through taxes allocated to Social Security). The others are *defined contribution* systems,

[5] Average labor productivity is measured by GDP at PPP in 2021 USD divided by the labor force (as estimated by the ILO).

where the amount to be allocated to finance the pension is known, but the value of the pension will depend on the returns of the fund's investments as well as life expectancy at retirement.

Defined benefit systems have historically been associated with pay-as-you-go schemes, since a mechanism of intergenerational transfers would not be politically viable without providing certainty about the future pensions of those who pay current benefits. Until a few decades ago, these systems were considered safe, because the State often assumed the role of collecting contributions and paying benefits, something that worked very well while economies and the working-age population were growing rapidly, as occurred in the post–World War II period, and governments used surpluses to create reserve funds and also to finance their deficits. The combination of entering the less favorable phase of the demographic transition, together with increased longevity, has put these systems under strain, as they are rapidly depleting their savings and facing growing deficits.

The political economy of adjusting benefits for a population whose political weight increases as its share of the total population grows is complicated, especially in democratic systems. Therefore, it is not surprising that this has been reflected in a sharp increase in pension spending, and the outlook is for further increases, even after the adjustments already made in recent years in the context of efforts to restore fiscal solvency in many European countries, as evidenced in Table 5.2. The levels of public spending are related to the parameters used to calculate pensions in each country, retirement ages, as well as the demographic characteristics of each country.

Funded systems, for their part, guarantee ownership of pension savings and their returns, but require individuals to bear the risks associated with changes in investment returns as well as changes in longevity.

To mitigate the first of these issues, authorities tend to regulate and strictly supervise the investments made by public or private entities that manage pension funds. These systems are more recent, but in most countries with pay-as-you-go systems, there are voluntary savings schemes that complement them. It was only from the 1980s onwards that general funded systems began to be designed, usually complemented by fiscal transfers to support the poorest pensioners through general tax revenues.

As has become evident since the mid-twentieth century, pay-as-you-go systems with defined benefits are highly vulnerable to demographic change, particularly population aging, as this leads to a declining ratio of contributors to pensioners over time. Growth in real wages helps to mitigate this problem if the value of pensions remains constant in real terms, but unless wages grow at a faster rate than the number of retirees, there will be a tendency toward deficits. These deficits could be financed by increasing current workers' contributions or by drawing on the fiscal budget when the system is managed by the state, which obviously has a limit determined by the capacity to reallocate public resources, raise taxes, or increase debt.

The path taken by most developed countries with mature pay-as-you-go systems has been to change the conditions for new generations of contributors and retirees, either by reducing retirees' benefits (postponing the retirement age or reducing the value of future pensions) or by increasing contributions.

This is no minor issue, as it involves breaking a fundamental element of the social contract that underpinned democratic societies during the twentieth century. In practice, countries have adapted through a combination of increased fiscal funding, higher contribution rates, reduced benefits, postponed retirement ages, and reductions in future pensions or other benefits.

Thus, pay-as-you-go (PAYG) systems have evolved and can now be divided into two variants: the traditional (NDB) model, which operates as a defined benefit system that (in theory) does not change over time and where the current generation of workers finances the pensions of retirees. There is another, more recent variant, which originated in Sweden and has since been adapted in various countries (NDC), where pensions are determined based on the entire personal contribution history to the system, recorded in so-called "notional accounts," but which maintains the pay-as-you-go logic, with pensions financed by current workers' contributions. This system also offers additional flexibility: workers can postpone retirement, thereby improving their contribution record to increase their pension. Unlike the traditional system, in this model pension rights are regularly recalculated to keep the system solvent,[6] which in practice makes it a pay-as-you-go system with defined contributions but with adjustable (possibly downward) benefits.

The main advantages of NDCs over NDBs are twofold: first, they explicitly incorporate all social security contributions in the calculation of pension rights, thereby reducing incentives for informality and the inflation of calculation bases in the final stages of working life, and they leave the decision of when to retire in the hands of individuals.

Second, they include automatic adjustment mechanisms for the value of each generation's pensions to ensure the system's financial solvency. Since the actuarial calculation of solvency involves several generations, this allows for a smoother adjustment, especially if, at the beginning of the process, contribution rates are high enough to accumulate a surplus.[7]

The issue of the economic impact of higher old-age financing costs in middle-income countries changes the global dimension of the problem, as these countries account for most of the world's population. Moreover, the risk is greater for them because they are at a very advanced stage of their demographic transition (DT), which is also occurring much more abruptly than in developed countries. In fact, China has just entered the stage where the working-age population begins to decline in absolute terms.

Some Latin American countries such as Brazil, Chile, and Mexico are a decade or slightly more away from that milestone. Brazil has a traditional, though segmented, pay-as-you-go defined benefit system. Chile and Mexico opted for individual capitalization account models with defined contributions (FDC). Their pension systems are very different and increases in old-age dependency ratios will have very different impacts on them.

[6] The acronyms NDB and NDC to refer to the two variants of pay-as-you-go (PAYG) systems are taken from Holzman and Palmer (2006).

[7] See Holzman and Palmer (2006) and Holzman et al. (2012).

A World Bank study published in 2011 shed light on the impact of population aging in selected Latin American countries, specifically identifying the pension component. It shows that countries with PAYG systems faced tremendous fiscal solvency challenges according to that study. Among them, some have introduced significant reforms to their pension systems, as in the case of Brazil, to mitigate their future fiscal impact.[8] Others, such as Chile, with an FDC system, introduced non-contributory pensions financed by the state to address the problem of insufficient self-financed pension benefits, while at the same time increasing contribution rates. Many adjustments are still pending to limit the impact resulting from the speed of their demographic transition.

It is important to note that some of these countries have very early retirement ages and extremely generous pension calculation mechanisms, which is possibly a reflection of the fact that when the parameters of their respective pension systems were defined, they reflected the prevailing reality in which old-age dependency ratios were very low. However, as the demographic transition has progressed and this ratio has increased, solvency problems have become evident.

Beyond the financial sustainability issues of pension systems arising from demographic changes, in developing countries the widespread prevalence of informal employment severely undermines the possibility of financing income that would allow individuals to maintain their standard of living in old age. Informality results in very few workers making full and regular contributions to Social Security, making impossible to finance adequate pensions, regardless of whether the system is pay-as-you-go or fully funded.[9]

Figure 5.7 shows the labor informality rates estimated by the ILO for a group of countries with per capita income equal to or lower than USD 40,000 per year

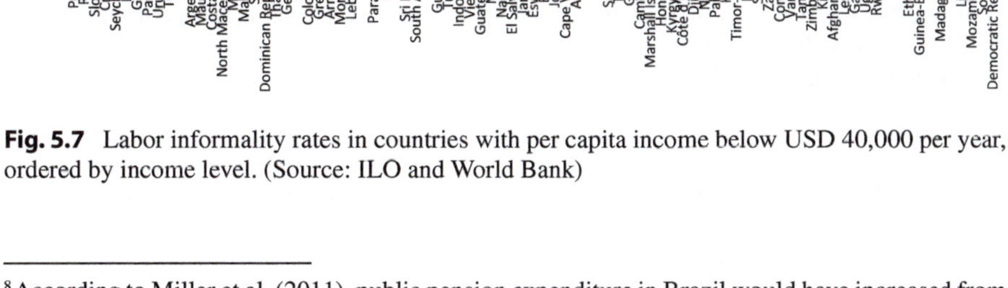

Fig. 5.7 Labor informality rates in countries with per capita income below USD 40,000 per year, ordered by income level. (Source: ILO and World Bank)

[8] According to Miller et al. (2011), public pension expenditure in Brazil would have increased from 7% of GDP to 25% of GDP by 2050, without a reform of the system.

[9] See, for example, Bosch M. et al. (2013).

according to World Bank statistics. It is important to highlight that the countries are ordered from left to right from the highest income (Poland) to the lowest income (Democratic Republic of the Congo). There is a very strong negative correlation between labor informality and per capita income, which is reflected in the dotted line.[10]

The chart also highlights in orange those countries whose informality rates are above the trend line, which, as expected, is approximately half. However, this distinction allows us to identify a particular feature of the region comprising Latin America and the Caribbean. Although almost all these countries have incomes above average, the vast majority are above the trend line in terms of informality. Of 16 countries in that region, only 5 are below the trend line and 11 are above it.

These countries are further along in their demographic transition and are aging rapidly, with a growing share of the population reaching retirement age, only to find that pensions are insufficient to maintain their standard of living, thereby putting significant pressure on public finances to address these issues. If this is not properly addressed, it could lead to future fiscal crises that may ultimately hinder development in these countries.

This is only one dimension of the challenges that population aging poses to social security systems worldwide, but which are already beginning to affect emerging countries. In addition to the issue of income security in old age, there is also the growing pressure on health systems, which until a few decades ago were heavily focused on childbirth and pregnancy care and on preventing infant mortality and malnutrition but now must increasingly focus on the treatment of chronic and age-related diseases. On the other hand, when many of these countries have not yet achieved adequate coverage in preschool education, they are already beginning to feel the pressure to provide intensive care for a growing share of the population that is no longer self-sufficient.

For emerging countries to have any chance of successfully addressing these problems, which until recently were typical of wealthy countries, they will necessarily have to increase their income, which will undoubtedly place greater pressure on the environment.

References

Arellano JP (1980) Sistemas alternativos de seguridad social: Un análisis de la experiencia chilena. Colección Estudios CIEPLAN N°4

Bar-On YM, Phillips R, Milo R (2018) The biomass distribution on Earth. Proc Natl Acad Sci 115(25):6506–6511

Bosch M, Melguizo Á, Pages C (2013) Better pensions, better jobs: towards universal coverage in Latin America and the Caribbean. IDB

Diamond J (1997) Guns, germs, and steel: the fates of human societies. W. W. Norton

[10] The choice of USD 40,000 per capita (PPP) threshold is not arbitrary. When all countries for which these data are available are included, it is found that, with only a couple of exceptions, informality rates are below 5% in all of them.

European Commission (2021) The 2021 ageing report: Economic & budgetary projections for the EU member states (2019–2070) (Institutional Paper No. 148)

Holzman R, Palmer R (eds) (2006) Pension reform: Issues and prospects for non-financial defined contribution (NDC) schemes. World Bank

Holzman R, Palmer E, Robalino D (eds) (2012) Non-financial defined contribution pension schemes in a changing pension world. World Bank

Miller T, Mason C, Holz M (2011) Impact of demographic change in ten Latin American countries: Projecting public expenditures in education, health and pensions. In: Cotlear D (ed) Population aging: Is Latin America ready? World Bank, pp 233–272. https://documents1.worldbank.org/curated/en/117291468299147625/pdf/588420PUB0Popu11public10BOX353816B0.pdf

OECD (2023) Pensions at a glance 2023: OECD and G20 indicators. OECD Publishing. https://doi.org/10.1787/678055dd-en

Ritchie H (2021) The world has lost one-third of its forest, but an end of deforestation is possible. Our world in data. https://ourworldindata.org/world-lost-one-third-forests

Rouzet D et al (2019) Fiscal challenges and inclusive growth in ageing societies. OECD Economic Policy Papers, No 27. https://doi.org/10.1787/c553d8d2-en

United Nations (2024) World population prospects 2024: Online edition

Part III

Inefficiencies and Excesses in the Use of Natural Resources

The combination of population growth and rising purchasing power is generating enormous pressures on the availability of natural resources, as well as on the capacity of ecosystems to absorb the direct impacts and waste generated by economic activity. This section will analyze the use of natural resources in the production of goods and services to meet consumer demand, their limitations, and their environmental and economic impacts.

Chapter 6 analyzes the evidence on limits to global economic growth arising from the potential scarcity of natural resources.

Chapter 7 examines the microeconomic challenges posed by the sustainable exploitation of renewable natural resources and the reasons why many of them are overexploited to the point of jeopardizing their future availability, with a specific focus on fisheries and forest resources.

Chapter 8 focuses on the challenges of sound management of exhaustible resources, with an emphasis on the capture of rents and the use of these to ensure sustainable flows over time.

Chapter 9 discusses the environmental impact of natural resource use and the potential constraints this could impose on sustainable economic development.

The final chapter of this section differs significantly from the previous ones, as it is devoted to the analysis of the macroeconomic challenges posed by the volatility of natural resource prices for countries that depend on them and on the policies to address them. While it could be argued that this issue lies outside the scope of the sustainable development of countries, the experience of many nations, particularly in Latin America and Africa, shows that if this is not properly managed, successive economic crises prevent the development of long-term visions and policies, including environmental concerns, and therefore it cannot be absent from the discussion in emerging countries rich in natural resources.

Natural Resources as a Constraint on Global Growth

6

Abstract

Historically, the availability of natural resources was a major constraint to economic progress and population growth, usually solved by people moving to other places. The Industrial Revolution removed that constraint and Nature disappeared from theories of economic growth. The issue reappeared in the early 70s, especially after the first oil shock. This chapter deals first with some distinct features of natural resources (heterogeneity, externalities, limited availability) that must be accounted for when analyzing their role in the economy. In what follows signals of scarcity, as reflected by prices, are compared with the use of exhaustible resources, concluding that, so far, there is little evidence that their availability has been a major restriction on economic growth. This, despite exponential growth of their use and consumption. Whenever there are prolonged price spikes, innovation coupled with investments have relaxed potential constraints. However, this works well only in cases in which prices reflect scarcity, something common with provisioning services, but absent in many ecosystem services that could eventually cause severe degradation for some critical variables for the economy, such as food production.

Keywords

Natural resources · Resources constraints · Malthus · Solow · Technological innovation · Economic rents · Commodity prices · Food · Metals · Oil

6.1 Heterogeneity of Natural Resources and Economic Rents

Natural resources differ from industrial products because their availability depends on nature and are limited. Since they are not evenly distributed across the planet, they are heterogeneous. As explained in Part I, this characteristic requires departing

J. Vial, *Sustainable Development and Natural Resources*, Integrated Science 39,
https://doi.org/10.1007/978-3-032-12263-6_6

from the standard theory of the firm, which in long-term equilibrium operates with constant returns to scale and can indefinitely increase production by expanding the use of capital and labor.

The heterogeneity of natural resources is reflected in their extraction costs. This means that the structures and levels of total and unit costs are highly diverse within the same sector. One way to represent this heterogeneity is through the unit cost curve of the activity that extracts a given resource. To illustrate this, suppose the activity is mining, where the industry cost curve is the envelope of the unit cost curves of the various deposits, ordered from lowest to highest cost.

Figure 6.1 represents equilibrium in the market for a natural resource. In this market, there are three firms: Firm 1 has the lowest unit costs, represented by the CF1 curve. Next, to the right, is the production of Firm 2 and its respective unit costs CF2, and finally Firm 3, whose unit costs (CF3) coincide with the point where market supply and demand are equal at the equilibrium price. In this equilibrium, Firm 1, which produces Q1, has a unit cost equal to CU1, which means it earns a rent equal to Rent F1. Firm 2 earns a rent equal to Rent F2, and the marginal firm (Firm 3) earns no rent.

A crucial point in this analysis is that the costs of each firm include both operational and capital costs, so even Firm 3 has incentives to produce, since the invested capital is earning what it would obtain in alternative investments. F1 and F2 earn rents, which originate from the natural advantages of each deposit: mineral content (grade) per ton of extracted material, water availability, distance to transportation infrastructure, etc.

The fact that these extraordinary profits are due to conditions not created by the company's management, and that the company would still find it worthwhile to operate even without them, creates the opportunity (and the incentive) for all stakeholders (company owners, the state, workers, local communities, etc.) to capture all or at least part of these rents.

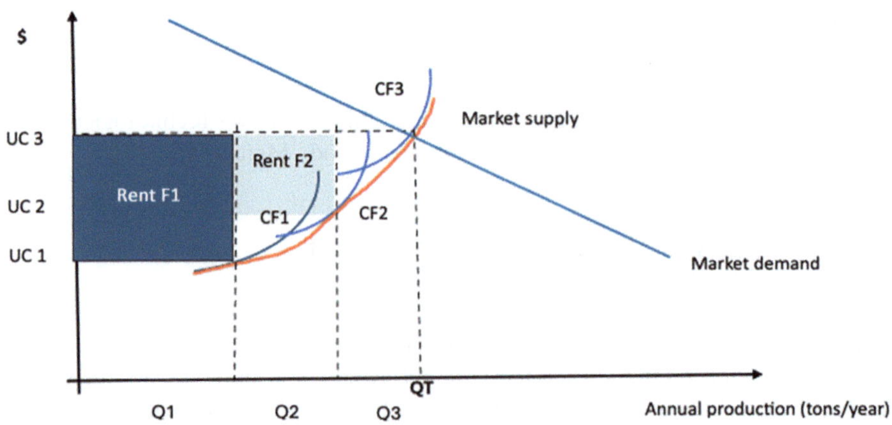

Fig. 6.1 Hypothetical cost curve of an industry exploiting natural resources

On the other hand, since the rent per unit of product is the portion of unit income the firm receives due to nature's contribution to the business, it is a key input for valuing natural capital services.

Since nature does not have a representative to exercise property rights, tension arises among the various actors involved in generating rents. In a free-market economy, the investor, as owner of the deposit, has the first claim to these rents. However, given that a country's natural resources are a common good, it is generally accepted that the state should receive at least a portion of the rents, either through taxes or royalties and, in some extreme cases, by acting as owner through a state-owned enterprise. Workers also typically seek to capture part of the rent through regular wage negotiations, so it is not surprising that unions in mining or oil companies are extraordinarily strong, partly because they have workers capable of contributing resources to hire good lawyers and to finance the basic needs of their members during prolonged strikes.

While from a theoretical standpoint it is relatively easy to justify a rule for distributing rents that addresses the tensions, putting it into practice is extremely complex. On the one hand, a general rule has the drawback of not considering the heterogeneity among firms mentioned earlier: within the same activity, there may be very different unit rents (even zero or negative in the short term) among the various companies involved.

Secondly, in a world with highly varied risks (geological, climatic, input availability, technological and political changes that can affect access to resources or even taxation itself, among others) and very long planning horizons, it is very difficult to calculate the true opportunity cost of capital. To some extent, it can be argued that rents are simply part of the reward for taking on the risks of activities such as mining or agriculture.

Finally, and this is particularly true in the exploitation of non-renewable resources, unit costs tend to increase over time precisely because previous extraction leaves less resources available for the future, which is obvious in the case of oil or a mineral. In the latter, this is reflected in the decline of the mineral "grade," that is, the proportion of mineral relative to the amount of the amount of earth and rock removed decreases. In addition, transportation and mineral processing costs also increase.

In other words, when calculating unit rents for extractive activities involving natural resources, it is necessary to account for cost differences between companies, as well as changes in these costs over time. A careful assessment of the risk-adjusted cost of capital specific to the activity and the characteristics of the resource being extracted is also required. This is not easy, and if rents are overestimated for the purpose of calculating specific taxes, also known as royalties, investment will be discouraged.

Furthermore, when resources are extracted in different jurisdictions and large companies operate in several of them simultaneously, it is also important to consider the advantages and costs in countries with competing projects, so as not to unnecessarily discourage investment in one's own country.

Heterogeneity in costs, their increase over time, and the existence of rents are distinctive features of economic activities related to the exploitation of natural resources, from agriculture to mining, and even aquatic resources.

Another challenge worth highlighting stems from the fact that these are natural resources, so their availability depends on nature and, therefore, determines their location. A garment manufacturing plant, for example, can be set up anywhere in the world where the availability of labor, capital, and access to transportation allow it to operate profitably. On the other hand, a banana plantation can only operate where the climate is suitable and there is access to soil and water with the required characteristics for cultivation. Whether that plantation is near urban centers or in an isolated jungle location makes a huge difference in terms of operating and processing costs.

Likewise, if demand for clothing increases, the owners of the manufacturing plant can invest, hire employees, and expand production without significantly affecting their unit costs. In the case of a mineral deposit, for example, increasing production also requires investment and hiring additional personnel, but unlike the previous case, unit costs will almost certainly be higher, either because the ore grade decreases, requiring more material to be processed to obtain the same amount of mineral, or because more water and energy are needed for processing.

For the same reason, a mine has a limited useful life, which can often be extended by adopting new technologies and making further investments, but at some point it will lose its cost advantages and begin to generate losses, forcing its closure.

In the case of a plantation, if the soil maintains its fertility, the plants can even be replaced as their productivity declines, but this is almost always accompanied by increased use of fertilizers to compensate for nutrient loss, as well as pesticides, since over time new species of insects or fungi adapt to this new food source. The industrial plant will require investment to replace worn-out capital and to keep up with technological advances, but it does not face a natural limit to its survival.

It is often thought that activities related to natural resource exploitation do not innovate because they sell relatively homogeneous, undifferentiated products, and therefore do not need to entice potential customers with something new. The reality is more complex, as innovation in production is essential to remain competitive in activities subject to increasing costs.

Fruit growing provides an interesting example in this regard. On the supply side, major innovations to increase productivity and remain competitive have come from the introduction of increasingly efficient automated irrigation techniques to save water, a resource that is becoming ever scarcer in this activity, as well as the incorporation of mechanization as rising labor costs make it necessary. Another way to increase productivity is to apply increasingly effective fertilizers and pesticides. However, this gives rise to problems such as soil and water contamination, as well as impacts on biodiversity and human health. In both cases, there are externalities that are not reflected in market prices, but which are increasingly attracting the attention of the public and authorities, especially at the local level. This has begun to be reflected in regulations and the threat of rejection by consumers at the international level.

These are risks, associated with impacts on nature, that are increasingly being incorporated into companies' decision-making processes. In practice, this is reflected in rising costs not only due to direct natural limitations, but also due to indirect constraints arising from growing public concern over environmental degradation. The counterpart to this is the possibility of product differentiation through certification of good environmental practices. The trend toward favoring organic crops, for example, can be a positive differentiating factor that could help maintain competitiveness, despite lower physical productivity.

But in the case of fruit, innovation to improve productivity is not enough. Since fruit is a relatively sophisticated consumer product, it is common to observe significant differences in preferences among consumers in different countries, as well as changes in preferences for certain attributes within a single country. This drives research to develop new varieties or improve aspects such as flavor, color, aroma, firmness, and shelf life. At the same time, as in other industries targeting sophisticated consumers, there are also those who learn from the successes of others and develop their own capabilities, leveraging local advantages (water, climate, labor costs, market access, etc.) to enter the market—often successfully—competing with those who were first to penetrate developed country markets.

On the other hand, there is also the threat of product substitution in end uses, a risk that is exacerbated as global exploitation of natural resources cannot keep pace with rapidly growing demand. This is particularly relevant for non-renewable and capital-intensive resources. In the short term, this may lead to price increases and be beneficial for countries and companies with cost advantages, allowing them to capture greater rents. However, over the longer term, this will incentivize the substitution of these materials in their end uses, or the development of new resource sources to meet demand, which will no longer depend on traditional ones.

Therefore, technological advances and innovation is critical in sectors based on natural resources, particularly non-renewable ones. This issue is complex because, on the one hand, it poses potential threats to current producers, but on the other, it is the very tool that has enabled production to keep pace with the expansion of demand associated with industrial uses of these resources. This can be well illustrated by the case of copper.

The more pessimistic view tends to emphasize the role of innovation as a threat due to the substitution of resources in their traditional uses. This has occurred in the case of copper with the introduction of fiber optics in telecommunications and PVC pipes in construction. However, the loss of these markets has been more than offset by the growing demand for copper in electrical energy transmission for all electronic applications.[1]

[1] This may change in the future due to the large volume of resources being devoted to developing synthetic materials with "superconductor" properties.

But innovation is not limited to demand. Innovations in production processes reduced costs and enabled the exploitation of deposits that were previously not economically viable. A current example is the reduction in the costs of seawater desalination, which has allowed the development of large desalination plants in the Atacama Desert to supply water to Chilean mining operations, once natural freshwater sources became insufficient. Along the same lines, electricity generation from solar energy has advanced to the point that it is now gaining importance in supplying mining operations in northern Chile, taking advantage of the high continuous solar radiation.[2]

Technological advances also pose potential threats to traditional producers. In the case of copper, one such threat arises from submarine exploration, which has identified significant deposits of nearly pure copper on the ocean floor. Extraction using available technologies is feasible, although not yet economically viable, and is further complicated by the regulations of the Law of the Sea Convention.[3] A longer-term review of copper production and prices reveals a slight downward trend in price. However, this trend disappears, and even marginally reverses, if data prior to 1920 are excluded. Until that time, only underground deposits could be exploited on a large scale, a situation that changed with the development of new machinery for large-scale earth and rock removal, enabling open-pit mining with significant cost reductions. One of the first open-pit mines in the world was Chuquicamata, which until a few years ago still held the world record for size in this type of operation.

Figure 6.2 also shows long-duration cycles, such as those in the mid-twentieth century, where the innovation was combined with the negative demand impact of the Great Depression. It also clearly shows the effect of increased demand from emerging countries following the rapid growth of the Asian Tigers and, especially, China at the end of the twentieth century and in the first decade of the present century.

Figure 6.3 shows that the proportion of copper use per unit of GDP at the global level declined significantly from the second half of the 1960s, explained both by copper substitution in certain uses and by the increasing weight of services in GDP, especially in developed countries. Toward the end of the last century, this downward trend eased, possibly reflecting the strong growth of emerging Asian economies. In simple terms, this means that as GDP increases, copper consumption also rises, but at a somewhat lower rate. This is a trend unlikely to reverse soon, but it is consistent with the fact that copper extraction has become more costly, especially after the period of strong demand expansion at the beginning of this century.

[2] Incidentally, it also helps reduce the carbon footprint of mining, another major challenge to keep it competitive in the future.

[3] This could also change, especially if copper prices rise and the profitability of traditional mining operations declines. In fact, some small Pacific island states are already granting exploration licenses to multinational companies, and some states in developed countries, including the United States, are authorizing such operations.

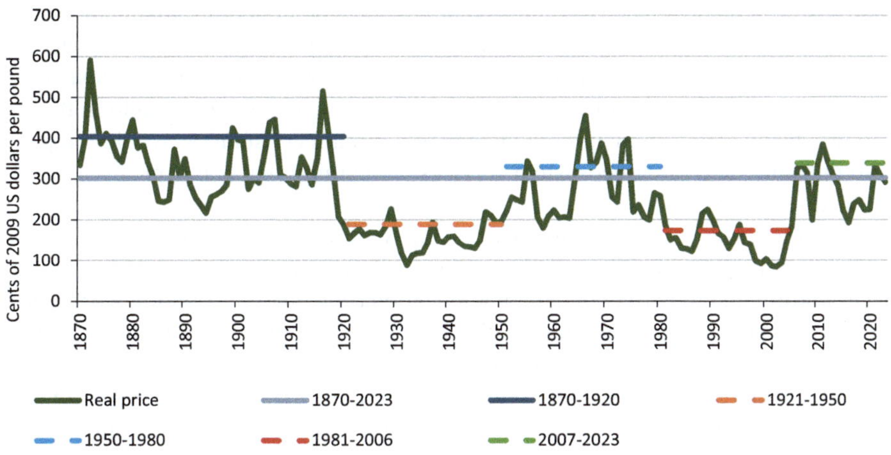

Fig. 6.2 Long-term evolution of the real price of copper (US$ 2009 cents/pound). (Source: Author's calculations based on LME and Díaz et al. (2007) data using the US Consumer Price Index)

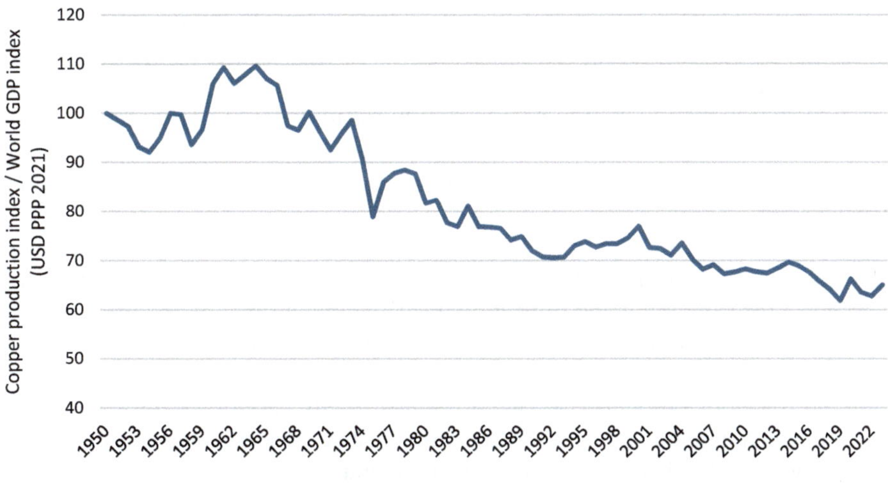

Fig. 6.3 Evolution of global copper production and its relationship to global GDP (Copper production index/Global GDP index, USD PPP 2021). (Source: World Bank and Maddison Project for global GDP, and USGS and Cochilco for global copper production)

6.2 Scarcity of Natural Resources as Constrains Global Economic Growth

The old issue of scarcity of natural resources as a limit to global growth re-emerged in the 1970s and comes up frequently. At the end of the day, the Earth is a closed ecosystem and there are absolute natural limits to provisioning services. Although

Malthus presented his ideas in a different context, the concern remains the same: with agricultural land being a fixed resource and population growing at exponential rates, food availability would adjust the population size to the carrying capacity of agricultural land. In this context, there is no economic development, and in the so-called "steady state," per capita income would remain constant, as would the pressure on nature and other ecosystem services (something that was typically not part of the discussion).

David Ricardo recognized the heterogeneity of soils (a fixed resource) in food production, both due to natural advantages (soil quality, water availability, market accessibility) and the effect of capital application (machinery, irrigation works, protection) to those soils. Thus, the profit obtained from sales, after deducting labor costs, must be divided between returns to capital and rent. The latter arises from natural advantages and does not correspond to payments required to mobilize production factors. In contemporary terms, rents correspond to the cost of using natural capital services. As these services become scarcer, the user cost increases and the unit rents of lower-cost producers rise.

In the case of industrial goods, operating with known technologies and competitive labor and capital markets, production can be expanded indefinitely without increasing marginal costs. If innovations allow for cost reductions, or if some differentiation valued by consumers is introduced, prices would change, giving rise to extraordinary but temporary profits (known as quasi-rents), since competitive forces will eventually eliminate them.

The Industrial Revolution and its aftermath led to the gradual disappearance of concerns about natural resource scarcity from economic discussions, relegating them mainly to analysis of the difficulties faced by poor countries in achieving development.

In modern economics textbooks, the issues of scarcity and heterogeneity of natural resources have virtually disappeared, and the Solow/Swan model[4] became the workhorse for explaining and projecting the economic growth of countries, conveniently based on the assumption of constant returns to scale and innovation rates that provided a floor for per capita income growth. For most of the twentieth century, the prices of raw materials and consumer goods based on natural resources showed no systematic increases that would justify concern over their scarcity, except for episodes that later proved to be temporary.

Nevertheless, there were important voices that raised concerns about the risk of resource depletion that could limit income and population growth. The best known of these was the work by Meadows et al. (1972), which was fortuitously published just a few months before the first oil crisis and the subsequent surge in energy prices that had a major economic impact worldwide.

In most studies that attempt to assess the scarcity of a resource, prices are used as a proxy. The idea is that if prices rise (relative to an index that captures general price increases), this should reflect scarcity, that is, a supply that responds to higher costs and has not been able to meet demand at the previous (real) price. However,

[4] See R. E. Solow (1956) and T. Swan (1956).

the true measure of scarcity should be the unit rent of products, which is the relevant variable for those making investment decisions. To illustrate with an example, the price of a product may fall because of the introduction of new production techniques that lower costs and increase the supply of that resource, and vice versa, for example, when prices rise because the cost of refining a metal increases due to higher energy costs and not because the mineral itself is scarcer. In the first case, rents increase, and in the second, they decrease. This is why it is important to focus on rents per unit of product and the evolution of unit rents over time. While these can be estimated for specific products in particular markets, the analysis becomes more complex when dealing with raw materials (commodities) traded globally by highly diverse producers operating in very different locations, in countries with varying levels of development and, therefore, with large differences in income and wages, currencies, and institutional arrangements (such as taxation). In these cases, it is very difficult to calculate rents per unit of product, as these will differ between companies and even between different plants within the same company.

Nevertheless, in practice, the scarcity of a natural resource is reflected first and foremost in increased costs, for example, due to a greater amount of material, inputs, and distance required to extract a unit of metal from a deposit. This, in turn, negatively affects unit rents, which will eventually be reflected in higher prices.

The most obvious candidates for scarcity are those natural resources that are depleted through use, such as oil, which, when used as fuel, is burned and its natural regeneration occurs over millions of years. The case of metals is similar, although some metals can be recovered and recycled after the product in which they were used becomes obsolete, a common practice in cases such as aluminum, copper, and steel. A third interesting case is agricultural land, which, although not necessarily exhausted by use if properly managed, fertilized, and with appropriate crop rotation practices, is globally fixed in area. In many cases, it can be expanded locally at the expense of replacing previous uses (forests, shrublands, natural grasslands, etc.), to the detriment of other services provided by the ecosystems that are destroyed; however, this also has a physical limit. Technology can also alleviate this constraint, as is currently the case with the cultivation of vegetables in hydroponic pools.

A particular feature of land is that, when dedicated to agriculture, there are usually several options for crop species, allowing for relatively rapid responses (from one planting season to the next) to adjust for market imbalances in specific markets. To identify trends indicating scarcity or abundance, it is necessary to control for this latter factor, which can be achieved by using an aggregate price index of various agricultural products.

Available evidence on price increases for non-renewable natural resources shows that, in general, they have not risen systematically, reflecting a sustained increase in productivity, whether through innovations (technology, management), increased use of capital and labor per unit of nature-based goods,[5] substitution by other materials, or product obsolescence, as occurred with the use of whale oil as fuel for public lighting when electric lighting became widespread, or as happened in the

[5] Nordhaus (1992), Krautkraemer (2005).

1970s and 1980s when copper was replaced by aluminum (which is lighter) in the manufacture of automobile radiators to save weight and improve fuel efficiency. In the 2000s, with the disappearance of landline telephony in households, a significant market for that same metal vanished. Of course, innovation works both ways: it can also increase demand for these products, as is the case with copper, where demand has been driven by the replacement of fossil fuels with electricity, given copper's significant advantage in conductivity.

The following discussion on evidence of scarcity will focus on three natural resources that are highly important for human activity: agricultural land, due to its relevance for feeding a growing and wealthier population; oil, as the main energy source for transportation and a wide range of other uses; and finally, copper, a metal often used as representative of this type of industrial raw material, characterized by a very active and competitive market.[6]

In the case of agricultural land, concerns about supply stem from the difficulties in expanding food production for a human population that is growing and consuming more as incomes rise. In this case, increasing scarcity should lead to systematically rising real prices. Prebisch and Singer,[7] for their part, argued that the income elasticity of food consumption is less than 1, so even if prices rise, they will do so at a slower rate than manufactured goods with income elasticities greater than one.

Agriculture depends on land, which is a limited resource and whose expansion faces increasing challenges. Moreover, it is a fragile resource that can be easily lost to erosion, contamination, or the expansion of urban and transport infrastructure. Data tend to show a downward trend in prices over time, but this seems to be supported more by a strong expansion of supply than by a "structural" weakness in demand. While there has been some expansion of cultivated areas, mainly thanks to technological advances and investments to capture and distribute water in regions previously lacking irrigation, increases in crop yields have outpaced population growth, to the point that today, global per capita cereal production is 50% higher than in 1970 as shown in Fig. 6.4. The key to this expansion has been the development of artificial fertilizers that supply crops with nitrogen in quantities unattainable directly from nature (Smil 2022).

Synthetic fertilizers and pesticides, along with the development of new, more resilient and productive varieties, are at the core of the so-called Green Revolution, which broke through food availability barriers in much of the developing world, eradicating endemic famines in India and other countries in Asia and Africa. Although the data are not very precise, it is worth noting that it is estimated that in 1980 about 700,000 people died each year from nutritional deficiencies, and most

[6] Steel has more uses and is traded in much larger quantities than copper, but it is a market dominated by long-term contracts, and transactions in competitive markets are less representative than in the case of copper.

[7] Although both Prebisch and Singer made their contributions (theoretical in the first case and an analysis of statistical evidence in the second) in working papers written in 1949, these were not formally published until later, by which time these ideas had already spread as the Prebisch–Singer Hypothesis. See Toye, J. and R. Toye (2003).

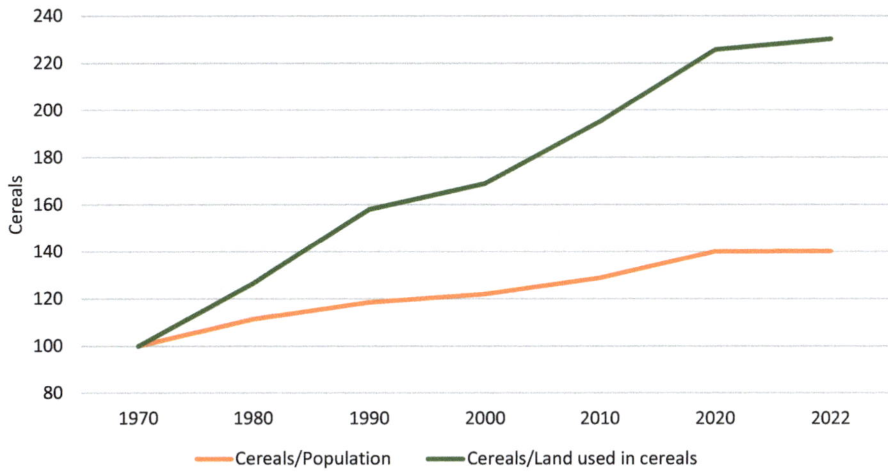

Fig. 6.4 Food availability and agricultural yields (Index, base 1970 = 100). (Source: World Bank, World Development Indicators)

of those deaths were children under 5 years old (approximately 500,000). By 2021, the total number had fallen to just over 200,000 deaths per year, and 75,000 for children under 5 years old, to 75,000.[8]

On the other hand, Fig. 6.5 shows that consumption of cereals has been falling steadily as a fraction of GDP, in line with the Prebish-Singer hypothesis, but given the expansion of global population and per capita income, innovation has been the critical factor to alleviate the natural constraints in the production of food.

Although there has been a systematic improvement in global food availability, this does not mean that famines have disappeared; however, instead of being relatively common phenomena as in ancient times, caused by climatic problems or armed conflicts, today they are isolated episodes, often related to institutional failures or conflicts. The data also show that the incidence of these crises has dropped dramatically since the last decades of the past century, as evidenced in Fig. 6.6.

In market economies, the signal of a product shortage is rising relative prices. Figure 6.7 shows the evolution of food prices, and they do not show an upward trend in real terms from a long-term perspective, which would be the clearest sign of scarcity. Although over the long run the trend in prices has been downward, it should be noted that this trend has stabilized in recent decades, coinciding with the period of rapid growth in highly populated countries such as China and other Asian nations.

Regardless of the value of the income elasticity of demand, what appears to be determining the evolution of cereal prices are technological changes on the supply side. Without these innovations, it would have been impossible to produce enough grain to feed the current eight billion people, and prices would have risen. The

[8] IHME, Global Burden of Disease (2024). OurWorldInData.org/causes-of-death | CC BY

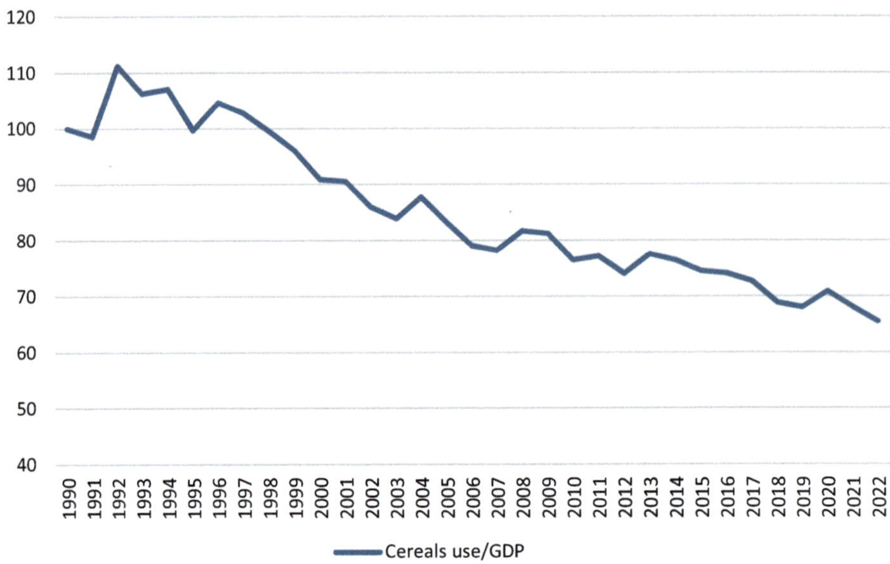

Fig. 6.5 Relationship between cereal consumption and global GDP (Index, base 1970 = 100). (Source: World Bank, World Development Indicators)

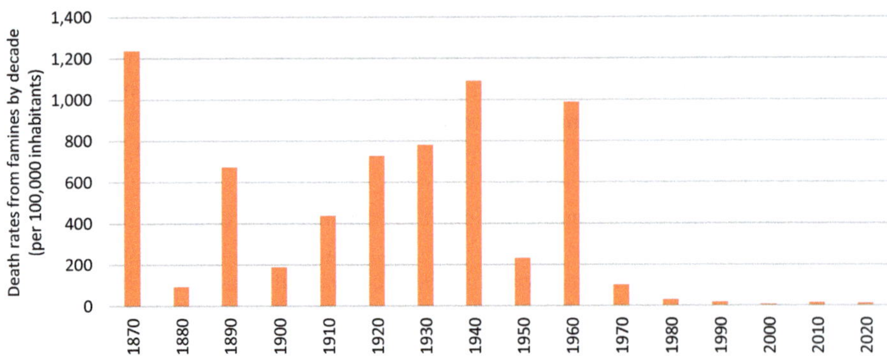

Fig. 6.6 Incidence of famines worldwide. (Source: Herre et al. 2017)

introduction of technologies that improve yields and reduce costs should be reflected in lower prices, but not necessarily in lower unit rents.

These results show that, at least so far, humanity has demonstrated the ability to successfully meet the challenge of feeding a growing population whose incomes have also multiplied, thereby increasing their consumption possibilities.

Scientific progress and its application to the development of new technologies, as well as the greater intensity of physical and human capital in food production, have played a central role in this outcome. This has increased confidence in human ingenuity and in market forces, as well as in the institutions created to promote their

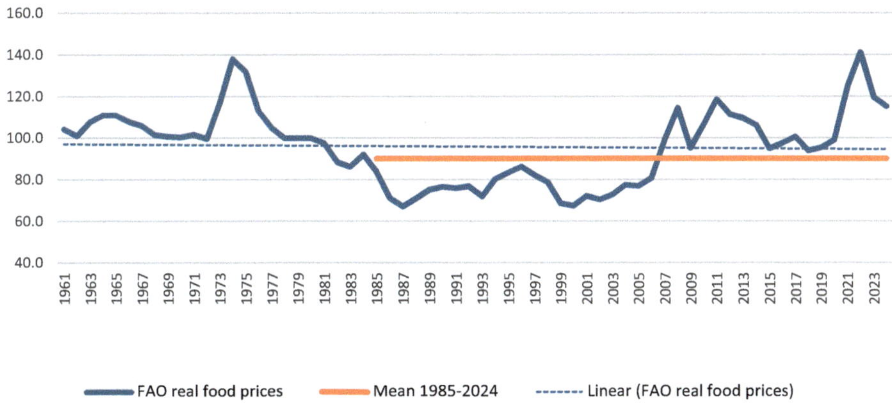

Fig. 6.7 Evolution of real food prices (FAO Index, 2014–2016 = 100). (Source: FAO)

optimal functioning and correct their failures, which, in principle, allows for relative optimism regarding the future, at least in this respect.

However, there are two caveats to this positive panorama: (a) Hunger and malnutrition persist in many parts of the world that remain very poor. (b) This increase in food production has not been without consequences: some aquifers are being depleted, fertilizers are degrading others, and the massive use of insecticides is threatening biodiversity in general, and the availability of natural pollinators in particular. Global warming may result in the loss of significant areas of agricultural land, either through desertification in regions experiencing higher temperatures and lower rainfall, or potentially through flooding of low-lying coastal areas. While it is true that currently uncultivated regions could be brought into production, it is unlikely that this would be sufficient to offset all the losses.

On the other hand, ocean acidification, an indirect effect of greenhouse gas emissions, is already putting highly productive marine ecosystems at risk. All this is occurring in a context where the loss of biomass in all major global fisheries (except for aquaculture) has reached alarming levels, causing the collapse of many fisheries and declining yields in others.

It is also important to note that these aggregate figures conceal significant differences in food consumption patterns between countries, even at similar income levels. For example, the typical North American consumer's diet contains a much higher proportion of beef than that of the typical Japanese consumer. These consumption patterns have very different impacts on resource pressure. Continuing with the previous example, on average, about 25 kilograms of grain are required to produce one kilogram of beef. For other types of meat, the grain-to-meat conversion rates are lower. At the opposite end of the spectrum is chicken, which requires about 3.3 kilograms of grain per kilogram of meat (Alexander et al. 2016).

This is relevant because, in the case of poor countries, it is not clear how much meat consumption is due to preferences or simply to financial constraints. In Latin American countries, for example, there have been sharp increases in meat

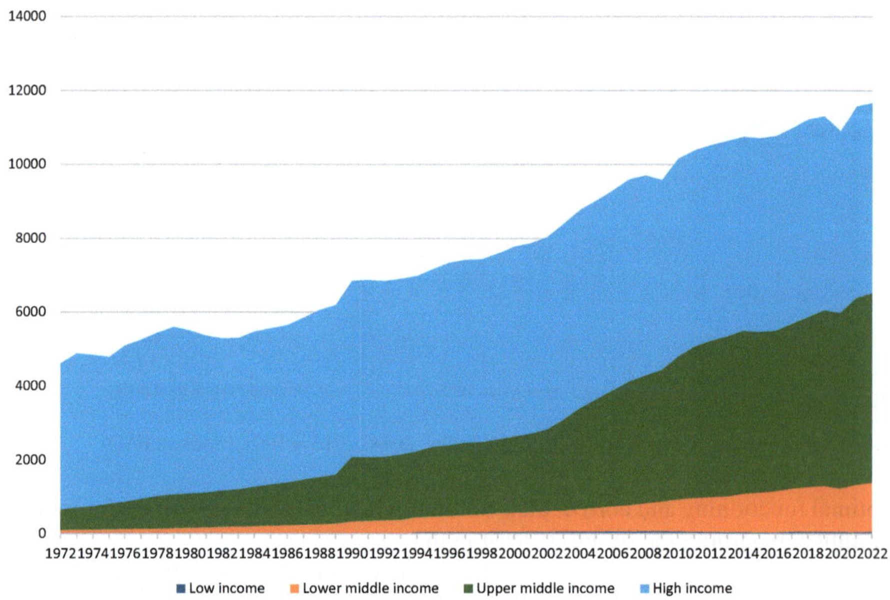

Fig. 6.8 Global consumption of fossil energy and composition by country income level (millions of tons of oil equivalent). (Source: World Bank, World Development Indicators)

consumption as incomes rise, something that has not yet been observed in some Asian countries. In China, there has been a significant increase in meat consumption in recent decades, but more concentrated in poultry and pork (which require about 6.5 kilograms of grain per kilogram of meat) than in beef.

Potential resource scarcity issues are not limited to food: Economic activity requires increasing amounts of energy and raw materials, many of which are available only in finite quantities globally. The Club of Rome drew attention to the potential scarcity of these resources, suggesting that they could become a constraint on future economic progress (Meadows et al. 1972).

To analyze this hypothesis, evidence from two products will be examined: oil and copper. Both are available in finite quantities on the planet, and while copper and other metals can be recycled, oil—at least when used as fuel—is depleted with use.

Figure 6.8 shows that in the last 45 years, fossil fuel consumption has grown almost threefold in absolute terms. Virtually all this increase has been concentrated in middle-income countries, especially those in the upper range of this income group. However, the market has remained well supplied, except for occasional disruptions, partly associated with political-military conflicts in the Middle East, a region that accounts for most of the world's oil exports. This period has also witnessed a profound transformation of these markets, where major multinational oil companies have lost power, first due to the nationalization of their assets in many emerging countries, as well as the development of new fossil energy sources such as natural gas and unconventional oil.

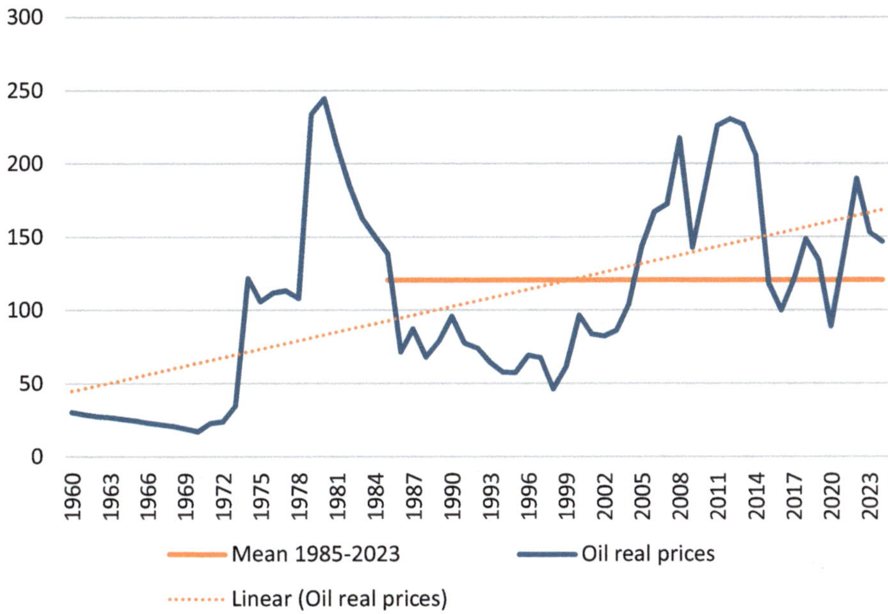

Fig. 6.9 Evolution of real oil prices (Crude oil price index deflated by US CPI, 2016 = 100). (Source: World Bank Commodity Price Data (The Pink Sheet), IMF)

When examining price trends, significant changes and an upward trend over time can be observed, as shown in Fig. 6.9. This trend persists even when the initial part of the period, dominated by the emergence of OPEC and the crises of the 1970s and early 1980s, is excluded from the calculation.

In this case at least, there is evidence of increasing scarcity, which was particularly pronounced between 2005 and 2012, coinciding with the so-called "commodity price boom," partly driven by China's rapid economic growth. This higher price level helped spur the strong development of unconventional energy sources, from shale gas and oil extraction to renewable sources such as wind and solar, which are growing at exponential rates, benefiting from efficiency gains and cost reductions associated with the scaling up of these operations.

As with food, technology is advancing rapidly, helping to contain the scarcity that would otherwise have emerged some time ago. In this case, these advances have materialized partly in response to the price increases during the "boom" period, which enabled the large-scale deployment of known technologies that had previously been too expensive to implement without government subsidies. A significant portion of these innovations resulted in a growing share of renewable energy sources in electricity generation, to the detriment of fossil fuels.

To conclude this section, the case of copper is analyzed, a metal used in multiple industrial applications and found in virtually all electrical devices and their connections, thanks to its excellent conductivity. Like many other metals, copper is also recyclable, with recycling now accounting for nearly 17% of the final supply of refined copper.

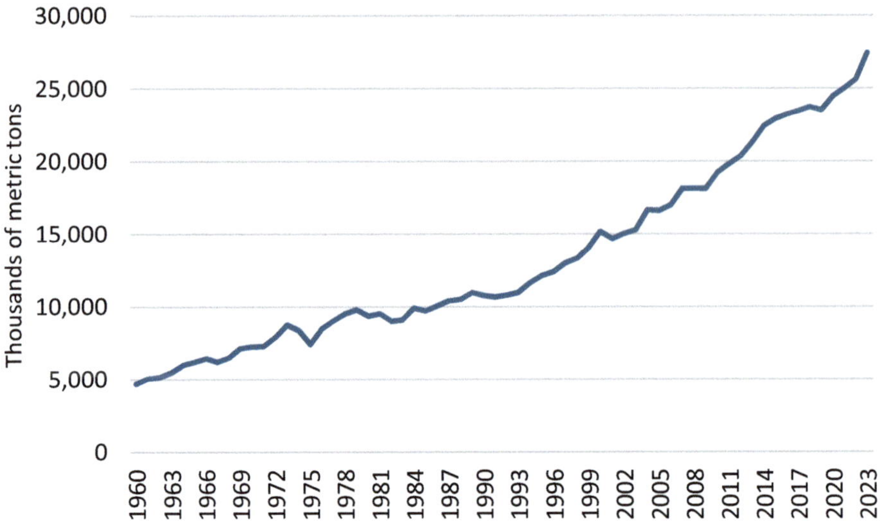

Fig. 6.10 World production of refined copper (thousands of metric tons). (Source: Cochilco)

As shown in Fig. 6.10, the growth in annual refined copper production has been even more impressive than that of food and oil, reaching almost five times the level achieved in 1960. This growth is remarkable both because it has occurred at exponential rates and because, over this period, copper has been displaced from many of its traditional applications, such as water pipes, telephone lines, automobile radiators, etc.

Figure 6.11 shows the evolution of real copper prices during this period, using, the US Consumer Price Index to correct for inflation.

In this case, a slight downward trend in price is observed. However, this calculation can be misleading because this market is characterized by overlapping cycles of different lengths. If a longer period is considered, from the late nineteenth century to the present, for example, no clear trend is observed. In contrast, if the comparison period is restricted to the years after 1985, the trend is upward. This is due, at least in part, to the fact that technological advances have been strongly operating on both sides of the market: on the demand side, facilitating the substitution of copper by other materials, and on the supply side, enabling substantial cost reductions that have made it possible to extract copper from deposits that, until a few years ago, were not technically or economically viable. In any case, both types of forces have combined in the past to alleviate potential mineral scarcity.

An interesting example is the price cycle between 2004 and 2011, also associated with the previously mentioned commodity boom. In this case, the price increase was caused by China's rapid economic growth, which drove a sharp rise in demand that led to shortages because mining production capacity takes a long time to respond. The short-term supply response comes with higher costs, so unit rents

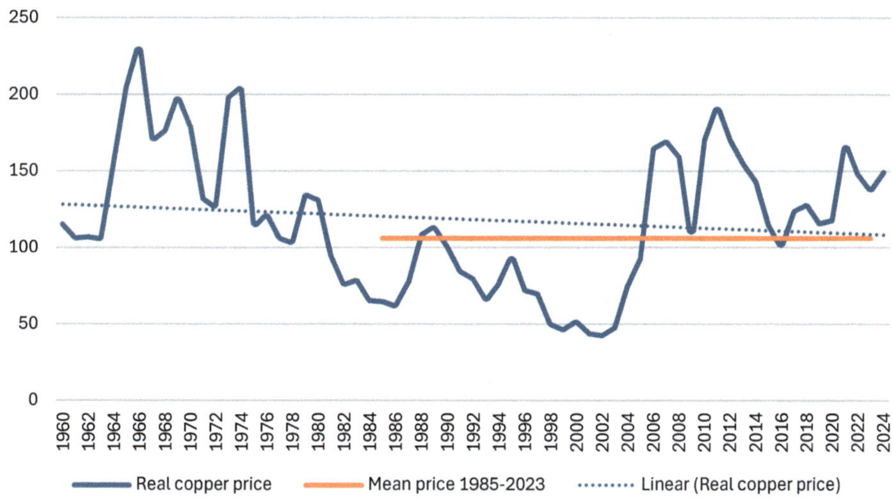

Fig. 6.11 Evolution of real copper prices (LME price, US cents per pound, deflated by US CPI, 2016 = 100). (Source: Cochilco, World Bank: World Development Indicators)

increase less than prices do. Part of these increases stem, for example, from lower ore grades, permanent wage increases, higher water usage costs, etc. However, the slowdown in global economic growth and the development of new mining projects, as well as cost reductions, allowed the price to stabilize at around USD 3 per pound for a time, but with profit margins not very different from those observed before the start of the cycle.

Looking ahead, some analysts point to a sharp increase in demand due to rising incomes in highly populated emerging countries, the substitution of fossil fuels in electricity generation, and the expansion of electricity into energy applications previously reserved for fossil fuels. However, we could also see copper being replaced in some of its current uses, or the development of new sources of supply that are not yet competitive with current technologies, such as deep-sea mining, for example. If the past is any guide, it is likely that both will occur.

In summary, considering the evidence, it is difficult not to be moderately optimistic about the power of science and technology, driven at least in part by market signals, to alleviate shortages of food and raw materials. Given that demand for these resources grows more or less in line with global GDP (exponential but moderate rates), while scientific knowledge and its applications are increasing at a much faster pace, it seems reasonable to expect that as long as prices signal scarcity, there will be responses that help to alleviate it, not always immediately, but within timeframes that avoid major economic disruptions.

What becomes clear is that technological advances, affecting either supply or demand, have been decisive in shaping the long-term trends of prices and unit profits, either by moderating demand or by increasing productivity and reducing costs.

References

Alexander P, Brown C, Arneth A, Finnigan J, Rounsevell MDA (2016) Human appropriation of land for food: the role of diet. Glob Environ Chang 41:88–98

Diaz J, Lüders R, Wagner G (2007) Economía Chilena 1810–2000. Producto Total y Sectorial. Una nueva mirada. Documento de Trabajo 315. Instituto de Economía UC

Food and Agriculture Organization of the United Nations (2022) The state of world fisheries and aquaculture 2022: Towards blue transformation. FAO

Herre B, Samborska V, Hasell J, Roser M (2017) Famines. Our world in data. https://ourworldindata.org/famines

Krautkraemer JA (2005) Economics of natural resource scarcity: The state of the debate (Discussion Paper 05-14). Resources for the Future. https://media.rff.org/documents/RFF-DP-05-14.pdf

Meadows D, Meadows D, Randers J (1972) The limits to growth. Potomac Associates

Nordhaus W (1992) Lethal model 2: The limits to growth revisited. Brookings Pap Econ Act 1992(2):1–59

Smil V (2022) How the world really works: the science behind how we got here and where we're going. Viking

Solow RM (1956) A contribution to the theory of economic growth. Q J Econ 70(1):65–94

Swan TW (1956) Economic growth and capital accumulation. Econ Rec 32(2):334–361

Toye J, Toye R (2003) The origins and interpretation of the Prebisch–Singer thesis. Hist Polit Econ 35(3):437–467

Challenges in the Sustainable Management of Renewable Resources

7

Abstract

When reviewing the situation of renewable resources, such as fisheries and forests, there is widespread evidence of overexploitation and useless destruction. They have in common the failure of market mechanisms to induce sustainable behavior, either because of absence of property rights, leading to the so-called "Tragedy of the Commons", or externalities that are not valued by markets, such as soil and water cycle regulation by forests, for example. In addition, many species are treated as exhaustible resources because their biological regeneration cycle is too long to compete with alternative use of the land, given the impatience signaled by interest rates. Different policy options to deal with these market failures are discussed, as well as the role of artificial production methods such as forests plantation and management, as well as aquaculture, to alleviate the pressures over natural assets.

Keywords

Renewable resources · Fishing · Market failures · Common property resources · Unsustainable extraction · Aquaculture · Regulation · Forestry · Externalities · Impatience · Interest rates · Plantations · Forest fires

Part I made it clear that developing countries exhibit a marked specialization in the production and export of goods based on natural resources, which poses challenges for achieving sustainable development, both at the microeconomic level and for the macroeconomy as a whole. This chapter examines the former.

The management of renewable natural resources faces the paradox that, in principle, efficient management should allow for a sustained flow of services over time. However, in practice, current data indicate that almost all renewable resources are being overexploited and the natural capital assets that provide them are in decline. This applies to the services provided by forests, marine species, soil, and, in many

J. Vial, *Sustainable Development and Natural Resources*, Integrated Science 39,
https://doi.org/10.1007/978-3-032-12263-6_7

areas, water. Much of these problems stem from the fact that the costs of obtaining provisioning services from natural capital are not properly reflected in the market prices of these services. In the case of fish and marine organisms in general, since they belong to no one until they are caught or collected, the cost of losing the potential offspring of captured fish is not considered. In the case of logging, the impacts on soil protection, water capture and circulation, loss of habitats for multiple species, and the release and loss of CO_2 capture capacity are also not considered.

The next two sections examine the cases of fisheries and forests to illustrate the challenges that must be addressed to achieve sustainable management of renewable resources.

7.1 Status of Marine Resources and the Causes of Their Overexploitation

Fishing activity worldwide is facing serious challenges: never has so much food been extracted from the seas and inland waters, but there is also evidence of declines in the stocks of most of aquatic resources. While demand for this type of food is growing at high rates, wild fish catches stagnated globally three decades ago, and fishing fleets must travel farther, often harvesting younger specimens (as evidenced by smaller sizes). Over these three decades, aquaculture (the farming of aquatic species, both at sea and inland water) has covered the increase in consumption. However, its future expansion is uncertain due to growing concerns about its environmental impact.

This situation is worrisome, because with a growing world population and greater purchasing power, it is virtually inevitable that demand for animal protein will continue to rise. This, in turn, puts additional pressure on livestock and the farming of other animals for human consumption, with potentially more severe environmental effects than those of aquaculture. Dietary shifts toward plant-based protein could ease the pressure on livestock, but not necessarily on land use and water consumption. In any case, it is very unlikely that such a change will be widespread, rather than a niche phenomenon as it is today, unless there is a drastic change in relative prices.

Seafood makes a very important contribution to human nutrition, especially in lower-middle-income countries, providing nearly half of the animal protein consumed globally.[1] In fact, almost 90% of seafood catch and production is destined for human consumption. Fishing is also an important economic activity, at least in some countries: although one third of global aquatic production is exported, this represents about US$150 billion per year. What is not often recognized is that by far most of production and catch of these products occurs in inland waters, and only one third comes from the sea.

The global powerhouse in aquatic products, both in terms of consumption and extraction, is China, which is also the leading exporter of seafood.

[1] FAO (2022).

Another noteworthy aspect is that aquaculture has gained significant importance in recent decades, to the point that, in terms of volume, it already matches the global capture and harvest of fish and algae, as shown in Fig. 7.1. The downside of this process is that, as reported by the FAO in its latest State of World Fisheries and Aquaculture Report (FAO 2022), this outcome results from stagnation in wild fish capture levels and a rapid increase in aquaculture production, both on land and at sea.

In its latest report, FAO notes that about 65% of fish stocks are at biologically sustainable levels, highlighting that this proportion continues to decline. The report emphasizes that fisheries management measures are proving effective and that over 80% of fish catches come from biologically sustainable stocks, indicating significant progress. However, this is not sufficient: 30% of fish stocks are exploited beyond their ecological sustainability, as shown in Fig. 7.2, and that there are any underexploited fish species.

When examining the status of fisheries by regions, it is evident that those bordering the South Pacific off the coast of South America exhibit the highest levels of overexploitation, as shown in Fig. 7.3. This area is rich in nutrients brought by the Humboldt Current, which has been utilized by Ecuador, Chile, and Peru, as well as by International fishing fleets that operate (at least in theory) outside the jurisdictional waters of these countries.

Overall, these are troubling developments, as total global fish consumption has increased steadily, partly because per capita consumption has doubled over the past 50 years, in addition to the doubling of the world population during the same period. Total fish consumption growth has slowed over the past two decades, with annual increases dropping from 3.5% to 4% at the end of the twentieth century to 2.5% in the last decade. This could be seen as positive news, were it not for the possibility that it signals some degree of scarcity, since FAO's general seafood price index has

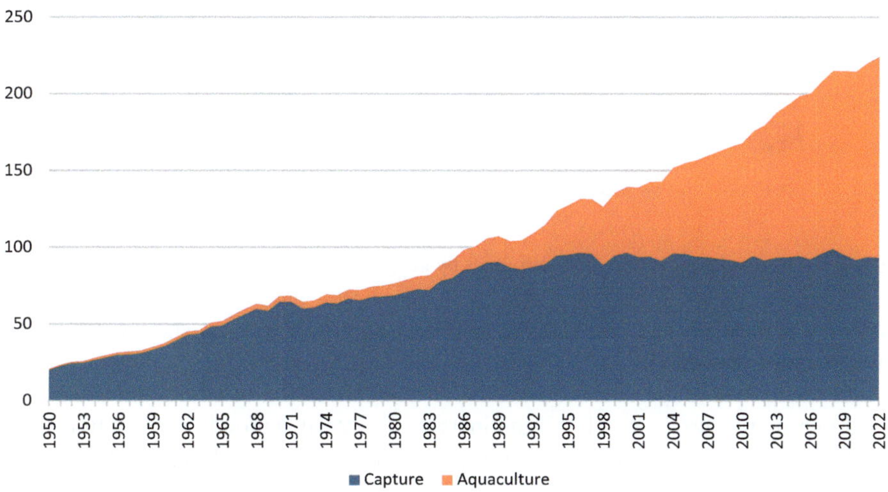

Fig. 7.1 Trends in global fish and algae capture (millions of metric tons per year). (Source: FAO Fishstat)

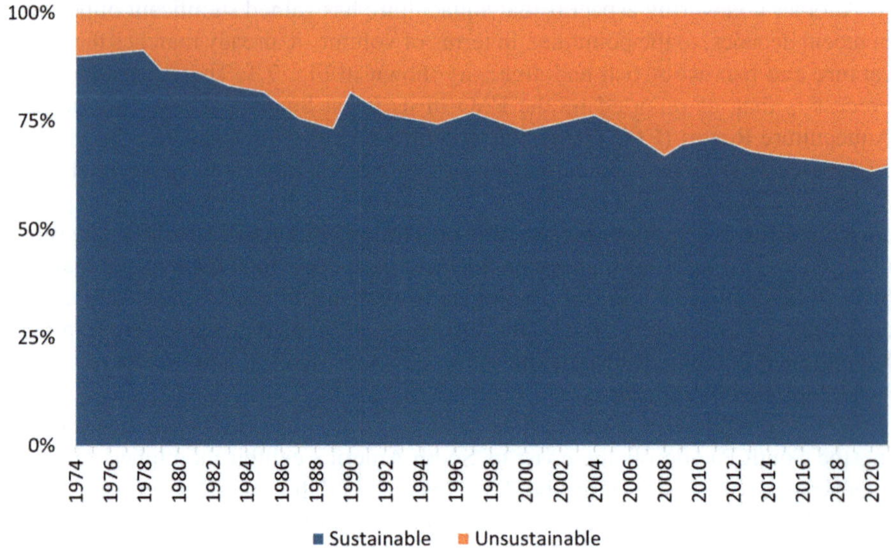

Fig. 7.2 Trends in the sustainability of global fish stocks. (Source: FAO 2022)

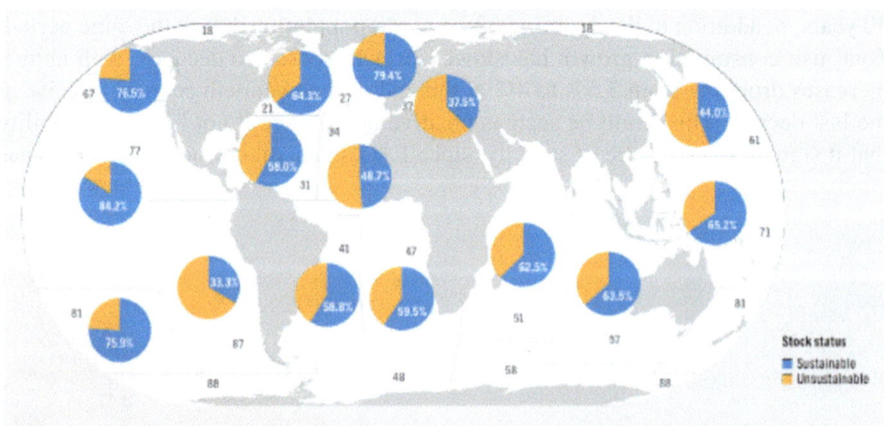

Fig. 7.3 Overexploitation of fishery resources by geographic area. (Source: FAO 2024)

shown a significant increase relative to other sources of food as shown in Fig. 7.4 over the past decade.[2]

As mentioned at the outset, given the virtual exhaustion of additional sources of wild marine resources, nearly all the increase in global seafood consumption since 1990 has been met by the growth of aquaculture production. However, this growth

[2] It should be noted that this is a relatively new index, based on foreign trade data and integrating a wide variety of goods. In fact, FAO has not yet incorporated it into its food price database.

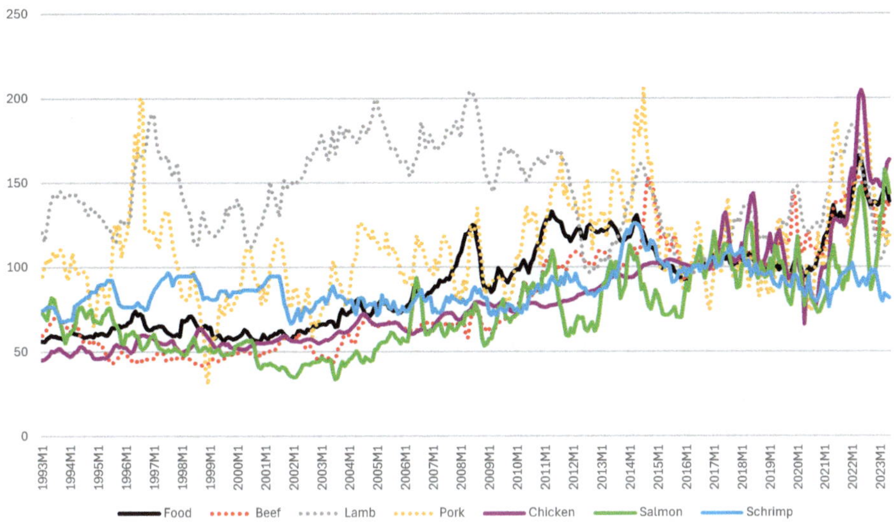

Fig. 7.4 Price indices for selected foods (current dollars, base 2016 = 100). (Source: IMF)

appears to follow a linear trend, which recalls Malthus's classic reasoning: with consumption growing at exponential rates, scarcity will eventually arise, which should begin to be reflected in prices—something that has not yet occurred, as shown in Fig. 7.4.

Aquaculture has undergone very dynamic development over the past 50 years as shown in Fig. 7.5. Initially, it focused on freshwater fish and other species, cultivated in ponds or lagoons for local consumption and mainly concentrated in Asia. This activity rapidly expanded to other regions as marine and coastal fish and crustacean farming developed. Advances in technology, especially in feed and physical systems for cultivation, also enabled large-scale industrial development of high-value species, such as salmonids, which in turn has fostered the development of international trade, further supported by advances in cold chain logistics.

Not all seafood, nor all aquaculture products, are used directly for human consumption. They include raw materials for animal feed (fishmeal) and others used as industrial inputs (algae), but this use has declined over time, with its share of total catch falling from nearly 30% in the 1990s to the current 20%.

The increase in the importance of aquaculture for human consumption is even more pronounced, rising from 10% of the volumes supplied by capture fisheries in 1970 to a figure that exceeds the latter by nearly 20% in 2020, as shown in Fig. 7.6.

Most seafood products, particularly those intended for human consumption, are destined for domestic markets, although in some cases the fishing fleets of these countries operate in regions far from their own coasts. International trade in these species is quite limited and is increasingly concentrated in high value farmed fish and shellfish compared to other species. In fact, the few international price indicators for seafood products refer specifically to aquaculture products: salmon and shrimp.

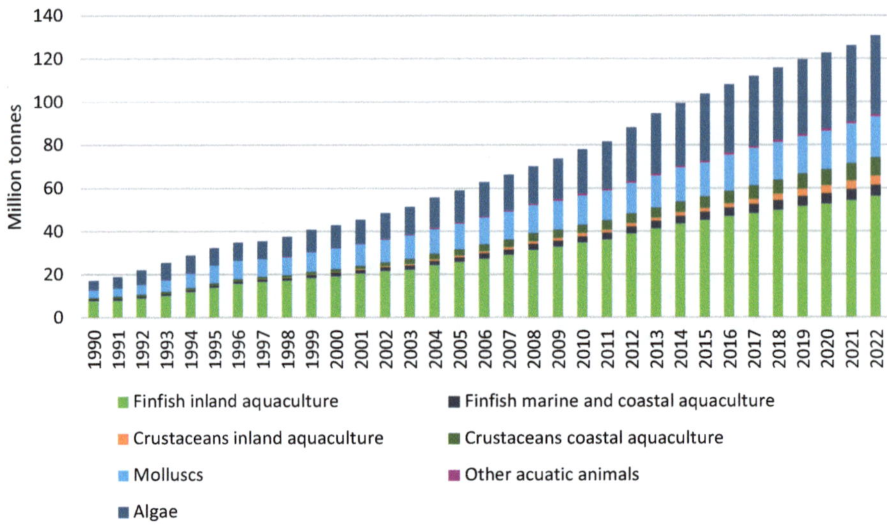

Fig. 7.5 Trends in aquaculture in the last 30 years. (Source: FAO 2022)

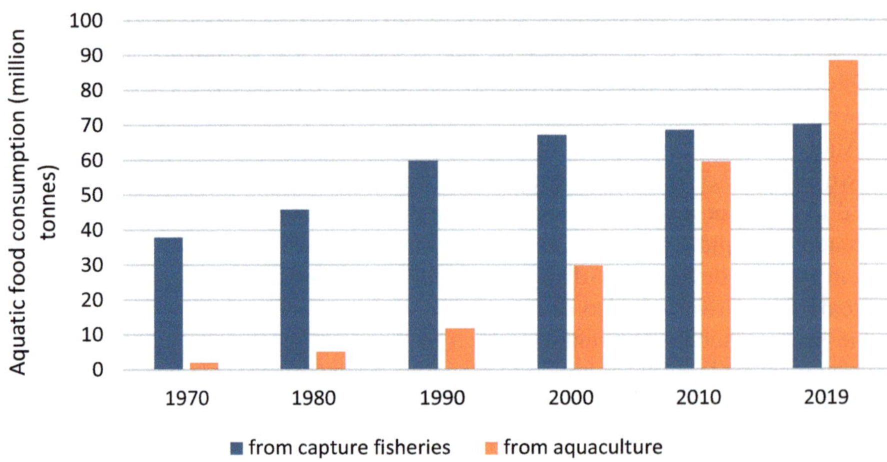

Fig. 7.6 Contributions of capture fisheries and aquaculture to human consumption. (Source: FAO 2022)

The impacts of human activity on marine ecosystems go far beyond the overfishing of species that support humanity. Other factors to consider include water pollution from human waste, most notably the discharge of organic matter and fertilizers, the increased presence of carbon in the sea and the associated changes in ocean acidity, as well as solid waste, especially new synthetic materials that are difficult to break down, such as plastics. To this list must be added climate change, which is affecting sea level, temperature, and salinity, and which already appears to be beginning to disrupt the proper functioning of the processes that regulate the operation of the oceans and the atmosphere.

Figures 7.2 and 7.3 clearly show that fisheries based on the capture of wild fish are being overexploited, in the sense that the annual harvest exceeds the natural regenerative capacity of the stocks. From the giant whales to the humble anchovies, there are numerous cases of species that once supported the rapid development of fishing fleets and industrial plants to then collapse due to overexploitation, dragging down communities and ports with them. This pattern is also repeated in the harvesting of mollusks and crustaceans when fishers begin to connect with larger markets beyond the inhabitants of nearby fishing villages.

In human history, this is a recent phenomenon. In the past, fishery collapses were limited to freshwater fisheries near cities and towns. The vastness of the sea and the very limited carrying capacity, and lack of refrigeration, of wooden vessels powered by oars or sails kept the great oceanic fisheries beyond the reach of human activity. Nevertheless, some highly valuable activities, such as whale oil extraction, caused such devastation that the Nantucket fishing fleets had to travel to the middle of the Pacific Ocean to hunt whales that had virtually disappeared from the North Atlantic. However, this was an anomaly and, moreover, with the advent of the Industrial Revolution and the replacement of whale oil by gas and electricity, the original business lost its economic appeal.

However, the same Industrial Revolution made it possible to expand the range of fishing vessels along with their extraction capacity and enabled the preservation of captured fish through refrigeration. This radically changed fishing after the start of large-scale industrial exploitation of all the oceans.

In addition to direct human action, some fisheries also collapse for environmental reasons such as temperature changes, infections, or other factors. However, most of the recorded collapses follow periods of intense extraction to meet human needs.

The reason for this widespread failure of the usual market-based scarcity regulation mechanisms is simple: for them to work, the traded objects must have a clear owner who limits extraction to preserve the business for the future. If this is not the case, everyone involved in extracting the resource will try to obtain as much as possible: if a fish is left in the water to reproduce and ensure the viability of the business in the future, it is very likely that another fisher will come and catch it for consumption today, removing that specimen and, with it, all its potential descendants.

This problem does not have an easy solution, since the law recognizes that wild fish have no owners and that ownership is acquired upon capture. In practice, this has meant that in international waters, fishing fleets face no restrictions on fish extraction, and vessels from the most varied (and distant) origins operate there.

This situation led to the recognition that countries can regulate fishing activities within their jurisdictional waters, which in the twentieth century led coastal countries to create exclusive economic zones, extending the areas under their regulation up to 200 miles from the coast, beyond the traditional 12 miles.

Initially, regulations were designed to protect the activities that sustained communities traditionally dependent on fishing and sought to limit access by new fishers to traditional fishing grounds. With the industrialization of the activity, this ceased

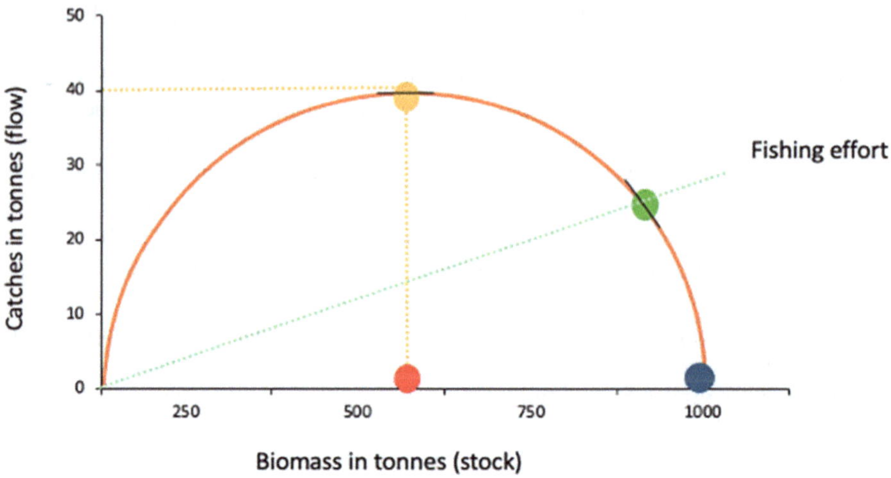

Fig. 7.7 Relationship between maximum catch levels for each sustainable biomass volume

to be effective, and other "rationing" mechanisms had to be sought, including those that limited fishing and transport capacity in certain fisheries, which soon proved ineffective. The next step was to incorporate scientific evidence to regulate fishing activity based on estimates of the biological regenerative capacities of species. To this, it was necessary to add the adjustment of incentives to optimize extraction.

Economics had to develop a conceptual framework to analyze these problems, incorporating the traditional principles of utility maximization by a "benevolent planner" to determine what would be an optimal extraction volume in a highly simplified model of a "representative" fishery, taking into account the interests of both current and future generations, as summarized below.[3]

The population dynamics of different marine species vary greatly among them, as well as within these depending on environmental conditions, making it difficult to generalize, and in the design and implementation of regulations, it must be acknowledged that there is considerable room for error.

Notwithstanding the above, the representation of this dynamic for a given species in a specific habitat is shown in Fig. 7.7. Starting from a no-fishing scenario, it can be assumed that there is a maximum capacity to sustain a given population, considering the availability of food, the presence of predators, the reproductive capacity, and the "life expectancy" of that species under "normal" natural conditions. This is represented by the blue point in the graph. The fish population can be expressed as a certain number of individuals, which is, however, quite imprecise because a normal population contains individuals of different ages and, therefore, different masses. For this reason, this capacity is generally referred to as a measure of the physical volume of the biomass of that species, expressed in tons.

[3] References on this subject can be found in Pearce and Turner (1990), Fisher (1981), and Tietenberg and Lewis (2015).

Fishing activity reduces the size of that biomass, and it is easy to represent the maximum catch capacity associated with different biomass sizes, as shown by the red line. When the total biomass is large, a small fishing effort can yield a large quantity of fish per unit of fishing time, as indicated by the green point. The fishing effort may correspond to the number of fishing hours by a certain number of vessels with a given hold capacity and is represented by the angle of the green dashed line in the graph. As this effort increases, that is, as the slope of the line originating from the origin increases, the additional yield from this effort (the slope of the red line at the green point) becomes progressively smaller as the fish density in the water decreases. Thus, as fishing effort increases, the volume caught continues to rise, but at a diminishing rate, until reaching a point where additional effort yields no further catch. This is the point of maximum biologically sustainable catch consistent with a stable biomass, corresponding to the yellow point, which, in the hypothetical numbers of the graph, represent an annual catch of 40 tons and a stable biomass of about 570 tons. Figure 7.7 depicts this situation in a highly schematic manner. It should be noted that there can be considerable variation around the curve and that its shape may be much more irregular.

The yellow point would represent something like a "technical optimum" consistent with the maximum extraction compatible with a stable biomass, if neither the fishing activity per se nor the biomass size of this fish species significantly affects the ecosystem.

This does not exhaust the subject, because this "technical optimum" in traditional microeconomic terminology does not necessarily coincide with the level of effort and extraction that maximizes the present value of profits derived from the exploitation of this fishery. In fact, under normal conditions, it never does.

To determine the economic optimum, the first step is to convert physical units into economic values. This, in turn, requires several assumptions. The first and simplest is to assume that the price at which the fish are sold is always the same, for example, one peso per ton, so that the axes now represent revenues from catch (vertical axis) and the economic value of the corresponding biomass (horizontal axis). Now, instead of tons, the axes measure values. For a given price level, the slope of the revenue curve represents marginal revenue.

The second important assumption is that each level of effort is associated with a cost level, and therefore, as effort increases, cost increases, and the slope of each line corresponds to the marginal cost associated with that level of effort. If it is assumed that there is only one vessel and that effort corresponds to the fishing hours of that vessel, where the only variables (in fixed proportions) are fuel and labor hours to be paid, then for each level of effort there is a cost level that changes only as a function of operating hours.

A third assumption—which will be relaxed shortly—is that there is only one period, so the only consideration in the fishing decision is the maximum utility that can be obtained from fishing in that single period. The first-order condition for this static optimization problem is the equality between marginal revenue and marginal cost, as shown in Fig. 7.8.

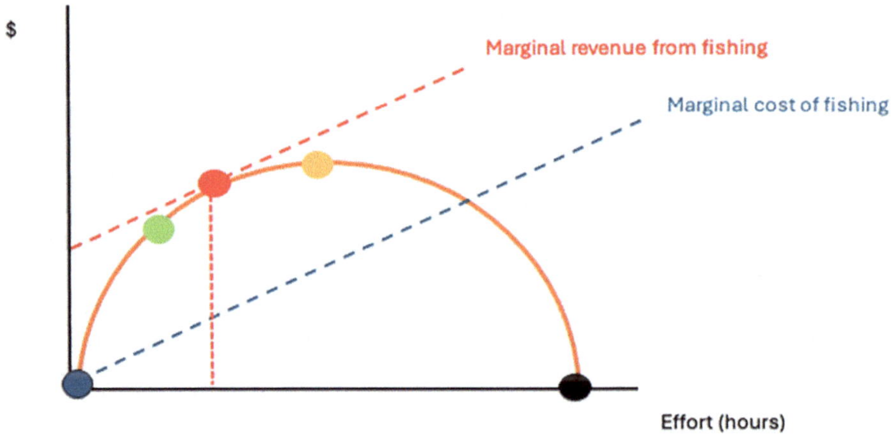

Fig. 7.8 The relationship between effort and economic returns

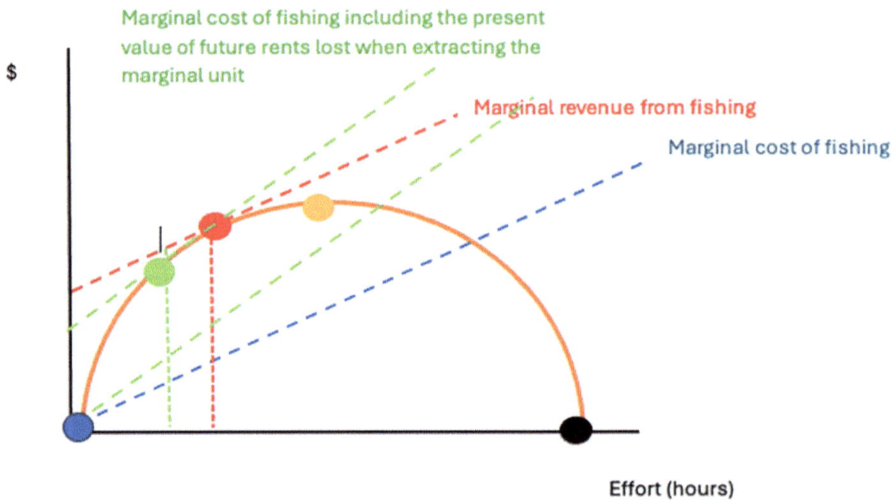

Fig. 7.9 Optimal equilibrium with more than one period

Starting from a point such as the blue one, which implies that this vessel begins by fishing one hour per day, for example, it obtains the corresponding revenues from the sale of those fish. As effort increases, the yield from additional fishing effort decreases until it reaches the orange point. From there, additional effort begins to yield less production in absolute terms due to the decline in biomass, and these returns decrease with further effort until they reach zero. Thus, based on these assumptions, a relationship can be graphed between economic returns and different levels of effort. This is shown in Fig. 7.8, where the minimum effort level corresponds to the blue point in Fig. 7.7, a higher level to the green point, the maximum yield effort to the orange point, and the maximum effort with zero yield corresponds to the black point.

Since the price is constant, the orange curve represents the revenues associated with each level of fishing effort. On the other hand, since the costs per unit of effort are also constant, they can be represented by a straight line starting at the origin, whose slope corresponds to the marginal cost of fishing effort. In turn, the slope of the revenue curve (orange) represents marginal revenue, which is determined by the relationship between biomass size and maximum sustainable yield. The red point, then, is the one that maximizes profits with a fishing effort that is sustainable over time.

In the real world, time exists, and it is crucial to incorporate it into the analysis, both because it allows for the introduction of population dynamics of species and the preferences of fishers for present versus future consumption. Both are central variables in any sustainability analysis.

When the temporal dimension is introduced, the fisher must decide how much to catch now and how much to leave in the sea to harvest in the future. If the fisher is very impatient and indifferent to future outcomes, he will try to extract the maximum amount that is economically optimal to exploit today, as shown in Fig. 7.8. However, if future income is valued to some extent, then limiting current effort to allow for greater future consumption has value. This can be represented either as a "reward" for leaving more fish for the future, which in equilibrium is reflected in an interest rate, or as a user cost for extracting an additional fish today (the lost offspring). This cost is reflected in the green dashed line in Fig. 7.9, which shows the user cost of natural capital (biomass) in the present, and the new equilibrium is at the green point, with a lower current catch and, therefore, greater future biomass. As shown in the graph, the fishing effort is lower than that which allows for the maximum sustainable annual yield (yellow point) if only the direct costs of fishing are considered. However, the optimal effort is even lower if the present value of lost income from extracting the future offspring of the fish caught today is included (green point).

Up to this point, the analysis has been conducted as if there were a single fisher or a "central planner," and it appears that, in principle, there would be no reason for massive overexploitation of fisheries, except in exceptional cases involving highly vulnerable species. In most cases, fishing is limited when the density of the species in the sea drops to the point where further effort is no longer profitable, which could allow time for partial recovery, unless other species begin to invade their ecological niche.

The underlying problem behind the overexploitation of marine resources is related to a peculiarity of these resources (and of all marine species): they have no owners, and ownership is acquired through capture. This leads to what is known as the "tragedy of the commons"[4] which is simply the fact that, as fish in the water are common property goods until the moment of capture, therefore no one has an incentive to conserve them, because "someone else" can catch them. Consequently, as

[4] A concept popularized by Hardin (1968) but with its formal origin in Gordon (1954).

long as revenues from fishing remain higher than costs, there will be incentives to enter into this activity, and the resource will be exploited to the point of the economic collapse of industry, and in many cases to the destruction of biomass and, consequently, consumption of natural capital.

Fisheries regulation must begin by limiting access to fisheries and then adopting the perspective of the "central planner" or single fisher to define a maximum extraction rate or annual catch quota for a given species. Secondly, a very robust system for monitoring and controlling both the species and fishing activity is required, as well as effective enforcement mechanisms in the event of violations. A challenge for regulators in developing countries is that the institutions responsible for these matters often lack resources of all kinds, and, very frequently, the companies potentially affected by regulation have experienced professionals and possess more information than the authorities.

The political economy of fisheries regulation is complex because, in practice, it means privatizing the capture of rents from a common property natural resource. This tends to be resisted, first and foremost, by incumbent fishers who see their investments and jobs at risk, as well as by public opinion, which perceives that access to wealth originating from the nation's natural heritage is being given away.

If the necessary technical capacity and political support exist, the most advisable solution is to define catch quotas for relevant species, which are then divided and distributed among a limited number of participants, allowing for economies of scale. The most politically challenging aspect is the allocation of individual quotas, for which the standard recommendation is to conduct well-designed and transparent auctions (Peña-Torres et al. 2022). Other contentious issues include the duration of assigned quotas and the possibility of trading them to facilitate adjustments within the industry to achieve maximum operational efficiency.

All of this can work well within territorial waters and exclusive economic zones, but the situation is different in international waters, where treaties are required to limit catches, something that has proven to be very complicated and difficult to implement in practice. The tremendous increase in the capture and storage capacities of oceanic fishing fleets makes this task increasingly difficult.

7.2 State of the World's Forests and Challenges for Their Preservation and Sustainable Use

Forests fulfill essential functions for sustaining life and the proper operation of ecosystems in general. It is no coincidence that biodiversity is greatest precisely in areas where forests abound: this occurs because these areas provide suitable climate, water availability, and food resources, and because of positive feedback between forests and other organisms.

The current stage in the history of life on the planet is characterized by particularly favorable climatic conditions for the development of the flora and fauna that prevailed over the past 100,000 years. Within this context, humans have gone from being a minor species, at severe risk of extinction, to dominating the Earth and leaving an unprecedented mark on ecosystems and landscapes. One of the clearest expressions of this is the change in land use distribution: at the dawn of agriculture about 15,000 years ago, nearly 60% of the planet's habitable surface was covered by forests, with the remainder consisting of grasslands, pastures, and shrubs. Today, forests cover slightly less than 40% of that area, and grasslands and shrublands are less than 15%. The remaining 45% is occupied by humanity for grazing (30%) and crops (15%).[5]

The change in land use has resulted in habitat loss for numerous species and, consequently, has contributed to massive biodiversity loss. In addition, it has led to the release of greenhouse gases, erosion and loss of soil vegetation cover, as well as severe disruptions to water regulatory systems.

In recent decades, the global trend of declining forest cover has continued, although the rate of destruction has slowed worldwide. In fact, in some areas of the Northern Hemisphere, particularly North America, Europe, and China, there has been a recovery of forest cover.[6] In contrast, in Central Africa, South America, and some Asia-Pacific countries, forest destruction continues at a high rate, although somewhat lower than in previous decades. Central America is a particular area of concern, as the proportion of forest loss relative to total area is the highest in the world. While this represents a small area globally, it is critical for biodiversity conservation, both because of its endemic species and because it also serves as a migratory corridor between the northern and southern hemispheres.

A first point to highlight from Table 7.1 is that the largest forest losses are concentrated in three areas: South America, Africa, and South and Southeast Asia, to which Central America should be added for the reasons mentioned above. These regions coincide with the last major reserves of biodiversity in the world, as shown in Fig. 7.10 below.[7]

A second point to note is that the correlation between deforestation and economic growth is not so clear, possibly due to the strong impact of China, which, as the largest country with the highest economic growth during the period, also shows a significant increase in forest cover. However, a positive relationship is observed between the level of development, measured by per capita income, and changes in forest cover. In this case, causality may be multiple, as birth rates also tend to correlate (negatively) with per capita income, so poverty and demographics may combine to exert pressure on natural areas.[8]

There are two areas of progress within these broader trends. On the one hand, the rate of forest destruction has decreased in all regions over the past decade, reflecting greater government efforts, supported by increased resources for protection, as well as advances in monitoring and control. On the other hand, there has also been a

[5] Urban land and infrastructure occupy about 1% of the habitable surface but are in areas critical for the development of life and biodiversity. Ritchie (2021) Our World in Data.

[6] This section is based on FAO (2021).

[7] For these figures, it is assumed that there is a strong association between deforestation and the release of soil carbon, something that is further supported later.

[8] Albagli and Vial (2022) present some statistical evidence that also points in this direction.

Table 7.1 Forest cover by region (thousands of hectares)

	1990	2020	Change	Var. %	Contribution
Africa	7,42,801	6,36,639	−1,06,162	−14.3%	59.8%
Eastern and Southern Africa	3,46,034	2,95,778	−50,255	−15%	28%
Northern Africa	39,926	35,151	−4,775	−12%	3%
Western and Central Africa	3,56,842	3,05,710	−51,132	−14%	29%
Asia	5,85,393	6,22,687	37,294	6.4%	−21.0%
East Asia	2,09,906	2,71,403	61,497	29%	−34.6%
South and Southeast Asia	3,26,511	2,96,047	−30,465	−9%	17.2%
Western and Central Asia	48,976	55,237	6,261	13%	−3.5%
Europe	9,94,319	10,17,461	23,142	2.3%	−13.0%
Russia	8,08,950	8,15,312	6,362	0.8%	−3.6%
Rest of Europe	1,85,369	2,02,150	16,780	9.1%	−9.5%
North America	7,21,317	7,22,417	1,100	0.2%	−0.6%
Central America and Caribbean	33,963	30,293	−3,669	−10.8%	2.1%
Central America	28,002	22,404	−5,598	−20.0%	3.2%
Caribbean	5,961	7,889	1,928	32.4%	−1.1%
South America	9,73,666	8,44,186	−1,29,480	−13.3%	72.9%
Oceania	1,84,974	1,85,248	274	0.1%	−0.2%
World	42,36,433	40,58,931	−1,77,503	−4.2%	

Source: FAO (2021)
Note: Contribution refers to the percentage of change attributable to each region or subregion

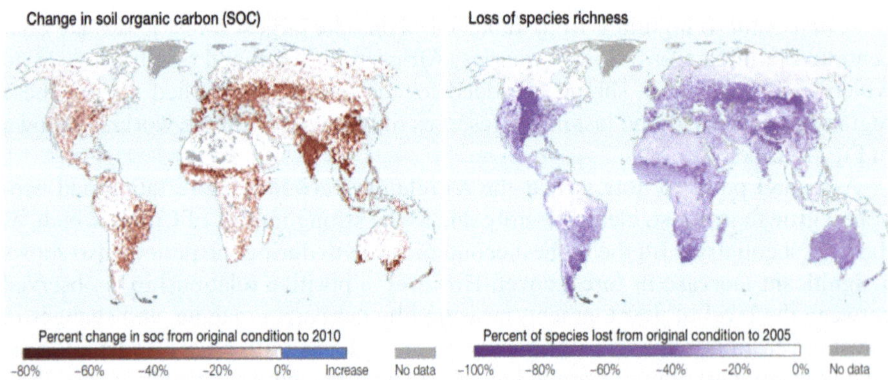

Fig. 7.10 Relationship between biodiversity loss and deforestation at the global level. (**a**) Changes in soil organic carbon. (**b**) Loss of biodiversity. (Source: IPBES 2019)

sustained increase in forest plantations, mainly to produce raw materials for industrial processes, which in turn relieve pressure on natural forests. Although forest plantations represent a relatively small fraction of global forest area (7%), they are concentrated in a small group of countries that, according to FAO estimates, supply about 20% of the world's timber needs.

At a time when concern about climate change is increasing, it is important to highlight the contribution of forests to carbon capture and storage. This occurs in three ways: carbon capture through tree growth, changes in forested areas, and

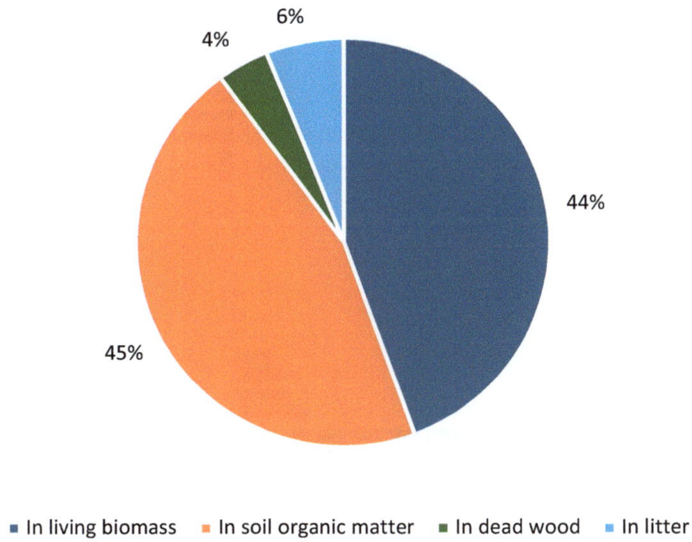

Fig. 7.11 Proportion of carbon stocks in forests in 2020. (Source: FAO 2021)

carbon retention in the soil. The first is a flow, while the second and third refer to stocks accumulated over time.

Figure 7.12 shows the extent to which changes in forest cover have contributed to the capture of atmospheric CO_2 (negative values indicate carbon release due to deforestation). Forests in Europe, Northern Asia (China), and North America have made significant contributions to reducing the carbon footprints of their respective regions. In contrast, deforestation has become the main source of greenhouse gas emissions in Africa, South America, and Pacific countries (IPCC 2022a, b, c). This occurs both through the direct release of carbon when forests or extracted firewood are burned, and through the decomposition of organic matter, as well as from soil preparation for crops and livestock. Figure 7.11 highlights that there is as much carbon stored in trees as in the surface layer of forest soils.

Figure 7.12 shows that the situation has improved in South America, but in the rest of the subregions that are net CO_2 emitters for this reason, emissions continues to increase.

It is worth noting that the FAO report also indicates that in developed countries (Europe and North America) there has been significant progress in the certification of sound forest asset management, which facilitates access to financing by taking advantage of potential compensation for emissions from industries required to meet reduction targets. In contrast, in other regions this is still in its infancy, with only a few exceptions.

The destruction of forests is closely linked to human activity for reasons that will be examined below. This occurs both directly (such as logging or fires) and indirectly (climate change, introduction of competing or predatory species). Among these causes, fires are becoming increasingly important, as climate change is already

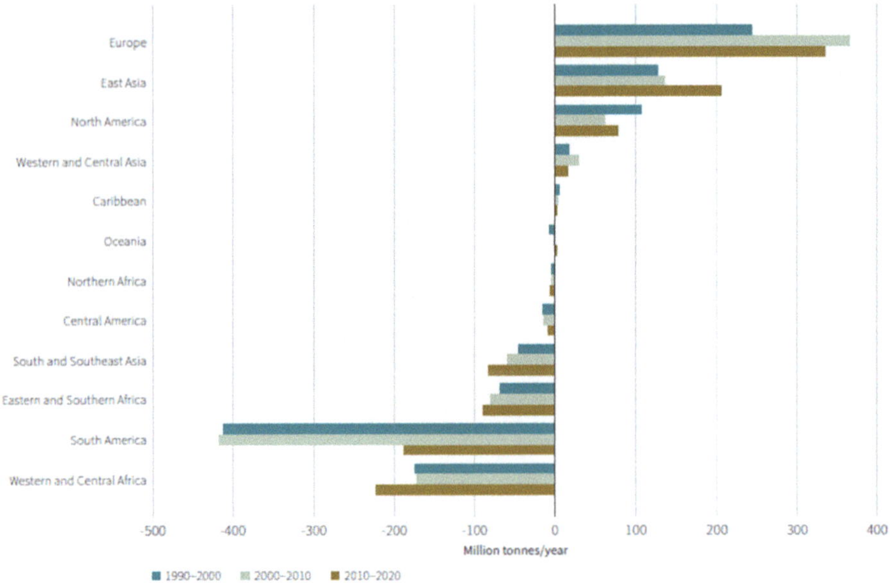

Fig. 7.12 Change in carbon stocks stored in forests 1990–2020 (million tons per year). (Source: FAO 2021)

creating conditions that facilitate their ignition and spread, especially in drier areas such as Mediterranean countries, Australia, and parts of South America.

Some forest fires have natural causes, such as lightning strikes, but the vast majority are due to human actions, often accidental, for example, when agricultural residues are burned or due to electrical discharges from power lines. However, the largest burned areas originate from fires set to "clear" land for crops or livestock. This, in turn, may result from small-scale actions associated with the need for higher incomes among low-income rural communities, or from operations to develop large-scale agriculture.

Data from Fig. 7.13 suggests that fires set to clear land to improve the incomes of poor rural families may be a significant factor behind the pressure to change land use. In these cases, part of the problem lies in the fact that agriculture and livestock farming are of very low productivity, and in the absence of capital, the high cost of fertilizers, and limited stable access to water, the only way to increase income is by expanding the cultivated area.

What are the economic factors behind deforestation? Forests provide a wide range of goods and services that are extremely valuable as inputs for economic activity, including plant fiber found in all forms of paper, wood for housing and furniture construction, and for energy generation, especially in the world's poorest regions. In addition, they protect soil from erosion, help maintain the integrity of watercourses, and sequester atmospheric carbon, offsetting emissions from other sources. They also serve as habitats for multiple species, helping to maintain

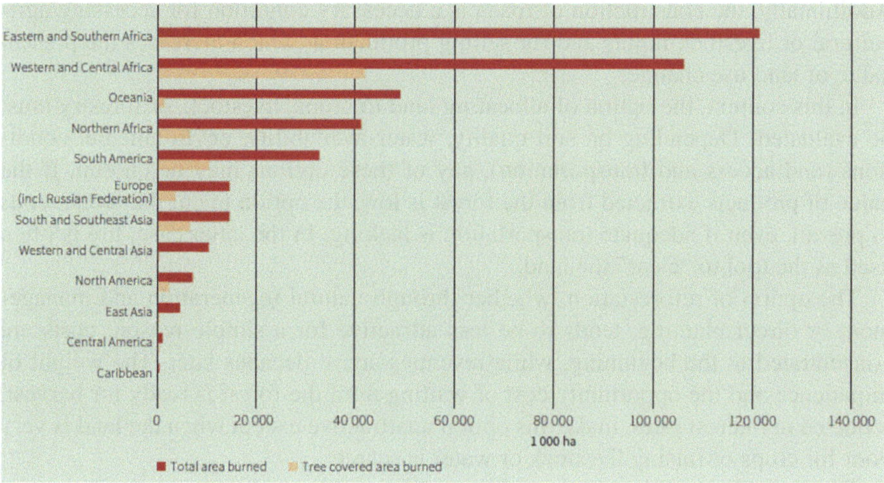

Fig. 7.13 Burned areas between 2001 and 2018 by region (Annual average, thousands of hectares). (Source: FAO 2021)

ecosystem balance. Given this, one might ask why humanity does not take better care of forests and prevent their destruction.

From an economic perspective, it is useful to start with the situation of an existing forest and then analyze the decision to plant a forest. In both cases, it must be recognized that the land on which the forest stands is a scarce resource with alternative uses. From a strictly economic standpoint, the landowner's decision involves finding the most profitable alternative over the relevant time horizon.

The decision to preserve a forest versus clearing it for an alternative land use depends on the value assigned to keeping it standing, compared to the value that can be obtained from harvesting the forest now, plus the present value of the flows that may be obtained from the new use of that land.

Except in some few cases, the private value of maintaining a forest standing without intervention is very low, as there are not many products that can be sold: limited quantities of timber and firewood, wild fruits, mushrooms, and, if conditions allow, some income from recreational uses.

In poorer societies, where there are no alternative uses for the land, the forest remains and provides some food and the main source of energy: firewood. However, as the population increases, pressure arises to expand arable land and extract more firewood, leading to more intensive use and gradual degradation of the forest.

The alternative of clearing the forest and converting the land to other uses becomes more valuable as access to transportation routes improves, allowing logs and other products to be taken to markets. The difficulty of access and the absence of transportation are the main factors protecting the existence of natural forests. For this reason, as soon as access roads are opened in these areas, deforestation quickly follows. Initially, this occurs on a small scale to sell firewood, and later, as access to urban areas improves, on a larger scale to sell logs for timber or fiber production.

Additionally, the construction of roads is a necessary condition for accessing agricultural or livestock inputs and for selling production, which increases the present value of land use change.

In this context, the option of allocating land to crops, livestock, or forestry must be evaluated. Depending on soil quality, water availability, environmental conditions (and access and transportation), any of these options may be chosen. If the value of products extracted from the forest is low, the option to cut the forest tends to prevail, even if adequate transportation is lacking. In the latter case, fire is often used as the tool to "clear" the land.

The option of reforestation, whether through natural regeneration and management or direct planting, tends to be less attractive for a simple reason: costs are concentrated at the beginning, while revenues arrive decades later. The weight of impatience and the opportunity cost of waiting until the forest is ready for harvest, reflected in interest rates, make this option unattractive except when the land is very poor for crops or raising livestock or water is scarce.

The exception to this rule are large corporations that industrialize the forest and add significant value along the way, and that have an almost infinite operational horizon. To make the project viable, they must invest large sums in industrial plants, and to justify that investment, they require a reliable flow of raw materials over a long-time horizon. These operations benefit from significant economies of scale, making them practically indivisible: a standard pulp mill costs several billion dollars and requires a large forest mass nearby, managed in a coordinated and sustainable manner to ensure a continuous supply of adequate raw materials.

Forest plantations are an integral part of this process, as they allow for planned supply and ensure a continuous flow of inputs of homogeneous quality. Furthermore, they optimize land use by selecting high-yield forest species with characteristics best suited for timber and fiber production, given the local environmental conditions (soil, climate, water).

Although natural forests can provide timber for industrial use, their exploitation is more complex due to their heterogeneity: different species of various ages, with a mix of individuals in different states—some with advanced decay and juveniles in early development stages—coexist in the same forest. All this increases the cost of harvesting operations, with highly variable yields over a given area. For this reason, when planting is chosen, usually only a few species are used—those best adapted to local conditions—and with well-planned densities to optimize growth and harvest.

All this requires a legal and regulatory environment that provides certainty of property rights over decades, something that does not always occur in developing countries. Moreover, investments of billions of dollars are only possible if there is access to credit and a deep capital market, which is rare in such countries. Not only they lack these markets, but local companies must also pay high risk premiums in international markets.

It is not surprising that in modern societies, natural forests are relegated to areas of difficult access, and that in more advanced countries, where low-productivity agriculture or livestock farming has ceased to be profitable, forest area is increasing—mostly due to land abandonment rather than forest plantations. Furthermore, in societies that have achieved high levels of basic needs satisfaction, where urban life with all its daily stress predominates, there is a growing appreciation for the other ecosystem services provided by forests and nature in general. This opens the possibility of generating income by preserving forests for recreational and spiritual purposes.

Historically, an adverse relationship was assumed between economic development and the preservation of natural areas, based on the correlation mentioned at the outset between land use change and the progress of human activity, both due to population growth and higher per capita consumption. What is now becoming apparent is that this correlation is changing, as consumer preferences shift from food to other goods and services, especially recreation in contact with nature, and as demographic pressure decreases with rising per capita income. If this trend continues, it could be expected that greater economic development, focused on poverty alleviation in those areas where the preservation of nature and biodiversity are priorities, help to reverse the correlation between economic development and environmental degradation.[9]

The existence of ecosystem services provided by forests that have no commercial value, combined with the impatience that drives deforestation for short-term income, as well as improved access that increases the opportunity cost of keeping forests standing and high interest rates, are economic factors that accelerate the destruction of natural forests. However, improvements in transportation infrastructure also open the possibility of industrializing timber, which could incentivize plantations. These plantations can maintain or even enhance some of the ecosystem services of forests, but they can hardly compensate for the loss of habitat for species that lived in the original forests, so biodiversity will not be the same.

References

Albagli E, Vial J (2022) Crecimiento económico y biodiversidad: Algo tiene que ceder (Documento de Política Económica No. 70), Banco Central de Chile

Collier P (2014) Resource revenue management: Three policy clocks. In: Caputo R, Chang R (eds) Commodity prices and macroeconomic policy. Central Bank of Chile

Fisher A (1981) Resource and environmental economics. Cambridge University Press

Food and Agriculture Organization of the United Nations (2021) Global Forest Resources Assessment 2020: Main report. https://doi.org/10.4060/ca9825es

Food and Agriculture Organization of the United Nations (2022) The state of world fisheries and aquaculture 2022: Towards blue transformation. FAO

Food and Agriculture Organization of the United Nations (2024) The state of world fisheries and aquaculture 2024. FAO

Gordon HS (1954) The economic theory of a common-property resource: the fishery. J Polit Econ 62(2):124–142

[9] In fact, Albagli and Vial (2022) find that the positive correlation between per capita income growth and biodiversity loss has tended to disappear, but the correlation between population growth and biodiversity loss persists.

Hardin G (1968) The tragedy of the commons: The population problem has no technical solution; it requires a fundamental extension in morality. Science 162(3859):1243–1248

Intergovernmental Panel on Climate Change (2022a) Agriculture, forestry and other land uses (AFOLU). In: Climate change 2022: Mitigation of climate change. Cambridge University Press, pp 747–860

Intergovernmental Panel on Climate Change (2022b) Summary for policymakers. In: Shukla PR et al (eds) Climate Change 2022: Mitigation of Climate Change. Cambridge University Press. https://doi.org/10.1017/9781009157926.001

Intergovernmental Panel on Climate Change (2022c) Summary for policymakers. In: Pörtner H-O et al (eds) Climate change 2022: impacts, adaptation and vulnerability. Cambridge University Press. https://doi.org/10.1017/9781009325844.001

IPBES (2019) Global assessment report on biodiversity and ecosystem services. IPBES Secretariat. https://doi.org/10.5281/zenodo.3831673

Pearce DW, Turner RK (1990) Economics of natural resources and the environment. Johns Hopkins University Press

Peña-Torres J, Muñoz R, Quezada F, Kaufmann C (2022) Entry deterrence and collusion at repeated multi-unit auctions of ITQs. Mar Resour Econ 37(4):417–442

Ritchie H (2021) The world has lost one-third of its forest, but an end of deforestation is possible. Our world in data. https://ourworldindata.org/world-lost-one-third-forests

Tietenberg TH, Lewis L (2015) Environmental & natural resource economics, 10th edn. Pearson

Microeconomic Challenges for the Sustainable Management of Exhaustible Resources

8

Abstract

Sustainable management of exhaustible resources requires that at least some meaningful proportion of their rents be saved to provide for the well-being of future generations. This is a key issue for emerging countries, rich in these types of resources, but facing social and political pressures to spend now to solve existing problems, and at the same time need to invest to sustain or increase production of these commodities. The estimation of rents from natural resources and their evolution over time, as well as the design of efficient mechanisms to capture the right amount of these rents, is central for these purposes. Current formulas go from taxation to outright nationalization. Then there is the issue of how to best save these resources for future generations. Investment in human capital and infrastructure to promote growth plays a role and they are usually allocated through the budget process. Special purpose funds have also been used, either to relieve the burden of Social Security for future generations, or just to provide financial wealth for them to dispose of.

Keywords

Economic rents · Weak sustainability · Taxation · Nationalization · Externalities · Wealth funds · Fiscal resources

As discussed in previous sections, the sustainable management of exhaustible resources presents significant challenges. The most obvious is precisely that they are exhaustible, which may be an easy fact to incorporate into the investment decisions of a company seeking to exploit them, but represents a challenge at the level of the country or the communities that must interact with the exploitation of the resource.

For the latter, the concept of "weak sustainability" becomes vital, with implications for participation in the rents generated by the activity, as well as for savings

J. Vial, *Sustainable Development and Natural Resources*, Integrated Science 39,
https://doi.org/10.1007/978-3-032-12263-6_8

and investment decisions. But this is not the only reason for state intervention: mining and oil extraction activities have significant environmental impacts, both through habitat destruction and severe contamination of the areas where operations are located, impacts that are normally not reflected in market prices and can be sources of conflict that may severely affect the economic and social outcomes of these activities.

Another dimension often associated with the exploitation of this type of resource is the impact on macroeconomic balances due to dependence on natural resources, but since this is common to all types of resources, including renewables, it is addressed in Chap. 10.

8.1 Taxation and Rent Capture in the Management of Exhaustible Resources

From the perspective of a country rich in such resources, sustainability in the extraction and use of exhaustible resources is directly linked to the concept of "weak sustainability" analyzed in the first part: the key is to find a way to ensure that the income stream from the rents generated by this activity can support an additional consumption flow that is sustainable over time, even after the resource is depleted. To achieve this, action must be taken in at least three areas:[1]

- Creation of an efficient mechanism for capturing a significant share of the rents for the State,
- Defining a rule to determine how much of these additional government revenues will be allocated to current consumption and how much to savings, to maintain these consumption flows for future generations,
- Defining how to invest the savings as efficiently as possible to optimize consumption flows over time.

Of all the above, the most complex in practice is precisely the design of a mechanism that allows the collection of a significant fraction of the rents, while maintaining incentives for efficient resource exploitation.

In principle, the idea of calculating a specific tax on resource rents is standard: once the rents are estimated, a proportional rate should be applied, which could range from 0% to 100%. In principle, even 100% would be acceptable, if costs indeed include the remuneration of capital, considering the time horizons of these investments and the associated risks.

However, as explained in previous chapters, the estimation of rents is not so simple, because they vary between companies and over time, and are also affected by both the depreciation of the natural resource and technological changes. Even

[1] There is a third factor to consider, regarding the amount of savings, which relates to the stabilization of available fiscal revenues during the different phases of the price cycle of the resource in question (Collier 2014).

risks are an endogenous variable: they depend not only on natural processes, but also on the reactions of the population, from the most general level related to public policy, to the most specific, such as relations with local communities affected by extractive activities.

A first consequence of this complexity in calculating rents is that such levies can only be applied in activities where a few large companies operate, which are usually highly capital-intensive activities, such as large-scale mining and hydrocarbon extraction. This is also politically easier to implement, since in most cases the entities taxed are multinational companies headquartered in other countries.[2]

This last point opens another dimension in the analysis: if the companies that can efficiently exploit the resource[3] are large multinationals, operating in various countries, with long time horizons for their decisions, they must allocate their capital to exploration and exploitation projects among the best available alternatives, which implies competition for investment resources that are scarce. This immediately means that rents cannot be taxed without affecting the competitiveness of resource development projects in the country that taxes rents more than others.[4]

All of this translates into the need to achieve a delicate balance between capturing the largest possible share of rents without undermining the attraction of investments necessary for resource development. This is not easy, as it requires public policy decisions capable of building credible agreements that span several decades with potential investors. In turn, this demands political systems that can support such agreements, which typically means that institutions must enjoy broad legitimacy.[5] Moreover, technologies and political environments are highly dynamic. Taking copper as an example, Chile has been the world's leading copper producer for 35 years, among other reasons, because even after all that time, it still holds more than a quarter of the world's known reserves. This situation was consolidated in the 1990s, when a successful transition to democracy reinforced the credibility of rules and institutions. This resulted in a massive inflow of investment into the country, allowing production to increase from 1.6 million tons in 1990 to 4.6 million in 2000, doubling its share of global copper supply to 35%. A similar process occurred in Peru, albeit on a smaller scale, in the 2000s, and in recent years, the Democratic Republic of the Congo broke the stagnation of copper mining in Africa, becoming

[2] There are some exceptions, such as the fishing industry, where access to fishing quotas can be auctioned as previously explained, which is an alternative way of capturing rent. This could also be applied in mining for new ("greenfield") projects, where access must be secured through government permits.

[3] In this context, this should be understood as maximizing the present value of rents over the useful life of the resource.

[4] It should be noted that these multinational companies typically extract various minerals, sometimes including energy resources. This means that even if the tax rates on rents imposed by other producing countries on a specific resource are compared, this may not be sufficient if it is more profitable to allocate investment funds to the extraction of other resources.

[5] Barma et al. (2012) analyze the experience of multiple resource-rich developing countries to draw some useful conclusions in this area.

the world's second-largest copper producer with 2.8 million tons in 2023, slightly surpassing Peru.

The evolution of the oil market clearly illustrates the effects of technological and market changes on the competitive position of countries. With the oil price rise at the beginning of the century, opportunities emerged for the development of shale gas and oil deposits, whose technology was known at an experimental level but became commercially viable with the new prices. This led to rapid progress in these technologies and substantial cost reductions, which in half a decade enabled the United States not only to achieve self-sufficiency in oil and natural gas, but also to export surpluses, thereby changing the geopolitical landscape of these markets.

From a practical standpoint, a crucial issue in defining the mechanism for capturing rents is the government's capacity to implement it, particularly that of tax and regulatory authorities. As a result, in most cases, special instruments are created that apply to only a few activities and only to relatively large companies, reducing the universe of potential special taxpayers to a handful of companies in a few sectors, typically mining and hydrocarbon extraction.

When faced with the decision to choose an instrument to capture rents from a natural resource-based activity, governments can opt among four main alternatives:

- Use the general corporate income tax regime,
- Direct negotiation with companies culminating in a contract, with appropriate legal backing, to agree on the terms of operation, including a payment regime to the government,
- A royalty, or special tax on companies exploiting the resource, which may substitute for the general corporate income tax or become a special surcharge for this activity,
- Partial or full (nationalization) participation in company ownership. In the case of a "greenfield" project, which is the simplest case, this may involve the creation of a dedicated state-owned company, or in the case of existing activities, through negotiation of a contract for the gradual transfer of a share or all the ownership to the state, or directly through expropriation.

The use of contracts with individual companies was very common in the 1960s and 1970s, but has since fallen out of favor, partly because it has lent itself to corruption and difficulties in the operation and realization of investments, but also because tax collection systems have improved and, at the same time, demands for transparency and accountability have increased in general. This has led to a gradual evolution toward other mechanisms, such as keeping a significant participation in the property of the company, as was the case of ZCCM in Zambia, or replacement by taxes.

When the tax system functions well, the general corporate income tax regime can be a simple and quite effective mechanism for capturing a significant share of rents, especially if the tax incorporates some progressivity in rates and consequently should not be dismissed "a priori."

The option of relying on the general income tax minimizes the potential adverse effect on investment and development of the activity, although it may generate political problems in the long term. This may be the best alternative for countries that lack strong monitoring capabilities for the detailed information required to collect a royalty. It is therefore not surprising that one of the most prominent recommendations by specialists in the field is to invest in strengthening supervisory capacity and specialized taxation.[6]

In the case of large multinational companies operating in mining, the main challenge for proper tax oversight lies in the accurate allocation of costs to correctly identify capital remuneration. This is often circumvented by indebting the subsidiary to the parent company to finance investments, subsequently transferring profits as interest payments to the parent, especially when these exceed market rates. Another method involves transfer pricing with suppliers related to the parent company or with processing companies, also linked to the parent, which pay for minerals sent for processing, smelting, or refining.

The move toward nationalization of the extraction and processing of natural resources, especially minerals and oil, began in Russia at the start of the twentieth century and then spread to Europe—first to the Scandinavian countries, then to the rest of Western Europe, and, after the Soviet occupation, to Eastern Europe. In the postwar period, and especially during the 1960s and 1970s, this trend rapidly extended throughout Latin America, Africa, and some Asian countries, particularly China. However, from the 1980s onward, the trend reversed, and a wave of privatizations began, which continues to this day. This is reflected in Fig. 8.1, which shows the share of state production of various minerals in the total global value of mineral production, excluding China. When examining the value of state production by

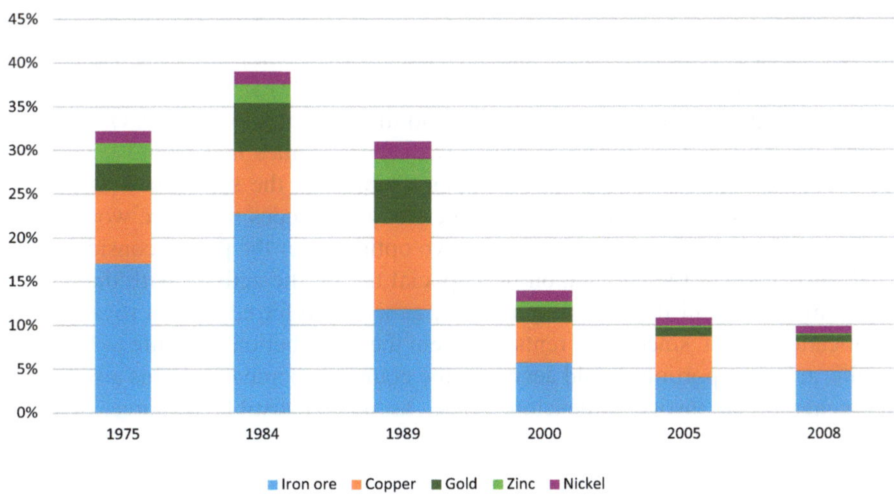

Fig. 8.1 Evolution of state participation in the value of global mineral production (percentage of production value). (Source: World Bank 2011)

[6] See Otto et al. (2006).

country relative to total mining output, Chile, India, and Indonesia, in addition to China, stand out in order of importance (World Bank 2011).

There are several reasons for the decline in state participation as a mechanism for capturing rents in mining activities. On the one hand, many state-owned enterprises have been subject to strong political pressures, with unclear and shifting objectives that failed to consider the need to focus on the company's long-term competitiveness. In the few successful cases cited in cross-sectional studies, the standout companies are precisely those that have built solid governance, where management teams possess skills and qualifications comparable to their private sector counterparts and where management is focused on long-term horizons.

However, even when these conditions are met, state-owned enterprises tend to decline if the Treasury does not allocate sufficient resources for investment to keep them competitive. The temptation for the owner (the State) to spend in the present and postpone investments that will yield returns 10 or 15 years later is too great, especially in emerging countries with relatively weak public finances. This, in turn, creates a vicious cycle: when production begins to decline and costs rise, contributions to the Treasury decrease and the state-owned company is authorized to take on debt to finance its investments. If these investments are delayed or prove less profitable than anticipated, the Treasury ultimately must forgo part of its revenues or even inject capital.

The problems are reflected in the fact that state-controlled companies generally have lower productivity, higher costs, and lower profits than their private sector peers (IMF 2020). In particular, the differences in profitability in mining are the greatest among the various sectors considered.

Considering this, the question arises as to which is the better option: to nationalize the extraction of a resource or to leave it in private hands and tax profits. There are few cases of reasonably well-managed state-owned companies coexisting with private firms, but a recent study by González et al. (2023) analyzes the experience of copper mining nationalization in Chile and the creation of CODELCO[7] by comparing its contributions to the Treasury with the revenues that would have been received had it remained a private company subject to the same taxes as private mining firms operating in the country. The study concludes that there would have been virtually no difference between the two options over the period considered.[8]

Finally, there is the application of a special tax on the activity with the aim of capturing part or all the rents it generates, generally referred to as a royalty. The justification is that, since these rents represent the contribution of nature to business profits, at least a portion should accrue to the country's population. This also has an intergenerational justice dimension, since, as an exhaustible resource, what is

[7] CODELCO is one of the few state-owned mining companies that the World Bank study classifies as successful. World Bank (2011).

[8] This is relevant because CODELCO has significantly reduced its contributions to the Treasury in recent years due to delays and cost overruns in major investment projects and the increasing amount of resources allocated to servicing its debt CESCO (2023).

extracted now will no longer be available for future generations, so investing part of these rents could compensate for the present consumption of natural capital.

There are many variants of royalties that can be applied, always considering the balance between capturing revenue now versus creating conditions for reinvestment in the activity so that the flow of societal consumption is optimized over time.

One characteristic of royalties is that they are generally not homogeneous (Otto 2017) and are often differentiated by activity. For example, within mining, different rates may be applied to different minerals and to different scales of operation. The latter is based on the understanding that there are significant economies of scale in this activity. Differentiation by type of expenditure is also common. For example, in the early stages of a project, payment is often waived, since operating expenses occur before the project generates revenue and the probability of success is low, and investment expenses are high, which also occur before generating revenue. In the case of investments, there may be exceptions to the general tax treatment of depreciation.

In practice, there are various ways to apply a royalty (Otto et al. 2006). The most traditional, but often very complex, especially when applied to different types of minerals, is the **charging of a fixed amount per unit produced** or sold, for example, 10 dollars per ton of iron produced. This becomes complicated when the operation extracts various minerals with different values (such as copper and gold), requiring different tax rates for each mineral. Moreover, a uniform fixed rate can be very burdensome for small producers, or excessively generous for larger operations.

A second option is to **charge a proportion of the value of mineral sales**, which simplifies the issue for polymetallic operations but introduces the challenge of determining the appropriate value of the products. In many cases, there are concerns about possible transfer pricing between the mineral extracted at one operation and the processor that smelts or refines it, especially when related companies operate in different countries. One way to address this issue is to use a reference price from a well-established market, such as the London Metal Exchange. In this case, different tax rates are often applied depending on the volume of sales or the capital of the companies.

Finally, there is the **imposition of a percentage rate on profits**, which is typically established as a surtax in addition to the general corporate income tax. This approach is closer to the idea of taxing rents and is also more closely aligned with the companies' ability to pay, since a fixed value on sales can exacerbate financial difficulties during prolonged periods of low prices. The main point of debate with this type of tax is the calculation of costs. In some jurisdictions, only operating costs are considered, although conceptually it is also correct to exclude the opportunity cost of capital, which, as previously discussed, presents significant challenges.

Regarding the level of rates or the fixed tax per unit produced, this depends primarily on the payment capacity of the affected companies. Therefore, regardless of the method used, the tax rate should be compared to the normal profit levels of the operation. A second factor to consider is whether it is a special tax that replaces the normal corporate income tax, or if it is in addition to it. Finally, the tax burden faced by companies in the sector in other countries competing to attract investment in similar activities should also be considered. Ultimately, the key is to capture the maximum possible rent without undermining the development prospects of the activity in the country. This is not an easy balance to achieve, as it must consider long-term effects that are difficult to predict. Maintaining a continuous, open, and transparent dialogue among all stakeholders, leading to an environment of mutual trust, is a factor that can help avoid costly mistakes for all parties.

8.2 Allocation of Fiscal Revenues from Rents and (Weak) Sustainability

The rents captured by the State from the exploitation of exhaustible natural resources should be allocated in a way that allows future generations to be compensated for the use of those resources. This does not happen overnight, and it is generally expected that, taking the resource's life cycle as a reference, a significant volume of savings will be generated in the early stages of exploitation, so that the income flow generated in this way gradually compensates for the eventual decline in the value of rents. This general principle is not without problems, because, as previously noted, over long-time horizons there is great uncertainty regarding technology and market conditions, making it very difficult to forecast the useful life of deposits and, even less so, prices.

On the other hand, in the short term, the public finances of countries rich in natural resources and dependent on revenues from them are exposed to high volatility in the prices of products derived from those resources.

This means that countries rich in exhaustible natural resources must implement fiscal rules that consider both the volatility of prices and the fact that these resources may diminish over time. These two issues operate on different time scales and probably require different rent-saving policies to smooth short-term cycles and to compensate for the eventual decline in rents as costs rise and the resource is depleted (Collier 2014).

The standard tool to address the loss of income due to resource depletion is the creation of a Sovereign Wealth Fund that manages long-term investments for this purpose. The case of Norway's Government Pension Fund (GPF) is often cited as a successful example in this regard. The fund was created in 1990, with the first contributions made in 1996. From the outset, it was designed with a dual mandate: to smooth the fiscal impact of oil price fluctuations and to accumulate savings to offset resource depletion. The operating rule is simple: fiscal surpluses are deposited in the fund, deficits are financed from it, and the Fund transfers to the Treasury the equivalent of the real return earned on its investments. The management of the Fund was entrusted to the Central Bank of Norway, with the resources kept entirely separate from the budget.

The GPF has been successful in achieving both of its stated objectives. A precondition for this has been a very prudent fiscal policy that has kept the government's non-oil deficit at around 9.5% of GDP in recent years, but with a Fund level that is equivalent to about three times the GDP level (and growing). This implies that the non-oil deficit is financed almost entirely by the transfer of the real returns from the GPF.

Norway's strong performance is explained by political consensus on the need to safeguard the pensions of current generations as well as the income of future ones. This is further reinforced by robust institutional strength, which has ensured the continuity of these policies over time. Such factors are uncommon in middle- and low-income countries.

Another option is to keep the resources within the budget but generate significant fiscal savings to directly finance investments that enhance the country's potential growth and, ideally, help diversify the productive base and reduce dependence on natural resource rents, as is the case in Malaysia, for example (Collier 2014). Achieving this is not easy, among other reasons because natural resource rents hinder such changes due to their impact on the real exchange rate (the so-called Dutch Disease).

One way to assess whether national savings are sufficient to finance investments that would allow for the gradual substitution of natural capital consumption—i.e., whether the principle of "weak sustainability" is being met—is to compare the group of countries that the IMF classifies as resource-dependent (IMF 2012) and that are systematically present in the World Bank's genuine savings database (19 countries). This comparison shows that, in general, these countries tend to maintain positive genuine savings rates, meaning they are covering capital depreciation and leaving a surplus to finance increases in other forms of capital. However, there is no systematic pattern of relationship between the natural resource rents they capture[9] and genuine savings rates,[10] nor are particularly high investment rates observed that would suggest they are driving a structural transformation of their economies.

References

Barma N, Kaiser K, Le TM, Viñuela L (2012) Rents to riches? The political economy of natural resource-led development. World Bank

Centro de Estudios del Cobre y la Minería (2023) Las interrogantes del presente y futuro de CODELCO. https://cesco.cl/wp-content/uploads/2024/10/CESCO_Las-interrogantes-del-presente-y-futuro-de-CODELCO.pdf

Collier P (2014) Resource revenue management: Three policy clocks. In: Caputo R, Chang R (eds) Commodity prices and macroeconomic policy. Central Bank of Chile

González A, Sanchez F, Castillo E (2023) The nationalization of the large-scale copper mines in Chile: successful investment or financial failure? Mineral Economics. Advance online publication. https://doi.org/10.1007/s13563-023-00412-z

International Monetary Fund (2012) Macroeconomic policy frameworks for resource-rich developing countries (Policy paper). https://www.imf.org/external/np/pp/eng/2012/082412.pdf

International Monetary Fund (2020) State-owned enterprises: the other government. Chapter 3, Fiscal monitor. April. https://www.imf.org/en/Publications/FM/Issues/2020/04/06/

[9] These fluctuate between 5% and 10% of GDP, although with considerable variance.

[10] Which are between 5% and 15% of GDP, also with significant dispersion.

fiscal-monitor-april-2020#:~:text=Chapter%203%20:%20State%2DOwned%20Enterprises,of%20transparency%20of%20their%20activities

Otto J (2017) The taxation of extractive industries: Mining (WIDER Working Paper 2017/75). United Nations University WIDER. https://www.wider.unu.edu/sites/default/files/wp2017-75.pdf

Otto J, Andrews C, Cawood F, Doggett M, Guj P, Stermole F, Stermole J, Tilton J (2006) Mining royalties: a global study of their impact on investors, government and civil society. World Bank

World Bank (2011) Overview of state ownership in the global minerals industry: long-term trends and future (Extractive Industries for Development Series No. 20). https://hdl.handle.net/10986/16968

Natural Resource Dependence and Global Sustainability

9

Abstract

This chapter takes stock of the global environmental impact of the use of different categories of materials provided by nature (biomass, fossil fuels, metal ores and non-metallic ores). Making use of recent research by the World Bank it shows the links between the use of these materials with the satisfaction of diverse sources of human demand for goods and services (food, housing, mobility, energy, etc.). There are some good news, such as the falling ratio of fossil fuels and biomass use per unit of GDP across all regions of the world. However, this is not the case for metals and minerals, especially in Latin America and Africa, where countries rely significantly on them for exports revenues. Since final demand for these commodities is coming both from high income countries as well as from fast developing nations, the reduction of their environmental impact will be challenging. In addition, market signals are distorted by significant externalities linked to greenhouse gases emissions and biodiversity in the places where they are extracted.

Keywords

Material resources · Environmental impacts · Intensity of use

The previous chapters presented a relatively optimistic view regarding the role of technology in alleviating potential natural resource scarcity that could constrain global economic growth. In the case of overexploited resources, such as marine resources, correcting market failures arising from externalities could enable the sustainable exploitation of natural resources over time. The intention has been to show that, in both cases, the knowledge and means exist to successfully address the challenge of scarcity.

The role of advances in knowledge and technology is most evident in the continuous increase in the production of metals and fuels from non-renewable resources.

J. Vial, *Sustainable Development and Natural Resources*, Integrated Science 39,
https://doi.org/10.1007/978-3-032-12263-6_9

In these cases, the incentive for innovation derived from the opportunity to capture rents from competitive advantages through cost reductions has enabled technological progress without significant cost increases. When prices have risen, as occurred with the oil price "shocks" at the beginning of this century, this made it possible to finance the development of new energy sources, both for traditional fuels such as shale oil, and alternative sources based on renewable resources such as solar and wind. Although those alternatives already existed, they still required high oil prices or subsidies to reach the scale needed for widespread adoption, which was achieved in less than two decades.

Although the scarcity of provisioning services has not been an obstacle to meet the material demands arising from the exponential increase in population and per capita consumption over the past century, this alone is not sufficient to achieve sustainability. It has already been shown that there are many areas where the presence of externalities and other market failures, combined with human impatience, lead to the overexploitation of renewable resources, the destruction of ecosystems, and even the excessive capture of non-renewable resource rents to finance current consumption at the expense of the well-being of future generations.

Moreover, even if resources are exploited rationally and in isolation from a broader context, the fact that the environmental impacts of human activity are putting at risk the support systems that regulate human activities and life in general require that the impact of resource extraction be examined from a global perspective. Following the approach promoted by the World Bank (Damania et al. 2023), it must be recognized that the world is no longer operating within what could be called an "efficient production frontier" defined by the trade-offs between the use of provisioning services and the supply of ecosystem services, as illustrated by point A in Fig. 9.1. The figure shows in red what would be the global production possibility frontier, considering the two major classes of natural capital: that which is linked to the production of goods and services to meet human consumption (horizontal axis)

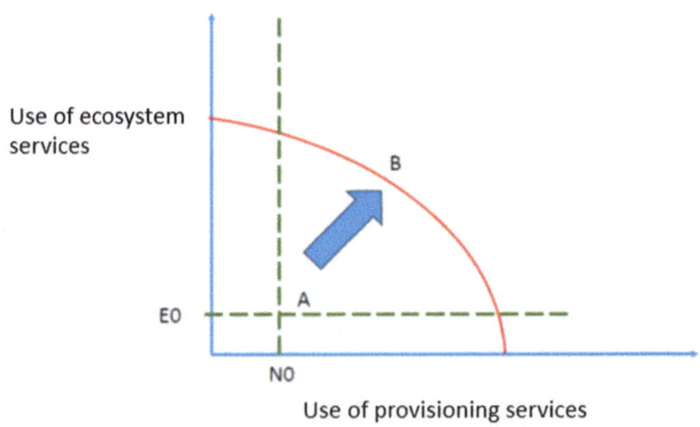

Fig. 9.1 Production possibility frontier considering natural capital constraints

and the associated combinations of ecosystem services relevant to life but are not inputs for human consumption (vertical axis).

This frontier shows the maximum levels of human consumption of goods and services for combinations of the different forms of capital. In the past, a point such as A, which lies within the frontier, may have been the maximum, but advances in knowledge and technology, international trade, and the application of greater physical and human capital have made it possible to expand the frontier so that a combination like point B, with much higher consumption than at A, becomes feasible. The fact that today these levels of natural capital use are reaching the point where the integrity of the biosphere is at risk implies that, unless very fundamental changes occur, in the coming decades the feasibility of continuing to push this frontier upward is becoming less viable and, on the contrary, it could well contract.

In the short term (which, at this scale, means decades rather than weeks as in traditional macroeconomic analysis), it must be considered that humanity will be operating along the frontier, so that increased extraction of provisioning services will come at the expense of the availability of other ecosystem services.

In the face of a retreating natural world, with severe ecosystem degradation and losses of biodiversity and wildlife in general, the relationship between the use of provisioning services provided by ecosystems and the preservation of the integrity of the ecosystems where economic activities take place is becoming increasingly complex. When natural spaces were abundant and there was a world to discover and conquer, as was the case throughout most of human history until the early twentieth century, the tension between the use of natural resources and the preservation of ecosystems was a local and secondary issue. This changed in the mid-twentieth century and has only worsened since then. The world still has ample reserves of copper, iron, coal, and other materials essential for the proper functioning of modern economies, but the growing scarcity of the natural services associated with the ecosystems where these resources are found is making their exploitation increasingly difficult.

In this context, the main concern is the impact that the extraction of natural resources is having on other ecosystem services. These tensions largely originate from externalities and the impatience of the human species, as discussed in previous chapters, which have led to the overexploitation of renewable resources and the generation of waste that threatens the sustainability of future development, both globally and at the level of individual countries or specific localities.

The latest report from the United Nations Environment Programme (U.N. Environment Programme 2024) addresses this issue, highlighting the significant environmental impact of increased demand for specific provisioning services associated with meeting the specific needs of consumers. The report identifies four major categories of material resources provided by nature (biomass, fossil fuels, metal ores, non-metallic minerals) that are used to meet the demand for seven major categories of services (food, mobility, construction, energy, communications, waste disposal and recovery, and others). In this way, it is possible to relate the effects of population growth and increased per capita consumption to meet these needs, with the requirements for the materials identified above.

As expected, and as noted in previous sections, the last half-century has seen a sharp increase in demand for all materials. However, this increase has not been uniform, and different trends are observed among materials and among countries grouped by income level.

Figure 9.2 shows the trend in the consumption of each of these materials, expressed in billions of tonnes per year. Mass is a relevant indicator of environmental impact, as it measures the volume extracted and, at the same time, the amount of material that must ultimately be disposed of at the end of its useful life. In the case of fossil fuels, the waste from their combustion takes the form of invisible gases, but they are no less real for being unseen.

In all cases, an upward trend is observed, but at very different rates: while the consumption of biomass and fuels has doubled over the past 50 years, that of metals has tripled, and that of non-metallic minerals (mainly sand and cement) has more than quadrupled in the same period, far outpacing the increase in the other categories.

When examining the uses to which these various materials are allocated, three stand out far above the rest: buildings and structures (excluding infrastructure associated with mobility such as roads, bridges, and ports), mobility, and food.

Buildings and structures, both for residential purposes and to produce goods and services (other than those associated with mobility, food production, and energy generation), currently require 30.6 billion tonnes of materials per year, of which 76% originate from non-metallic minerals, mainly sand and cement. Anyone over the age of 50 who has had the opportunity to look out of the window of an airplane and compare landscape changes over this period can attest to their impact on the territory. One of the greatest environmental impacts of this activity is the high

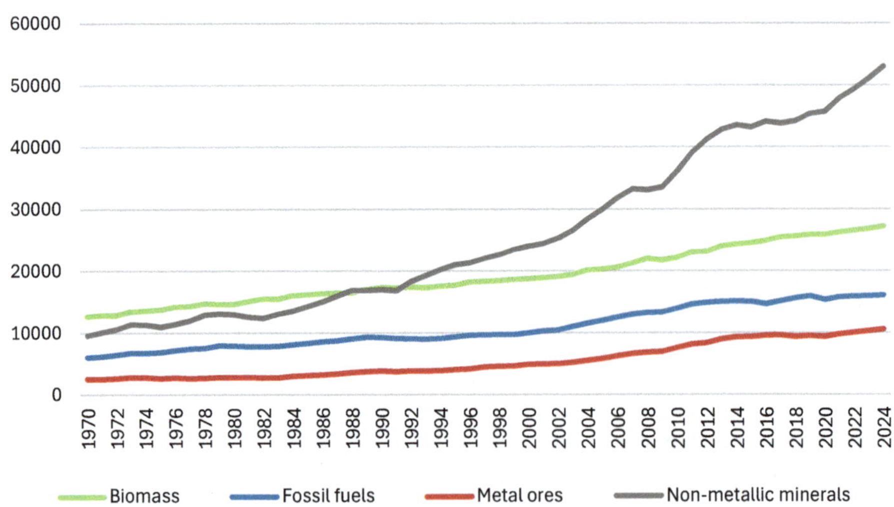

Fig. 9.2 Trends in global material use (million tonnes). (Source: UNEP-IRP (2023). Global Material Flow and Resource Productivity Database. Global Material Flows Database | Resource Panel)

carbon footprint of cement production, the main use of these materials. When comparing regions by their level of development, it is evident that upper-middle- and high-income countries account for most of the use of these materials, not only because of their higher consumption levels, but also because in both groups this represents about one third of total material consumption.

The second most important use of materials is associated with mobility, at 28.6 billion tonnes per year, of which 64% also corresponds to non-metallic minerals and 19% to fossil fuels. Here, the difference in material consumption is greater between high- and middle-income countries compared to low-income countries, where only 12% of material consumption is allocated to this purpose.

Finally, the third major factor driving material use is food, with 23.6 billion tons per year, of which 84% consists of biomass. This has a significant impact on land use and land-use change, despite the substantial increase in productivity per hectare previously mentioned. In fact, in middle- and low-income countries in South and Central America, as well as in Africa and South Asia and the Pacific, deforestation and land-use change are by far the largest sources of greenhouse gas emissions and are also the main drivers of the massive losses of biodiversity as a result of habitat destruction. In the case of biomass, it accounts for more than half of material use in low-income countries and more than a third in lower-middle-income countries. In upper-middle- and high-income countries, it represents around 20% of total material consumption.

In the first part of the book, the need to drastically reduce material use to achieve sustainable development at the global level was raised. This, among other things, requires a substantial reduction in the materials intensity of GDP generation, if a major decline in quality of life is to be avoided. Statistics show that the world is still far from this goal, at least in two cases (non-metallic minerals and metals), as can be seen in Fig. 9.3, which shows the evolution of the materials intensity of use over the past 50 years.

While the use of biomass and fuels per unit of GDP has decreased significantly, this has not been the case for metals and, of particular concern, for non-metallic minerals.

To better understand the origin of these trends, it is necessary to distinguish how the use of these materials is distributed across different regions of the world, with varying levels and trajectories of development. In the absence of data disaggregated by country income groups, in this case it was necessary to use geographic regions which, although they include countries with different levels of development, are dominated by the majority group. For example, the weight of high-income countries in Europe and North America is large enough to consider these regions representative of high-income countries. In Latin America and the Caribbean, as well as in Asia-Pacific, middle-income countries predominate, and in Africa, lowh-income countries.

An important initial point is the relative weight of these regions in the use of each of the four major material groups and how this has changed over time. This is shown in panels A, B, C, and D of Fig. 9.4.

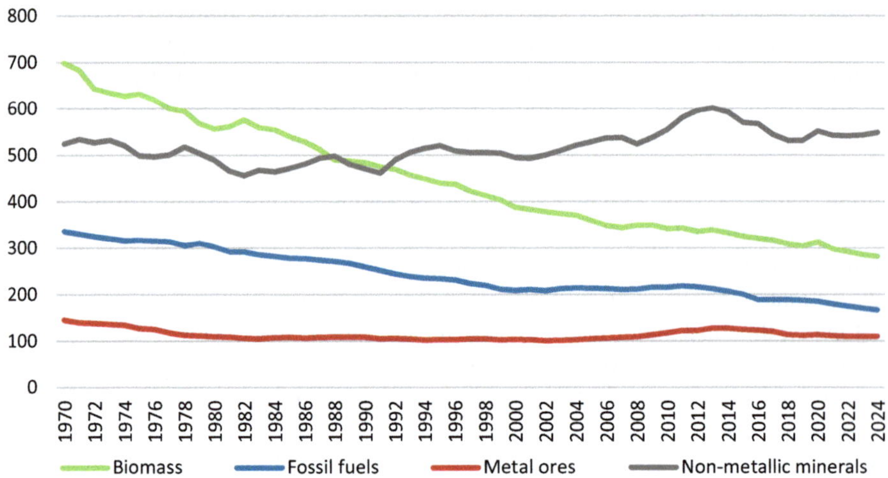

Fig. 9.3 Evolution of global materials use intensity (tons per dollar of GDP). (Source: World Bank and UNEP-IRP)

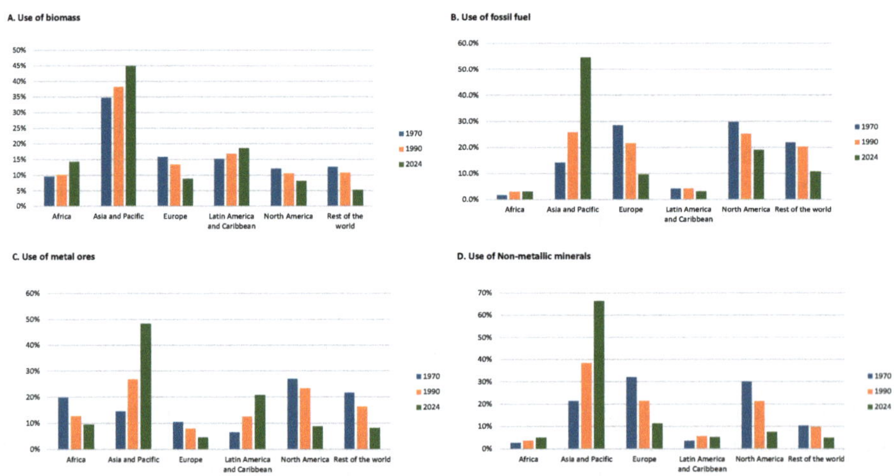

Fig. 9.4 Evolution of the relative importance of five major world regions in the use of each of the main material groups. (Source: UNEP-IRP (2023). Global Material Flow and Resource Productivity Database. Global Material Flows Database | Resource Panel)

The most notable aspect of the information on the relative weight of regions in global material use is the sharp increase in the importance of Asia and the Pacific in all categories, except for biomass, rising from levels between 12% and 15% in 1970 to figures between 50% and 60% today. Even in biomass use, the region stands out, although its share reaches "only" 45%.

It is important to note that biomass use is mainly for food, so it is not surprising that the most populous region in the world has the highest share in the use of materials for this purpose.

Another point to consider in the analysis is that minerals and fuels are mainly used as inputs in the production of final goods, and international trade plays an important role because final demand often comes from high-income countries. For example, iron ore produced in Brazil can be converted into steel or exported as ore to be processed in Japan or China, and then that steel becomes part of a car sold in the United States. If the steel is produced in Brazil and then re-exported, statistics report Brazil as the user of iron. If, on the other hand, the ore is exported, the final user reported is the country that transformed the iron into steel, but not the country where the car was sold. This explains, for example, the sharp increase in the use of metallic minerals in Latin America.

As Asia became the world's main manufacturing center during this period, it is not surprising that it appears as the main user of raw materials, but it should be remembered that a very large fraction of these resources is re-exported as manufactured products, mainly to developed countries and, increasingly, to middle-income countries.

Figures 9.2 and 9.3 highlight the sharp increase in global material use but also show that in three of the four material categories there has been a decrease in the ratio between their use and GDP: that is, there is some (partial and slow) decoupling between economic activity growth and materials use. This is a minimum requirement for restoring the sustainability of global economic growth. However, this adjustment is still insufficient and, moreover, is not occurring in the case of non-metallic minerals, an activity closely linked to the construction of housing, mobility infrastructure, and habitable spaces. In these uses, the tremendous size and growth of middle-income consumer groups in emerging countries is the main driver of global demand for these materials, confirming what was stated in the first part of this book: the greatest challenge for global sustainability is to adjust the economic growth patterns of developing countries, especially those with middle incomes.

The charts in Fig. 9.5 show very different trends among materials and regions. In the case of biomass use, which is closely linked to food and therefore constrained by population size, there are significant downward trends, although these are more recent and modest in Latin America and Africa, regions that are poorer than the others. The most recent demographic trends point to an acceleration of the decline in biomass use intensity in Latin America. On the other hand, the fact that countries such as Brazil and Argentina are major food exporters points in the opposite direction.

Regarding fossil fuels, in developed countries and in Latin America there is a declining trend in their intensity of use, which halved over the past 50 years. However, Asia-Pacific, by far the region with the highest consumption of these fuels, shows a steady trend in the ratio of fuel consumption to GDP. This is very bad news, as this is also the region with the highest GDP growth rates in recent decades, a trend expected to continue in the future. This is one of the main sources of tension for achieving sustainability goals. Fortunately, China, Japan, and the rest of the region are taking this challenge seriously, in part because, as will be seen later, these countries are highly vulnerable to climate change.

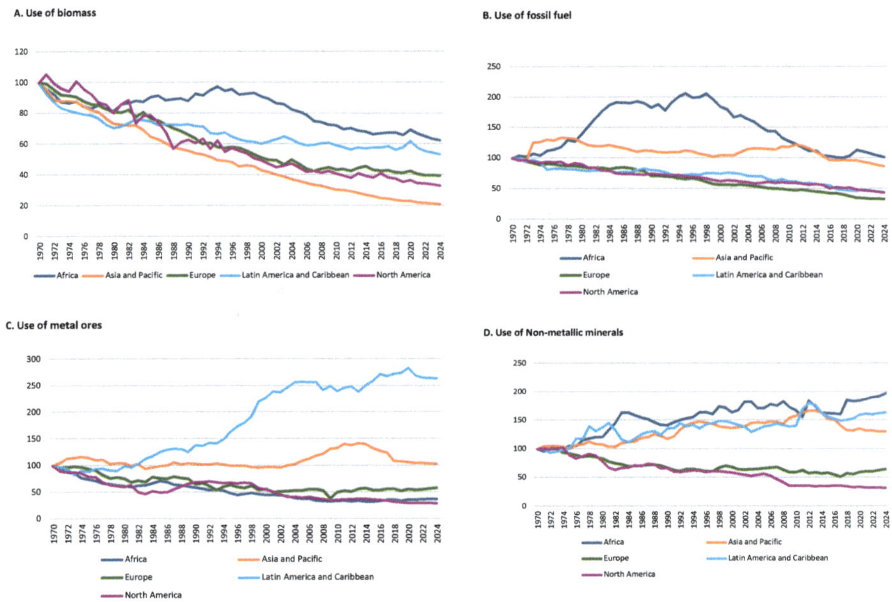

Fig. 9.5 Evolution of material use intensity relative to GDP by region. (Source: UNEP-IRP (2023). Global Material Flow and Resource Productivity Database. Global Material Flows Database | Resource Panel and World Bank, World Development Indicators)

As for metal minerals, regional trends are largely determined by the geographic location of deposits, with Latin America having a disproportionately large share relative to its economic size, something that is also beginning to occur in Africa. What really matters here is the intensity of final use of metals, which is determined by the levels of final consumption of goods and services worldwide, as shown in Fig. 9.2, which indicates a flat trend in global use of metal minerals. Among these, iron is by far the dominant due to its use in steel, a material indispensable for a multitude of human activities (Smil 2022) and with few substitution possibilities. Another important material is copper, used mainly for its electrical conductivity properties, making it an essential element for the energy transition, so its demand is expected to continue rising in the future. A similar situation occurs with aluminum, although its industrial applications are more varied than those of copper. All these metals require high energy inputs for their production.

Finally, in the case of non-metallic minerals, dominated by sand and cement, which are widely used in the construction of structures for human habitation, mobility, and the production of all kinds of goods and services, the global trend in their use intensity is increasing and driven by developments in emerging countries across all continents. Only in Europe and North America are there declining trends, but these are rather mild and even show some reversal in Europe. The rapid demographic transition will be an important factor in breaking the upward trend in middle-income developing countries, but it will also have to compete with strong demand for infrastructure construction, especially those needed to facilitate the movement of goods and people. This will be another focal point of tensions between growth and sustainability at the global level.

What has been shown in the previous chapters of this section is that the conceptual frameworks to successfully address these challenges have been developed. However, applying what has been learned requires political will, supported by strong institutions, ranging from the law-making process to the building of fiscal and sectoral oversight institutions. This also entails that technical criteria based on scientific and economic evidence prevail in decision-making. This remains an outstanding task in many countries, and even in those that have done relatively well so far, there is always the risk of setbacks. The current global environment of polarization and populist policy decisions that emerged recently does not help to achieve these objectives in emerging countries.

One conclusion from the previous analysis is that the scarcity of natural resources per se has not been a significant constraint on global economic growth so far, nor does it appear likely to be so in the coming decades, if prices reflect any potential scarcity. However, what has become evident in recent decades is the growing tension between the demand for provisioning services from nature and its capacity to provide the ecosystem services required to maintain the proper functioning of the biosphere. This is reflected daily and in the most diverse places on the planet, in the increasing resistance to developing mining, oil extraction, real estate, and many other investment projects, for reasons that ultimately can be traced to concerns about the preservation of the environment and the ecosystem services it provides, including cultural and aesthetic services.

It is very likely that in the coming years and decades these environmental constraints and distributive conflicts will become the main obstacle to obtaining the material resources needed to produce the goods demanded by consumers worldwide.[1]

References

Damania R, Polasky A, Ruckelshaus M, Russ J, Amann M, Chaplin-Kramer R, Gerber J, Hawthorne P, Martin P, Saleh M, Ruta G, Schmitt R, Smith J, Vogl A, Wagner F, Esha Z (2023) Natures frontiers – achieving sustainability, efficiency, and prosperity with natural capital: main report (English). Washington, D.C.: World Bank Group. https://documents.worldbank.org/curated/en/099062723135519167

Gazzani A, Herrera V, Vicondoa A (2025) The asymmetric effects of commodity price shocks in emerging economies [Unpublished manuscript]. https://drive.google.com/file/d/13DdP7pVRSrYwyfWH7WEyrfEPsZm0XdYs/view

Smil V (2022) How the world really works: the science behind how we got here and where we're going. Viking

United Nations Environment Programme, International Resource Panel (2024) Bend the trend: Pathways to a livable planet as resource use spikes (Global Resources Outlook 2024). UNEP. https://wedocs.unep.org/20.500.11822/44901

[1] In all figures containing data on the use of non-metallic minerals in North America, this variable had to be replaced by non-metallic mineral extraction in that region because the series is not available in the database. Since international trade in this category of materials between different world regions is low, the estimated error should be small.

Macroeconomic Challenges of Natural Resource Dependence

10

Abstract

Natural resources provide a large fraction of foreign exchange and fiscal revenues in many Middle- and Low-Income Countries. Some of them have been able to channel those funds to improve their living conditions. However, they have also been a source of macroeconomic instability and corruption, to the point that some authors talk of a "Natural Resource Curse". This chapter deals with the macroeconomic challenges, and it shows significant progress by many countries in the last decades. Many countries learned a lot on how to manage volatility in prices and incomes from natural resources since the days of the Debt Crisis. The combination of stronger fiscal and monetary institutions, the development of financial markets and the opening of these economies to international trade, coupled with inflation targeting, exchange rate flexibility and fiscal rules, helped most of Latin America to weather the Global Financial Crisis. However, it is too early to tell if these reforms will survive the political and social sequels of the Covid Crisis. These countries now know how to handle the challenges posed by natural resources price volatility, but the political will is in doubt.

Keywords

Commodity prices volatility · Macroeconomic imbalances · Macroeconomic policies · Fiscal rules · Stabilization Funds · Monetary policy · Exchange rates · International Financial Flows · Financial Development

Specialization in the export of natural resource-based products poses several major macroeconomic challenges that, if not successfully addressed, become sources of instability. This instability arises from the difficult-to-manage volatility in fiscal revenues as well as for the impact on the foreign exchange market, due to the significant role of these products in exports and the capital flows associated with

J. Vial, *Sustainable Development and Natural Resources*, Integrated Science 39,
https://doi.org/10.1007/978-3-032-12263-6_10

investments and profit remittances. In recent decades, it has also become evident that prolonged commodity booms can trigger large-scale real investments in the sector, which may become an additional source of macroeconomic instability.

Price volatility is related to the fact that the supply of natural resource-based products is relatively inelastic to prices in the short term and is subject to natural constraints that limit its ability to respond quickly to demand expansions. Moreover, it is also subject to unexpected shocks: a copper mine may have its operations disrupted by landslides or flooding, agricultural harvests vary depending on weather conditions, and oil prices can change abruptly in response to conflicts in the Middle East, among other factors. However, it should be noted that these elasticities differ significantly over different time horizons. In the case of agricultural production, price shocks generate signals to which producers in different hemispheres can respond quickly, so supply elasticity can be quite high, even within a single year if production is well diversified between hemispheres. In contrast, developing a new mineral deposit or expanding an existing one usually takes years. The case of oil and fuels is also different because, in principle, if spare capacity exists, it is relatively easy to adjust supply flows, and there are countries such as Saudi Arabia and the United States that can adjust their supply quite rapidly. However, this is also a market where supply decisions are heavily influenced by geopolitical factors.

On the demand side, there is considerable inertia, although the price elasticities of demand for consumer goods such as food and, to some extent, fuels like natural gas and gasoline, are responsive to price changes. In contrast, demand for industrial raw materials generally exhibits lower price elasticity, especially if these raw materials represent a small share of the final cost of the goods in which they are used.

One buffer against these fluctuations is inventories, which play a very important role in markets for storable products such as copper or oil. This is not the case for perishable agricultural products, which cannot be stored for long periods.

In general, however, as shown in Fig. 10.1, the volatility of these prices is high, except for food products. This volatility has significant macroeconomic effects on the foreign exchange earnings of countries that depend on these exports, as well as on public finances when these products provide substantial revenues to the Treasury.

These effects are often amplified by financial markets, as they factor in their credit and investment decisions the strong correlation between domestic cycles and fluctuations in terms of trade. Regarding this point, it is worth noting recent evidence showing that the effects are asymmetric; that is, declines in terms of trade tend to be more detrimental to the growth of emerging economies than improvements, partly due to this amplifying effect of financial markets.[1]

The volatility of export prices for raw materials is a recurring topic in any discussion of macroeconomics in Latin American, African, and some Asian countries. Many of the economic crises in Latin America during the twentieth century, as well as the almost inevitable recurrence of inflation, were linked to the difficulty of adequately managing this problem. Suffice it to say that, after countless crises and

[1] See: Gazzani et al. (2025) and the references to the specialized literature on this topic cited therein.

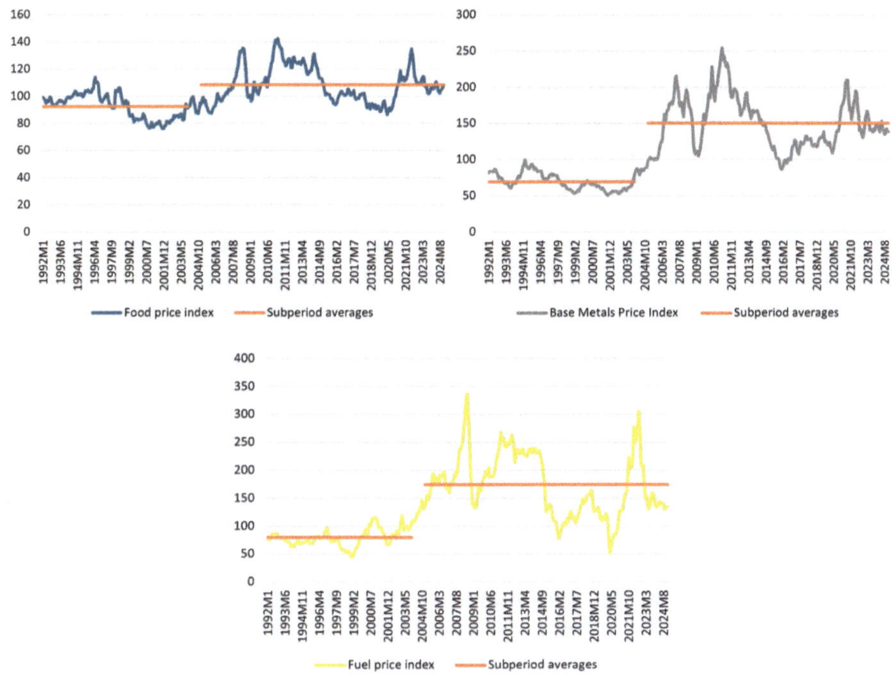

Fig. 10.1 Trends and fluctuations in real price indices of natural resources (indices base 2016 = 100, based on 2016 US dollar figures, deflated by US CPI). (**a**) Evolution and variability of food prices. (**b**) Evolution and variability of metal prices. (**c**) Evolution and variability of fuel prices. (Source: IMF and Federal Reserve Bank of St. Louis)

failed attempts, by the end of the last century a consensus emerged around the need to shield the economy from fluctuations in the prices of these exports. This involves allowing the exchange rate to fluctuate in order to absorb the impact on foreign exchange availability[2] and decoupling public spending from the cash flows provided by these products, whether derived from state ownership or from taxes or royalties levied on them.[3]

Evidence from many countries in Latin America, as well as from Africa and some Asian countries, attests to the challenges they face in dealing with the price volatility of these products. The good news is that, over the years and decades, a wealth of valuable experience has accumulated, with concrete policy recommendations that enabled various countries to improve their macroeconomic management.

[2] This requires, among other things, having sufficient international reserves, as well as a deep financial system without currency mismatches, where the local currency is the main means of payment.

[3] Well-designed fiscal rules or stabilization funds, within a framework of fiscal policy determined by macroeconomic objectives rather than cash availability, are key instruments. This presupposes governments with public savings or access to credit that is both expeditious and reasonably priced IMF (2012).

The main channels through which the prices of natural resource-based products affect macroeconomic developments, when these products account for a large share of GDP and exports are threefold:

The availability of foreign exchange, which originates from exports, as well as from foreign investment—including associated loans—and profit remittances. Available evidence shows that the more rigid the economy, particularly in the foreign exchange and labor markets, the more costly it becomes to adjust to such shocks. It is worth noting that one of the reasons why many countries maintain Central Bank intervention in the foreign exchange market is precisely because, when exports are concentrated on a single product, exporters of that product wield significant market power over a key economic variable. Of course, this means that Central Banks lose control over monetary policy, so they tend to add interventions in financial markets, piling up distortions in the economy. As exports diversify, capital markets expand, and these markets become more integrated internationally, this argument begins to lose relevance.

The second transmission channel operates through public finances, as the government captures part of the revenues from these exports, either through general income taxes, specific royalties on the sales of such products, transfers of profits from state-owned enterprises operating in these markets, or a combination thereof. Naturally, the greater the weight of these products in the economy, the larger the impact on public finances.

Finally, the third transmission channel operates through the macroeconomic impacts of investment in these activities. In this case, the effect is not immediate but occurs with a lag and only operates if it results in a more persistent cycle over time that affects long-term price expectations. For example, the so-called "super cycle" in commodity prices, which lasted about a decade between approximately 2005 and 2013, gave a strong boost to global investment in mining and oil during that period, which was a key driver of the economic cycle in resource-rich countries, both during the expansionary phase and its subsequent reversal. This is illustrated in Fig. 10.2 with the case of Chile, the world's largest copper producer. Investment in copper mining in Chile, which typically fluctuated between 2% and 3% of GDP, began to gain importance in response to the upward price cycle, reaching record highs of over 7% of GDP in 2011 and 2012, before falling (both in relative and absolute terms) from 2014 onwards.

Unlike the previous channels, which are mainly reflected in financial variables, this is a real shock, with strong impacts on the labor market and productive activities in general. As a result, the tools designed to address shocks to foreign exchange availability or fiscal revenues are not necessarily effective in dealing with a real economic stimulus of this magnitude and duration, since the development of mining projects takes several years.

To better understand the transmission of commodity price fluctuations in short-term macroeconomic management, it is useful to focus on the first two transmission channels, as investment booms in natural resources are less frequent and are generally associated with more permanent (actual or perceived) structural changes that play out over longer periods.

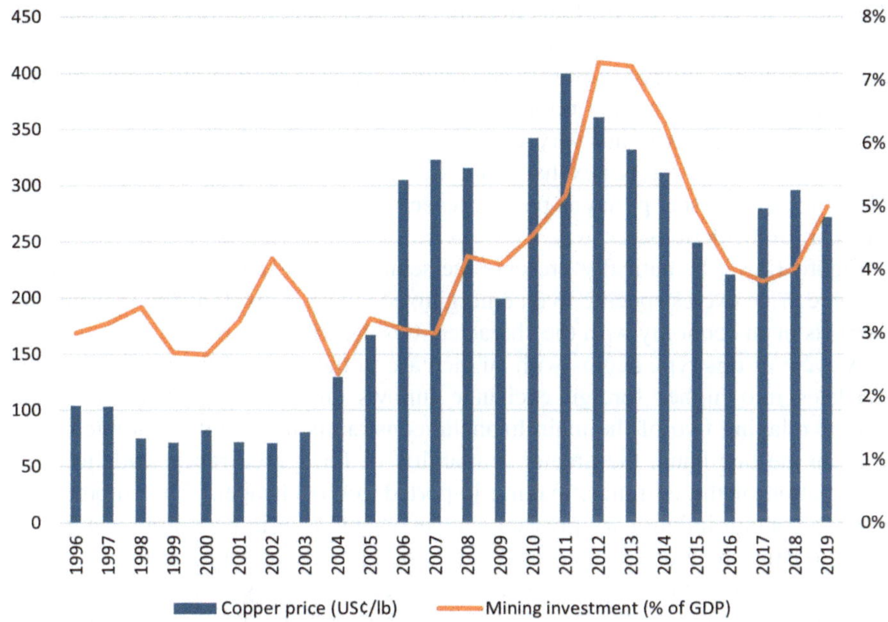

Fig. 10.2 Copper price and mining investment in Chile during the commodity "super cycle." (Source: Cochilco, Central Bank of Chile)

An additional factor complicating macroeconomic management is the absence of a deep local capital market to finance temporary deficits, both for the government and for companies, especially if they do not have ready access to external financing. This general description closely matches the macroeconomic environment prevailing in many emerging countries at the end of the last century. Although in the 1990s and the early decades of this century many countries undertook reforms that increased the flexibility of their economies and developed deeper financial markets more connected to the outside world, many countries still face similar circumstances. Moreover, the economic setbacks caused by the COVID-19 pandemic, the sharp increase in global fiscal indebtedness, and rising protectionism are restricting governments' room for maneuver, making it worthwhile to recall traditional cyclical impacts.

An economy with these characteristics, which also depends on resources such as copper or oil, faces considerable difficulties in coping with external shocks arising from fluctuations in the prices of these products. Such fluctuations can be abrupt and prolonged, as can be seen in the charts shown earlier. A common dilemma faced by economic authorities when the change has been large and has already persisted for several years is to identify whether what is being observed is part of a transitory but prolonged cycle, or whether it reflects a new trend that requires a more permanent adjustment in the exchange rate and public finances.

To illustrate the effect of these variables on the macroeconomy, it is worth to examine the effects of changes in the prices of a dominant product, such as oil in

Venezuela, Russia, or Middle Eastern countries, or copper in Chile or the Democratic Republic of the Congo, for example, when there are rigidities in the foreign exchange market and fiscal policy is severely constrained by cash availability. Some of these countries have implemented reforms that have enabled them to better cope with such shocks, but this is not very common, and there are also many examples of setbacks, where the foreign exchange market remains subject to heavy government intervention and fiscal policy is passive, even resorting to monetary financing of at least part of the deficits.

Figure 10.3 schematically presents the sequence and interaction of the natural resource price cycles specific to an emerging economy, as well as the transmission channels in an economy with the characteristics described above.

A cycle of this type begins with an increase in the price of the resource, which translates into higher foreign exchange inflows and increased fiscal revenues, thereby relaxing two of the main financial constraints that limit economic expansion: on the one hand, the greater availability of foreign currency leads to a real appreciation of the currency, making imported goods cheaper.[4] This appreciation helps to contain inflationary pressures, a persistent issue in economies with these characteristics, resulting in increased purchasing power for wage earners.

On the other hand, the government can increase spending, which translates into higher investment, as well as increases in transfers and operating expenditures. The

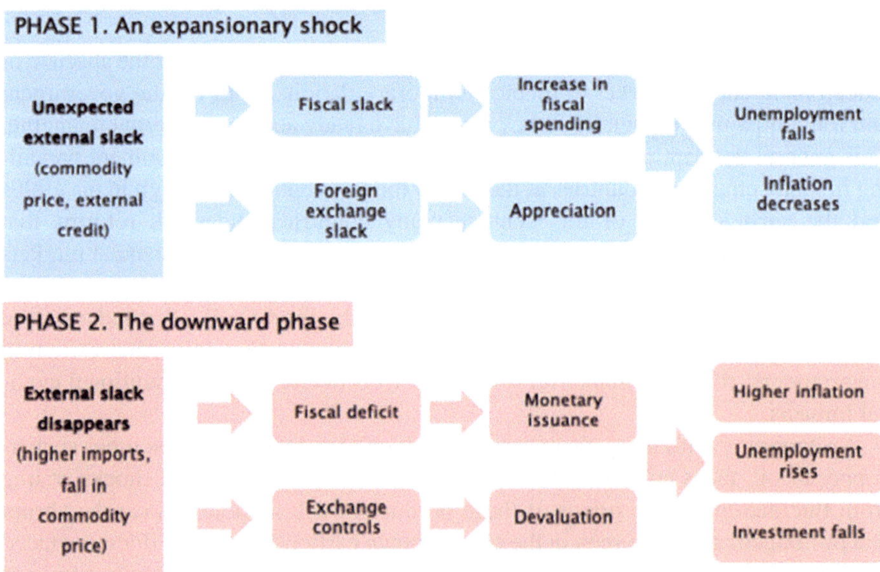

Fig. 10.3 Macroeconomic transmission channels of resource price cycles in a dependent economy

[4] It also reduces the profits of other producers of tradable goods, who may lobby for subsidies or protection against foreign competition.

economy then enters an expansionary phase with low inflationary pressures. The well-known "Phillips Curve",[5] which has served as the basis for monetary policy since the mid-twentieth century, operates in the opposite way to what is typical in these cases.

Indeed, when foreign exchange and fiscal constraints coexist with idle productive capacity, relaxing these constraints allows for accelerated growth and reduced inflationary pressures simultaneously. If the period of high commodity prices is prolonged, explanations begin to emerge that justify an even longer duration because of a trend change, which in turn leads fiscal authorities to make more permanent spending commitments. Unfortunately, even when there are changes in price trends, these do not occur without fluctuations, so almost inevitably the cycle reverses direction.

In the downward phase of the cycle, the initial constraints reappear in a more pressing form: the government faces deficits that cannot be eliminated in the short term and, without access to financial markets, has no choice but to turn to the Central Bank, while cutting investment and other expenditures wherever possible. At the same time, foreign currency becomes scarce to cover the current account deficit that arises from the drop in commodity prices, and inflation expectations rise, which, after a time, are confirmed when the authorities are forced to devalue the currency. The economy slows down and eventually enters a recession in response to the decline in public investment and lower consumer spending as purchasing power erodes, as well as tighter credit conditions as part of the Central Bank's efforts to curb inflation.

Thus, the economy dependent on natural resources moves to the rhythm of commodity prices, a pattern accentuated by public policies constrained by the availability of foreign exchange and fiscal resources.

What factors determine these outcomes? At least three very important ones can be mentioned[6]:

The absence of a sufficiently deep and developed domestic capital market to cushion fluctuations in the incomes of government, businesses, and households.

Weak fiscal and monetary institutions, where the finance minister cannot curb spending initiatives (often permanent), and where the Central Bank is required to finance government deficits.

An inadequate policy framework, which resorts to fixing the exchange rate to control inflation, and where fiscal policy is not formulated on the basis of sustainability of public finances, nor does it have the flexibility to act countercyclically.

Reforms based on the Washington Consensus helped to mitigate all these problems. On the one hand, many countries liberalized trade and financial markets, and most emerging economies have become more and better integrated into the global economy, with greater diversification of their exports. In most middle-income countries, governments improved their fiscal position and are applying policy rules to accumulate savings during periods of high temporary revenues and vice versa,

[5] The Phillips Curve represents a (positive) relationship between growth and inflation.

[6] See IMF (2012).

making fiscal policy independent of cash availability. The gradual development of local financial markets has also contributed to this, and in some cases, has been accompanied by better performance and greater depth of capital markets. Monetary institutions have also progressed, and today many central banks enjoy autonomy in their monetary policy decisions, with numerous cases where government financing is prohibited. All this has been accompanied by a strengthening of technical capacities, to the point that, in much of the emerging world, senior technical staff in economic teams are composed of professional economists with postgraduate studies from internationally prestigious universities, something that was rare at the end of the twentieth century.

10.1 Fiscal Policy: Institutions and Instruments

An effective fiscal policy, capable of eliminating or mitigating the procyclical bias induced by dependence on a particular resource, is essential for sound macroeconomic management. However, such a policy runs counter to the instincts of political authorities, who are always overwhelmed by demands they cannot meet due to a lack of resources. Unlike monetary policy decisions, which also face similar tensions, fiscal authorities perform the most political of all technical functions: allocating scarce resources to purposes consistent with government priorities, which in turn respond to the people.

A suitable and robust institutional framework is an essential condition for maintaining sound macroeconomic management with a medium- and long-term perspective. There are no universal formulas for this, and experience shows that even in successful cases, political conditions can change, thereby altering the circumstances that enabled those achievements.[7] Nevertheless, there are some empirical regularities that recur and are worth considering. One of these is the importance of the primacy of the Fiscal Authority in matters with budgetary impact within the Executive, and, in turn, the limitation of Congress's powers to introduce motions with budgetary implications,[8] at least in presidential systems.

One way to address this tension has been the establishment of fiscal policy rules to insulate the budgetary process. Originally, these rules were very simple, such as maintaining a balanced budget, which in practice did not work because it ended up exacerbating the cycle in the presence of transitory shocks. Attention then shifted to setting fiscal targets adjusted for the position in the cycle, a practice that became common after its incorporation into the agreements preceding the European Union. However, this is of secondary relevance in countries dependent on natural resources, where most revenue volatility originates from fluctuations in the prices of those resources.

[7] A more in-depth discussion of the relationship between institutions, policy decision-making processes, and their effectiveness can be found in Stein et al. (2005) and Scartascini et al. (2010).

[8] There is extensive literature on this topic, especially in Latin America. See Alesina and Perotti (1995), Alesina et al. (1999).

In the case of countries dependent on natural resources with highly volatile prices, there are basically two alternatives: one is to use market instruments to hedge exchange rate and fiscal risks arising from price fluctuations. The other is to create "self-insurance" mechanisms, such as funds built up from savings accumulated during periods of high prices, which are then used during the downturns of the cycle.

Price fluctuation risk hedges have gained importance as these markets have developed and reached sufficient depth to cover large volumes of money for periods that typically range from two to even three years. Some oil-exporting countries have successfully used, at least partially, such instruments to hedge revenues for periods of one or two years. Where state-owned enterprises exist, as in Mexico, Venezuela, Russia, and various Arab countries, governments often use these companies to contract such hedges and thus avoid assuming political risks when prices unexpectedly rise, for example. The same occurs in the case of metals with large transaction volumes, such as copper. These types of instruments could generate substantial gains for countries to the extent that declines in export product prices are magnified by capital outflows (Caballero and Panageas 2008).

The other option, which was the only one available until a couple of decades ago, was self-insurance, established through Sovereign Funds specifically designed to cover income fluctuations, or as savings to compensate for the present consumption of natural capital, or mixed funds to cover both purposes.

Experience with these funds has not been successful (IMF 2012), except in cases where there is strong institutional capacity, so that savings are effectively protected from the temptation to spend them, especially in pre-election periods.

For these funds to function well, it is essential that they be integrated with the Fiscal Budget, both in their design and execution. In addition, they must have a clear rule for determining the long-term price, around which the mechanisms for fund accumulation and withdrawals pivot.

A case that is often cited as among the most successful in this area is Chile, which created a Copper Revenue Stabilization Fund (FEC) after the major debt crisis of the early 1980s, as part of a Structural Adjustment Loan (SAL) agreement granted by the World Bank in the mid-1980s.[9] This fund operated reasonably well for almost 15 years, until it was merged into a Structural Balance Rule in 2001 that adjusted for both cyclical fluctuations in copper prices and GDP. During that period, the funds were used primarily to reduce external debt, including prepayments. When the Asian crisis occurred, it was accompanied by a sharp drop in copper prices, the operation of this fund helped offset part of the revenue decline, and the government was able to maintain a very ambitious (and costly) reform program in the areas of school education and the criminal justice system.

With the establishment of the Structural Balance Rule in 2001, the FEC's accumulation/deaccumulation rule was replaced by a cyclical adjustment focused on a revenue calculation corrected for deviations of the market price of copper from its ten-year trend projection, made by a group of specialists independent of the government. A cyclical adjustment of tax revenues was also added.

[9] For further background, see IMF (2023), Frankel et al. (2013), Arellano (2015).

This rule was extraordinarily successful in addressing the commodity supercycle, as it allowed for the accumulation of large amounts of financial assets in the funds that supported the fiscal rule. These served to cushion the impact of the global financial crisis in 2008–2009 and the major earthquake at the beginning of 2010. However, these events left a legacy of higher spending levels that gradually eroded Chile's fiscal position.[10] This was reflected in the increase of public debt from around 5% of GDP in 2007 to 25% of GDP in 2018. Following the political crisis at the end of 2019 and the COVID-19 pandemic, there was a severe deterioration in political and fiscal institutions, which led to the depletion of accumulated resources in the funds and a sharp increase in public debt to 42% of GDP—a level very close to what is estimated to be the maximum prudent debt threshold.

This was not an isolated episode, nor was it limited to emerging economies. Globally, a significant fiscal deterioration took place since the pandemic, exacerbated by military conflicts in Europe and geopolitical tensions in Asia. In Latin America, the post-global financial crisis optimism that led many to believe that most countries in the region had finally "graduated" from fiscal weakness has unfortunately been called into question by the severe deterioration of their respective fiscal positions.

10.2 Exchange Rate and Monetary Policy

The other key component for mitigating the macroeconomic problems arising from commodity price fluctuations is the framework of monetary and exchange rate policy in emerging economies that depend on these resources.

Until the beginning of this century, almost all Central Banks in Latin America and other developing regions followed monetary policies with dual objectives: controlling inflation and stabilizing the exchange rate. This was partly a legacy from the era when local capital markets were very small and access to international financing was limited. The dual objective sought to prevent an "excessive" appreciation of the exchange rate from undermining export diversification, or a depreciation from jeopardizing inflation control. This situation began to change in the early 1990s, when short-term capital flows returned to these countries after the reforms that followed the Debt Crisis of the early 1980s.[11] Among these reforms, notable developments included the increasing autonomy of Central Banks, the adoption of legal restrictions on monetary financing of fiscal deficits, as well as significant improvements in fiscal institutions and, consequently, in the solvency of public accounts.

[10] See IMF (2016). Another factor contributing to this deterioration was systematic errors in the projection of the output gap, caused in part by a similarly systematic overestimation of total factor productivity growth. See Vial and Isea (2025).

[11] That decade was labeled the "lost decade" for Latin America in terms of growth, although it was also a period in which significant economic reforms were implemented, leading to substantial improvements in the economic performance of many countries in the region from the 1990s onward.

Table 10.1 Average annual inflation by decade in selected Latin American countries (percentage of average annual CPI variation)

	1980–89	1990–99	2000–09	2010–19	2020–24
Bolivia	1383	11	5	4	2
Brazil	328	843	7	6	6
Chile	21	12	3	3	6
Colombia	23	22	6	4	7
Ecuador	34	39	18	3	1
Mexico	69	20	5	4	5
Paraguay	20	16	8	4	5
Peru	481	808	3	3	4
Uruguay	58	49	9	8	7
Venezuela	23	47	21	8627	903
Simple average inflation of countries	244	187	8	867	95
Median inflation of countries	46	31	7	4	6

Source: IMF

The tension between the objectives of controlling inflation and avoiding real currency appreciation in a context of abundant foreign exchange led some countries to resort to capital controls, which generally achieved only partial and temporary success. The "Asian Crisis" exposed these contradictions, leading to the gradual abandonment of exchange controls and fixed exchange rates by Central Banks, although the vast majority continued to intervene in these markets beyond what would be warranted by prudent management of international reserves.

By the late 2000s, most Central Banks had adopted inflation targeting and used the policy rate at which they provide short-term liquidity to banks as their main monetary policy tool. The combination of all these factors led to a significant reduction in inflation in most South American countries, a region characterized by persistently high inflation rates.

As shown in Table 10.1, progress in controlling inflation has been remarkable, except for Venezuela and Argentina.[12] Since the first decade of this century, inflation in most Latin American countries became indistinguishable from that of other emerging economies and is very similar to that of developed countries.

One concern in establishing this policy framework was that the real exchange rate could fluctuate significantly and, in some cases, deviate excessively from long-term equilibrium values, distorting resource allocation and negatively affecting growth prospects. The experience of the past two decades does not show evidence of such problems, even though the region's economies have gone through periods of strong positive exogenous shocks (very high commodity prices during the "super cycle") and negative shocks (the crises of 2008–09 and 2019–20). Figure 10.4 compares the evolution of the real exchange rate in five Latin American economies that, while operating with exchange rate flexibility, differ in the frequency, scale, and

[12] Argentina is not included in Table 10.1 because the IMF has not published continuous inflation data for any of the decades considered. In Venezuela, the time series are also incomplete, so the averages for recent decades likely underestimate inflation in that country.

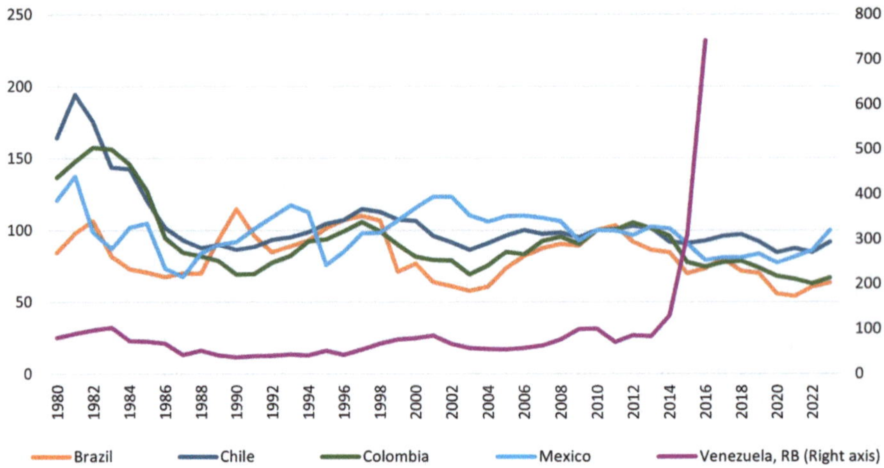

Fig. 10.4 Evolution of the Real Effective Exchange Rate in five Latin American economies with floating exchange rates and varying degrees of exchange rate intervention (See, among others, Sachs and Warner (2001), Lederman and Maloney (2007), and He et al. (2024) (Index base 2010 = 100). (Source: World Bank, World Development Indicators)

form of their exchange market interventions. Chile and Peru exhibit the smallest deviations from the average.[13]

Among these countries, Venezuela, Brazil, and Colombia have intervened the most in foreign exchange markets, while Chile and Mexico have only done so occasionally since the beginning of the present century. A common trend across all countries was the currency appreciation that took place in the second half of the 1980s, which merely reflects the extreme real depreciation they previously experienced because of the external debt crisis. Since the start of the century, all countries have shown fluctuations around the average and, paradoxically, Chile and Mexico exhibit the least variability in their real effective exchange rates.

10.3 Financial Development: An Enabling Factor

A key element for the effective functioning of monetary policy based on relatively sophisticated instruments, such as the management of interest rates that provide short-term liquidity to banks, is the existence of a developed, deep, and competitive financial market. Conversely, in an economy where this is not the case, the macroeconomic impact of providing liquidity to banks will be weak, which would require very drastic movements in policy interest rates, resulting in high volatility in the prices of various financial assets.

[13] The data are annual averages. Naudon and Vial (2016) report similar results using monthly averages with BIS data.

In such cases, there is no alternative but to use simpler instruments, such as attempting to control the money supply, which is practically impossible without exchange rate flexibility. In a context where the supply of foreign currency is closely linked to the highly volatile prices of a few export products, the effectiveness of monetary policy is left at the mercy of fluctuations in those prices. This closely resembles the situation described in Fig. 10.1. Certainly, if the domestic financial market is small and shallow, the government will be forced to finance its deficits abroad, which increases the risk of macroeconomic volatility stemming from the prices of the commodities on which both the trade balance and fiscal financing depend.

The foundation for the development of a deep financial market is the domestic savings rate. If this is low, financial development will depend excessively on external financing, which in turn can be highly sensitive to the volatility of export product prices.

There is extensive literature on the so-called "natural resource curse"[14] and among the various hypotheses proposed to explain a possible transmission channel from dependence on natural resources to low sustained economic growth are institutional weakness, instability arising from competition over rent distribution, and, of course, macroeconomic instability. The latter is closely related to the difficulty developing economies face in effectively managing the volatility of their export prices when domestic savings are low and the financial market has not reached sufficient depth.

Figure 10.5 shows the evolution of the domestic savings rate in various groups of emerging countries (10.5a) and for the group of the largest countries in Latin America and the Caribbean (10.5b).

Figure 10.5a highlights the large differences in savings rates between regions, which are not related to relative income levels. East Asian countries tend to maintain domestic savings rates above 40% of GDP, and those in South Asia are close to 30% of GDP, both much higher than those in Latin America and the Caribbean, with similar levels of development, where rates fluctuate around 20% of GDP, barely above the domestic savings rates of Africa. The savings rates of the Middle East and North Africa are heavily influenced by the savings from oil revenues in the Middle East, which simply do not have the capacity to be spent in the short term.

In the case of Latin American countries, their domestic savings rates have generally fluctuated between 15% and 25% of GDP, with Chile maintaining rates above the rest since the second half of the 1980s and Peru also diverging from the rest since 2005; these are the only countries with savings rates close to those of South Asia.

Domestic savings help stabilize the economy to the extent that they are channeled through financial and capital markets, as this makes them available for the government, businesses, and households to access local currency financing when there are declines in terms of trade or abrupt interruptions in capital flows from abroad.

[14] See, among others, Sachs and Warner (2011), Lederman and Maloney (2007), and He et al. (2024).

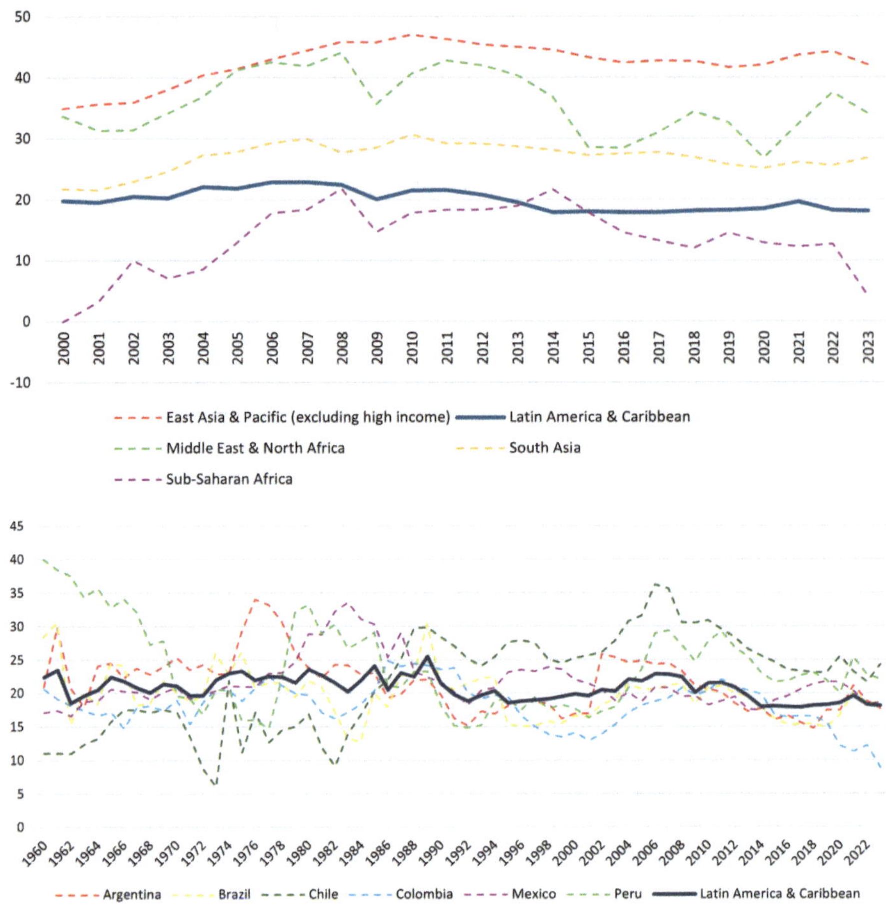

Fig. 10.5 (**a**) Evolution of domestic savings rates in regions composed of emerging countries (percentages of GDP). (Source: World Bank, World Development Indicators). (**b**) Domestic savings rates of selected Latin American countries (percentage of GDP). (Source: World Bank, World Development Indicators)

Figure 10.6 show the evolution of two important components of financial markets available to domestic firms: bank credit and the market capitalization traded on stock exchanges. To this should be added the size of the long-term debt market (bonds).

Consistent with what occurs with savings rates, the bank credit markets in East Asia are much more developed than in the other country groups, which have tended to stabilize at figures around 40% of GDP, except for Africa.

Figure 10.6b also shows that Latin America exhibits a stock market capitalization close to half that of the emerging regions of Asia and the lowest compared to

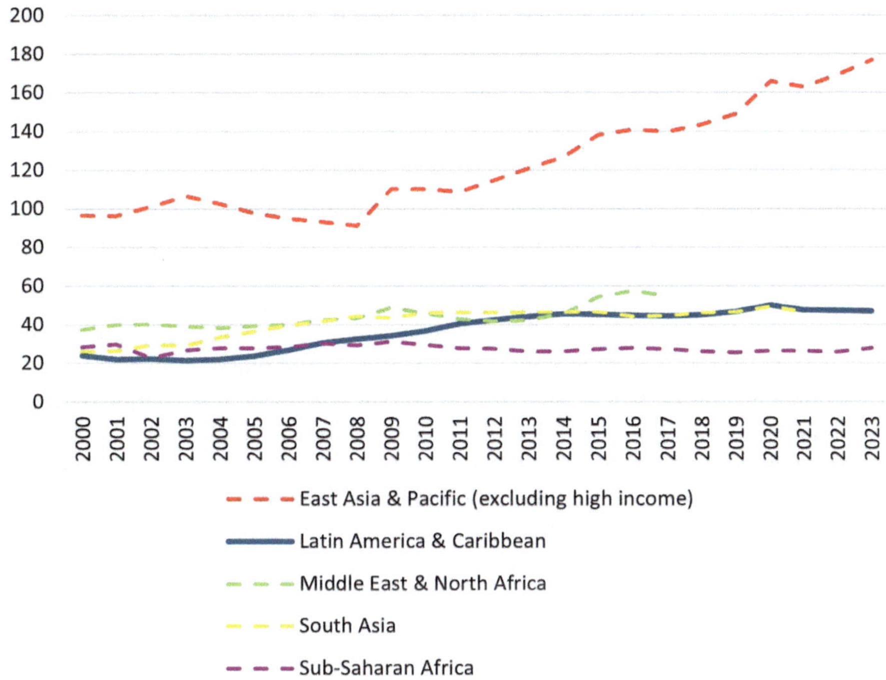

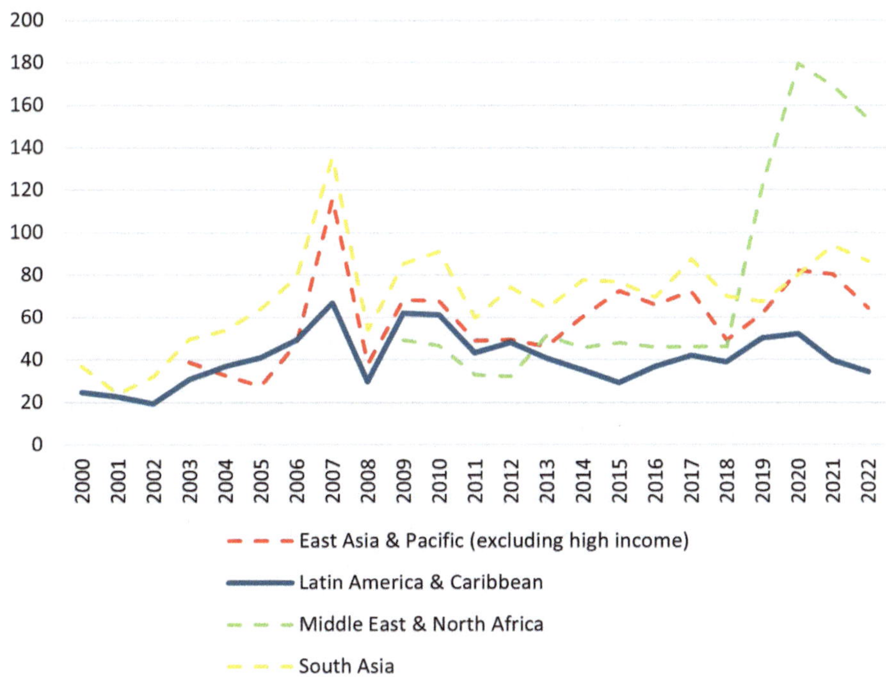

Fig. 10.6 (**a**) Evolution of bank credit to the private sector by region in emerging countries (Percentages of GDP). (Source: World Bank, World Development Indicators). (**b**) Evolution of stock market capitalization in regions of emerging countries (Percentages of GDP). (Source: World Bank, World Development Indicators)

the other regions, except for Africa, for which there are no aggregate data. The anomaly in Middle Eastern countries is due to the IPO of ARAMCO, the Saudi oil company, which took place in 2019 and, due to its size, greatly affects the regional average.

These figures are consistent with the fact that historically Latin America has faced great difficulties in neutralizing the volatility of commodity prices. This is partly because the region depends more on natural resource exports than other emerging regions, possibly with the exception of Africa, but also due to its low savings rates, which have limited the development of deep financial markets, leaving these countries at the mercy of the availability of external financing.

To complete this overview, Fig. 10.7 present disaggregated information by country on the evolution of bank credit to the private sector and stock market capitalization in the largest countries of Latin America.

In terms of banking penetration, Chile and, more recently, Brazil stand out, while Argentina and Colombia consistently maintain a low level.

On the other hand, it is also evident that, regarding capital market development, Chile clearly differs from the rest, which, in addition to stock market capitalization, is also reflected in the existence of a deep long-term bond market, as well as access to external financing by relatively large Chilean companies that have expanded abroad.

The above data tend to support the idea that Chile's better performance in mitigating commodity price swings is closely related to sound public financial management, combined with independent monetary policy focused on inflation control and minimal intervention in foreign exchange markets since the early 2000s. This, in turn, has been supported by the development of a diversified[15] and relatively deep financial market, which enabled government, companies, and households to access local currency financing to cope with the volatility of their terms of trade.

This framework has become the paradigm for addressing the macroeconomic challenges of small economies that are open to international trade and finance and dependent on natural resources. In this framework, the exchange rate is the first line of defense against external shocks, just as monetary policy plays that same role in response to domestic shocks that threaten the medium-term inflation target. Fiscal policy, in turn, can act in concert with the others in the face of major shocks, as was the case during the Global Financial Crisis of 2008–09 or the COVID-19 crisis. In more limited cyclical fluctuations, the rules help avoid amplifying the shocks. The combination of these elements, in turn, creates a virtuous circle, as it builds confidence and credibility in institutions and the commitments of authorities, resulting in lower costs for accessing both domestic and external financing, which not only benefits the state but is also passed on to the private sector.

[15] This development is largely due to pension system reform, which shifted from pay-as-you-go systems to one based on savings and capitalization in the early 1980s, and which, prior to the Covid crisis, had accumulated long-term savings equivalent to about 80% of GDP.

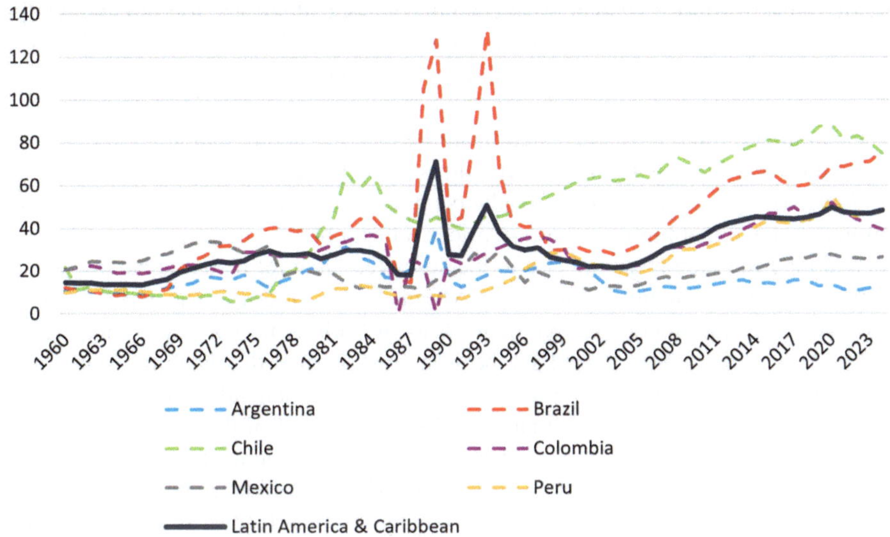

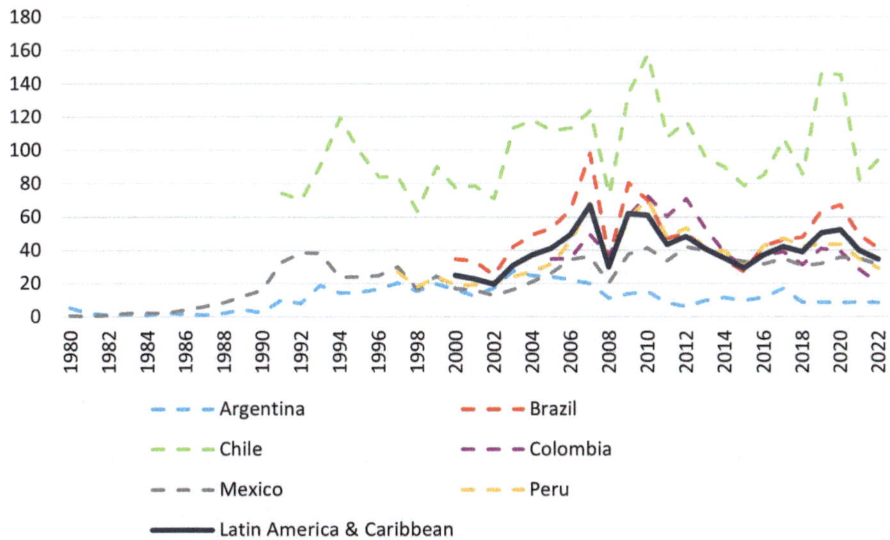

Fig. 10.7 (**a**) Evolution of bank credit to the private sector in Latin American countries (percentages of GDP). (**b**) Evolution of stock market capitalization in Latin American countries (percentages of GDP). (Source: World Bank, World Development Indicators)

The main conclusion of this analysis is that, through an appropriate institutional framework and macroeconomic policies, it is possible to mitigate the impacts of commodity price volatility. However, this is not easy and requires prioritizing long- and medium-term considerations in public policy decisions. On the other hand, even the strongest institutions can be weakened in complex situations. It is not enough for these principles to be enshrined in the Constitution and laws: there must also be a conviction among authorities and the general population that these precepts are necessary.

In summary, there have been significant lessons learned in managing the macroeconomic volatility arising from fluctuations in the prices of natural resource-based products. The great vulnerability to adverse price shocks and the ever-present temptation to believe that high commodity prices reflect changes in previous trends have been among the factors giving rise to the so-called "natural resource curse," which to this day afflicts many countries in Africa, Latin America, and parts of Asia. However, after the major external debt crises in Latin America in the early 1980s and in Asia in the 1990s, there have been lessons learned that have translated into substantial improvements in the fiscal, monetary, and financial institutions in most of these countries. Unlike the situation in the second half of the twentieth century, when the causes of inflation were debated, the development of financial markets was viewed with disdain, and fiscal deficits were considered structural and irremediable, today professional economists and many political leaders in these countries have learned from the mistakes and excesses of the past. This represents an addition to the human and social capital that offers hope for better macroeconomic management of revenues from natural resources in the future.

References

Alesina A, Perotti R (1995) The political economy of budget deficits. IMF Staff Pap 42(1):1–31

Alesina A, Hommes R, Hausmann R, Stein E (1999) Budget institutions and fiscal performance in Latin America. J Dev Econ 59(2):253–273

Arellano JP (2015) ¿Qué hacemos ahora que terminó el super ciclo del cobre? Programa CIEPLAN—Universidad de Talca

Caballero RJ, Panageas S (2008) Hedging sudden stops and precautionary contractions. J Dev Econ 85(1–2):28–57

Frankel JA, Végh CA, Vuletin G (2013) On graduation from fiscal procyclicality. J Dev Econ 100(1):32–47

He K, Gan Y, Wu H (2024) Natural resources, decentralized system, financial inclusion and sustainable development: Evidence from top emerging economies with resources abundance. Resour Policy 89:104674. https://doi.org/10.1016/j.resourpol.2023.104674

International Monetary Fund (2012) Macroeconomic policy frameworks for resource-rich developing countries (Policy paper). https://www.imf.org/external/np/pp/eng/2012/082412.pdf

International Monetary Fund (2023) Chile: Technical Assistance Report—Fiscal considerations in managing stabilization funds. Country Report. https://www.imf.org/en/Publications/CR/Issues/2023/07/07/Chile-Technical-Assistance-Report-Fiscal-Considerations-in-Managing-Stabilization-Funds-535833

Lederman D, Maloney WF (2007) Neither curse nor destiny: Introduction to natural resources and development. In: Lederman D, Maloney WF (eds) Natural resources: Neither curse nor destiny. World Bank, pp 1–14

Naudon A, Vial J (2016) The evolution of inflation in Chile since 2000. BIS Papers, 89g

Sachs J, Warner AM (2001) The curse of natural resources. Eur Econ Rev 45(4–6):827–838
Scartascini, C., Spiller, P., Stein, E., & Tommasi, M. (Eds.). (2010). The political game in Latin America: How are public policies decided? Inter-American Development Bank
Stein, E., Tommasi, M., Echeverría, K., Lora, E., & Payne, M. (Eds.). (2005). The politics of policies: economic and social progress in Latin America (2006 Report). Inter-American Development Bank
Vial J (2019) Reserves management and FX intervention: Chile's experience. BIS Papers, 104f
Vial J, Isea M (2025) Errores en el cálculo de la brecha del PIB y sus impactos en la aplicación de la regla fiscal en Chile. Clapes UC

Part IV

Challenges for Economic Development Arising from Planetary Boundaries

In Part I, the relationship between the explosive growth of global economic activity since the Industrial Revolution and the state of the environment was analyzed. The concept of "planetary boundaries" was introduced to describe the thresholds within which the systems regulating the biosphere can continue to absorb the impacts of increased human activity while maintaining acceptable risk margins. Evidence shows that in several dimensions these boundaries have already been exceeded, due both to population growth and increased consumption, especially of goods.

Part II examined the determinants of population growth and found that it is highly likely that the world population will reach its maximum size in the second half of this century, after which it will begin to decline. Thus, one of the main drivers of environmental degradation is already losing momentum. This is good news from the perspective of the sustainability of human living standards, but it will pose major challenges in adapting to this new reality, which will not be an easy process.

Part III focused on the pressure of human consumption on the provisioning services provided by natural capital, concluding that in cases where markets function well, the endogenous response of technological progress has made it possible to avoid scarcity. However, where markets fail, either due to the presence of externalities or because the goods are common property, human impatience and greed have led to the overexploitation of renewable resources and the destruction of habitats worldwide. The positive note here is that the problem is now recognized, and solutions are known, but there is still a long and difficult road ahead to convince the public and their leaders of the need to act effectively to prevent environmental damage, as well as to build national institutions capable of addressing the problem.

This section returns the focus to global problems because, as noted in Part I, it is at this level that serious issues are already emerging, threatening the planet's habitability for species such as humans, but which transcend the scope of action of individual nations—the largest units with the capacity to make and enforce decisions. Although multiple global organizations have been established since the end of World War II, with the most important being those linked to the United Nations system, they still lack the regulatory and executive capacities needed to enforce the agreements reached.

Among the various dimensions of the global environmental crisis, this section will focus on just two which share common elements and interact with each other but also have very different characteristics. Chapter 11 will analyze the problem of biodiversity loss, mainly resulting from the destruction or degradation of natural habitats. This is a global problem, as it is a phenomenon that has already occurred or is occurring worldwide, for the same reasons discussed in Part III. However, it is distinctive in that, since it concerns individual habitats and ecosystems, this crisis has a very specific spatial expression and must be addressed by national and local authorities.

Chapter 12 addresses the issue of climate change, its causes and consequences, the actions needed to prevent it, and what must be done to protect against the adverse impacts that are already known to be inevitable. This is a genuinely global problem because it affects the atmosphere, and no country can hope to isolate itself. On the positive side, this dimension of the global environmental crisis is one of the most manageable, because there is sufficient knowledge of its causes and predictive capacity to estimate its impacts globally and by region. Given this body of knowledge, it is possible to define actions to address it and even monitor progress toward achieving objectives.

Human Activity and Degradation of the Biosphere

11

Abstract

One critical area of concern is the degradation of Nature and its impact on the integrity of the Biosphere and, particularly, on the massive loss of biodiversity occurring during the last century. This chapter reviews the evidence on the state of Nature and Biodiversity, as well as the main reasons behind their decline. Externalities and the absence of properties rights play an important role in this crisis, especially for the destruction of ecosystems due to changes in land use, mostly to support the expansion of food production. This pressure is especially important in poor regions, where people depend on Nature for their sustenance, and human fertility is still very high. This also presents an opportunity as carbon markets develop and eventually might be used to finance land conservation, at least for natural forest rich in carbon. In any case, there is a major role for governments to preserve and protect critical ecosystems, most of whom are in remote and poor areas in the world. Given the importance of local action, international cooperation will be needed to stop current trends in those regions.

Keywords

Biodiversity · Externalities · Ecosystem services · Integrity of ecosystems · Land use · Environmental regulations · Preservation · Protected areas

At the beginning of the twenty-first century, there were virtually no ecosystems on Earth that were not experiencing alterations because of human activity and its "environmental footprint." Over the past century, humanity pushed to the limit the biblical mandate to grow, multiply, and dominate the Earth, even though it may still not fully grasp the consequences and responsibilities this entails. Advances in science and technology have provided a false sense of security that obscures the fragility of current lifestyles.

J. Vial, *Sustainable Development and Natural Resources*, Integrated Science 39,
https://doi.org/10.1007/978-3-032-12263-6_11

11.1 The Degradation of Nature and the Impact on Biodiversity

In recent decades, advances in scientific knowledge and the recognition of market failures in addressing common goods,[1] especially on a global scale, are sounding alarm bells, prompting various governments to propose actions to address these problems and to create the necessary institutions. This is no easy task, as these issues transcend national boundaries and require governments with different interests and perspectives to reach agreements.

Nature, biodiversity, and ecosystems are interrelated concepts. The following definition (Johnson et al. 2021) provides a useful starting point for defining the objects by which the state of nature can be monitored: "Nature refers to the ensemble of living organisms and the functions of the biosphere. The symbiosis between living organisms and the abiotic (physical and inorganic chemical) environment gives rise to ecosystems that regulate the flows of water, carbon, energy, and nitrogen, among others. Biodiversity is the variability of genes, species, and ecosystems." However, our understanding of the interrelationships among living organisms, as well as between them and their abiotic environment, remains limited.

The concept of biodiversity encompasses several dimensions that must be considered. Starting from the most basic level, it refers to the **diversity among individuals within a population.** Although to the human eye all individuals in a flock of birds may appear identical, each one is different, just as all humans riding a bus may look the same to a bird flying nearby. This diversity translates into genetic differences that result in variations in susceptibility to infections, differences in appearance, intelligence, singing ability, and so on. The greater the genetic diversity, the higher the probability that the population can adapt to and survive potentially catastrophic events.

A second dimension is the **diversity of populations** within a single species. Here, genetic diversity also plays a role, but it is complemented by the ability to adapt to and inhabit different locations or ecosystems. Thus, if a drought wipes out an amphibian population in one area, the species is more likely to withstand local catastrophes if it exists in many populations elsewhere. As the ecosystems they inhabit differ, these populations may, in turn, evolve toward greater genetic diversity among populations and eventually give rise to new subspecies or even distinct species.

Third, the **diversity of species** is also important, as it reflects the richness of an ecosystem and, at the same time, strengthens it through specialization in various functions, from water regulation to pollination and the control of species whose unchecked growth could compromise the stability of the ecosystem.

[1] See Sachs (2008) and especially Sachs (2015).

There is also **diversity at the ecosystem level** within a given area. The greater this diversity, the richer and more resilient the collective system is. This also results in greater complexity and interdependence, with all components mutually reinforcing one another.

Finally, it is important to consider **functional diversity within ecosystems.** This is highly relevant because it enhances resilience: if, for example, there is only a single pollinator species, a pest specific to that species could degrade this function, endangering the surrounding vegetation and, consequently, the integrity of the entire ecosystem. The same applies to predators: if there are few species fulfilling the role of population control and they are highly specialized, the ecosystem becomes very vulnerable to the emergence of invasive species.

This is why it is crucial to consider biodiversity in its entirety, which is often overlooked when attention is focused on a single flagship species, forgetting that it performs functions that complement or compete with those of other species. Protecting one species may harm others, or it may be forgotten that the "flagship" species depends on others that may also be at risk. At the same time, highly targeted protections for one specie that do not consider the state of their habitat may fail in their purpose, incurring significant costs in terms of lost opportunities by restricting human activities in that environment.

The protection of nature has intrinsic value, as a minimal respect for the lives of other organisms with which humanity share the habitat. However, from an economic perspective, which is necessarily anthropocentric, it is important to identify the contributions that nature makes to human well-being. These are summarized in Table 11.1, taken from the IPBES (2019) report.

Although the table does not specify the relative importance of each of these services, nor how critical they are for sustaining current living standards worldwide, these services are indispensable. Therefore, it is essential to preserve the capacity of ecosystems to continue providing them.

While so-called provisioning services are the easiest to identify and relate to human consumption, it is also important to highlight ecosystem regulatory services: these include the absorption and degradation of all types of waste, which was relatively easy, but extremely necessary, when most human-generated waste was organic and therefore biodegradable. With the massive emergence of synthetic materials in the twentieth century, waste absorption has become one major environmental problems, now reaching planetary dimensions. First came the ozone layer depletion caused by CFCs, then climate change due to greenhouse gas (GHG) emissions and already by the end of the twenty-first century, the pollution of oceans and land near urban areas with plastic waste and all kinds of liquid and solid garbage, whose degradation may take decades or centuries.

The destruction or degradation of natural habitats weakens the capacity of ecosystems to provide services that regulate air, water, and soil quality, resulting in a loss of biodiversity as well.

A healthy ecosystem requires many species that perform complementary functions to maintain its integrity. These range from bacteria that decompose dead biological material, insects and birds that pollinate, plants that filter and purify water,

Table 11.1 Services provided by nature to humanity

	Contribution of nature to people	Global trend over the last 50 years	Selected indicator
Regulation of environmental processes	1. Creation and maintenance of habitats	↓	Extent of suitable habitat Integrity of biological diversity
	2. Pollination and dispersal of seeds and other propagules	↓	Diversity of pollinators Extent of natural habitat in agricultural areas
	3. Regulation of air quality	↘	Retention and prevention of atmospheric pollutant emissions by ecosystems
	4. Regulation of climate	↘	Prevention of emissions and absorption of greenhouse gases by ecosystems
	5. Regulation of ocean acidification	→	Capacity of marine and terrestrial systems to sequester carbon
	6. Regulation of the quantity, location and temporal distribution of freshwater	↘	Ecosystem effects on the distribution of water among the atmosphere, surface and soil
	7. Regulation of the quality of freshwater and coastal water	↘	Extent of ecosystems that filter or add constituent elements to water
	8. Formation, protection and decomposition of soils and sediments	↘	Soil organic carbon
	9. Regulation of risks and extreme events	↘	Capacity of ecosystems to absorb and buffer hazards
	10. Regulation of harmful organisms and biological processes	↓	Extent of natural habitat in agricultural areas
		↘	Diversity of competent hosts of vector-borne diseases

(continued)

Table 11.1 (continued)

	Contribution of nature to people	Global trend over the last 50 years	Selected indicator
Material and assistance	11. Energy	↗	Extent of agricultural lands and lands for possible bioenergy production
		↘	Extent of forest lands
	12. Food and feed	↗	Extent of agricultural lands and lands for possible food and feed production
		↓	Abundance of marine fish populations
	13. Materials and assistance	↗	Extent of agricultural lands and lands for possible material production
		↘	Extent of forest lands
	14. Medicinal, biochemical and genetic resources	↘	Fraction of local species with known medicinal properties
		↓	Phylogenetic diversity
Non material	15. Learning and inspiration	↓	Number of people in proximity to nature
			Diversity of views from which to learn
	16. Physical and psychological experience	↘	Area of natural and traditional terrestrial and marine landscapes
	17. Support identities	↘	Stability of land use and occupation
	18. Maintenance of options	↓	Probability of species survival

Source: IPBES (2019)

plants and trees that fix and transform atmospheric gases and provide food for various species, as well as predators that limit the excessive growth of certain species. This interaction among highly diverse species means that these systems are constantly changing. They are living systems that grow, shrink, and change in composition, even if they are not directly interacting with humans.

The greatest problem with wildlife loss is that Nature operates as a highly complex system, with multiple interactions among species and habitats, many of which remain unknown. For example, for years there has been discussion about the problems caused by the mass die-off of bees, due to their impact on the pollination of agricultural crops, especially fruit plantations. However, it is often forgotten that bees are just one among many species of insects and birds that perform this function, with the only difference being that bees are the only pollinator species that has been "domesticated." The mass disappearance of other insects and birds is also affecting the reproductive capacity of many plant species, both cultivated and wild. Under conditions of climate stress, a decline in the reproductive capacity of plant species can hinder nature's ability to regenerate.

11.2 State of Nature and Biodiversity

For diagnostic purposes, it is possible to identify the area covered by various ecosystems and to assess their condition, at least partially, by observing increases or decreases in species of all kinds, as well as changes in the number of individuals within a species. It is also possible to track their status by monitoring the flows of ecosystem services they provide, for example, through the characteristics of water flows passing through them, soil loss, landscape changes, etc. However, all these measurements are limited by our ability to observe natural phenomena, much of which occurs at the level of microscopic organisms, often underground or inside other living organisms, such as trees and their roots, for example.

Biodiversity loss is reflected in two dimensions: the risk of species extinction and the decline in populations of a given species. Certainly, extinction usually follows a severe drop in the number of individuals, but a population decline does not always lead to extinction and, consequently, the loss of all the "genetic capital" of that species.

On both fronts, there has been a severe deterioration of biodiversity. Regarding species extinction, it is worth clarifying that throughout the history of life on the planet, there have been species extinctions due to natural causes, whether from completely exogenous factors, such as the asteroid impact that led to the extinction of the dinosaurs, or from imbalances in the systems that regulate the atmosphere, whether due to volcanic eruptions or the excessive proliferation of species that altered the presence of carbon in the atmosphere.[2] While there is academic debate about whether we are currently experiencing the sixth "Great Extinction," there is abundant evidence of significant increases in the rate of species extinction, as shown in Fig. 11.1.

This graph shows only part of the story, as it excludes all plant and insect species, for example. Although the slopes of these curves show a sharp acceleration coinciding with the demographic and economic explosion following the Industrial Revolution, the cumulative percentage of lost species may seem small, but all indications suggest that the worst is yet to come, especially if climate change intensifies, as all projections suggest. This danger is better reflected in the assessment of extinction risk by species families, also described in IPBES (2019).

[2] The largest of these occurred approximately 250 million years ago, when it is estimated that more than 95% of the species inhabiting the planet disappeared. For further background, see Kolbert (2014), Brannen (2017).

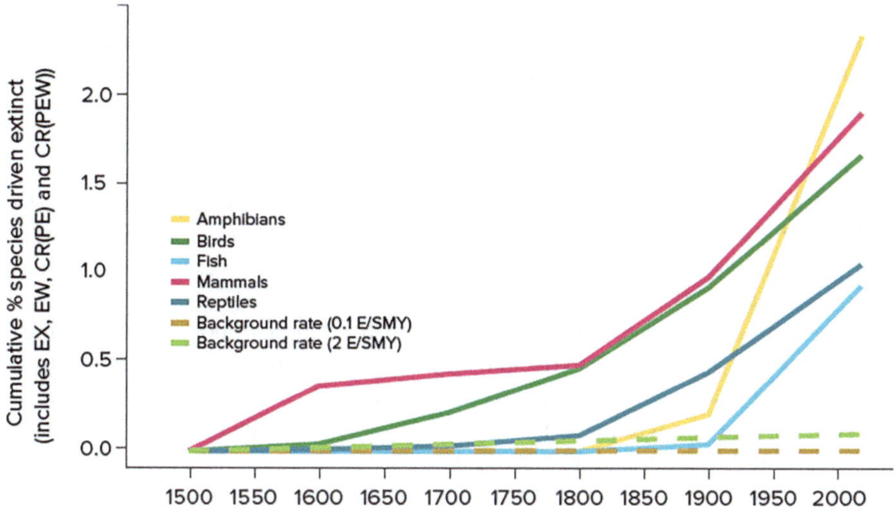

Fig. 11.1 Species extinctions by family from the year 1500 to the present (proportion of total species in each family). (Source: WWF (2024))

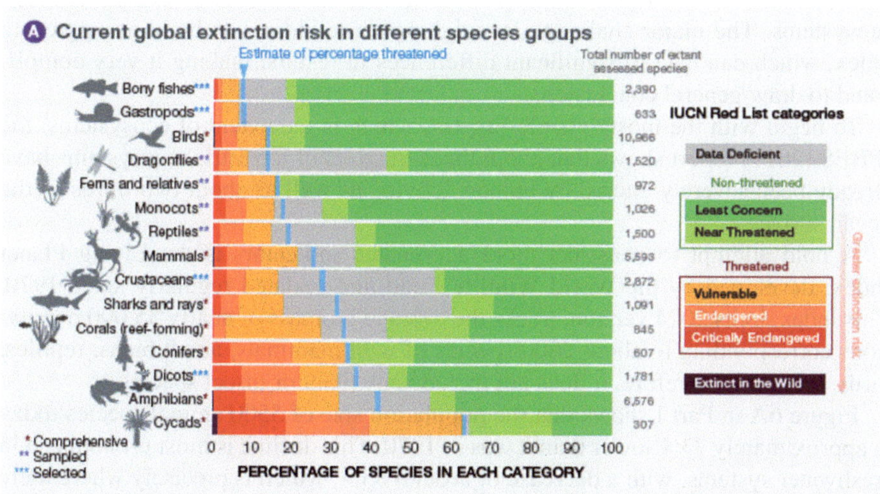

Fig. 11.2 Current extinction risk of species groups (percentage of total species in each group, by risk category). (Source: IPBES (2019))

When examining the percentage of threatened species for each group, as shown in Fig. 11.2, it becomes clear that if current trends continue, the number of extinct species will continue to rise in the coming decades, which is especially serious for certain tree species (cycads and conifers), amphibians, herbs (dicotyledons), and corals.

One of the most dramatic cases is the loss of insect populations and species. It is estimated that there are currently between 10 and 30 million **insect species**, but only about 10% of them are known and have been described, so many species are still being reported. Insects are among the oldest and most diverse forms of life on the planet, having survived and evolved by adapting to major environmental changes that led to mass extinctions, which have contributed to their enormous diversity. They play a vital role for other species, forming the base of the food pyramid for birds, amphibians, reptiles, and aquatic species, and, as if that were not enough, they contribute to the pollination of approximately 80% of flowering plant species (Gebremariam 2024).

More than 60 years ago, Rachel Carson, in her famous book "Silent Spring," raised the alarm about the rapid decline of insects. Although the information is fragmented, since studies on insect species refer to individual ecosystems and observations of a few species within them over several years, subsequent evidence confirms her fears, and it is estimated that nearly 40% of insect species in Europe and North America could face extinction in the coming decades (Gebremariam 2024).

Measuring the state of nature is very difficult. Unlike the economy, there are no units of measurement that allow us to add apples and oranges by converting them to monetary values. However, it is possible to construct indices and maps based on observations of the number of species and the abundance of individuals in various ecosystems. The major challenge is weighing these indices to build an aggregate index, which can lead to significant differences in results, making it very complicated to draw general conclusions.

To begin with the most "visible" aspect, which is the extent of ecosystems, the IPBES (2019) report shows that about three-quarters of terrestrial ecosystems have already been severely altered by human activity, as well as about two-thirds of the ocean surface.

A bold attempt to construct more aggregated indicators is the Living Planet Index, developed by the World Wildlife Fund and updated regularly since 1970. This index, in its 2024 version, tracks the average biomass of nearly 35,000 populations corresponding to about 5600 species of wild mammals, amphibians, reptiles, birds, and fish. Therefore, it does not include any plant or insect species.

Figure 6A in Part I shows that the population size of 5500 animal species today is approximately 73% lower than it was in 1970. This decline is most pronounced in freshwater systems, with a decrease of around 85%, which is precisely where there is the greatest presence and interaction with humanity. This is followed by terrestrial systems (69%) and marine systems (56%). In all of them, relative stabilization has been observed in the last decade, even though the first symptoms of climate change are already beginning to appear.

In terms of regions, the recent deterioration of nature is concentrated in emerging countries, led by Latin America and the Caribbean (−95%), Africa (−76%), and Asia and the Pacific (−60%). Developed countries that led the industrial revolution and population growth in previous centuries now show smaller declines in biodiversity, partly because the major losses occurred before 1970 and because in some of them, the combination of higher incomes, which no longer depend as much on

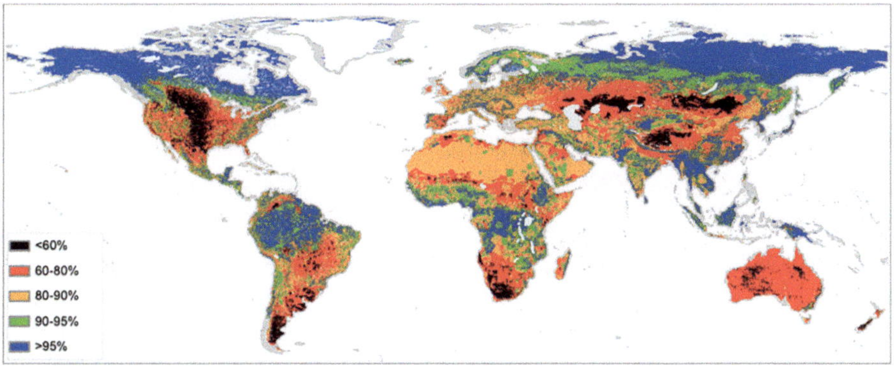

Fig. 11.3 Map of terrestrial biodiversity by potential level. (Source: Newbold et al. (2016))

nature, and slower population growth has allowed for the reversal of some previous trends. In any case, in aggregate, they are still losing biodiversity: North America (−39%), Europe and Central Asia (−35%), as shown in Fig. 11.3.

11.3 Causes of Ecosystem Degradation and Biodiversity Loss

With the explosion of human activity documented in Part I, there has been enormous pressure on natural areas and the state of nature. The main factors that have directly caused biodiversity loss are land use change, the exploitation and direct use of natural resources, climate change—which is already beginning to make its effects felt—pollution of land, water, and air, and the introduction of invasive species, which, lacking natural predators, occupy ecological niches previously filled by endemic species. Recently, the emergence of new diseases affecting species living in natural environments has also been added, which in turn is related to the effects of climate change and the movement of species (both animal and plant) facilitated by humans.[3] Figure 11.4 shows the relative importance of each of these factors in the global degradation of terrestrial ecosystems at the global level in the main groups of developing countries.

These graphs show that habitat destruction, linked directly to land use change and the destruction of forests and wetlands is the main cause of biodiversity loss in terrestrial ecosystems, followed by the overexploitation of resources, particularly in the case of fish and mammals. Climate change also plays a significant role in the decline of amphibians, along with diseases, which may also be linked to rising temperatures.

In the case of insects, which are not reported in the LPR, the main causes of extinction are identified as habitat destruction (grasslands and wetlands are especially important), the use of pesticides in agricultural activities, and climate change,

[3] See IPBES (2019) and UN (2019).

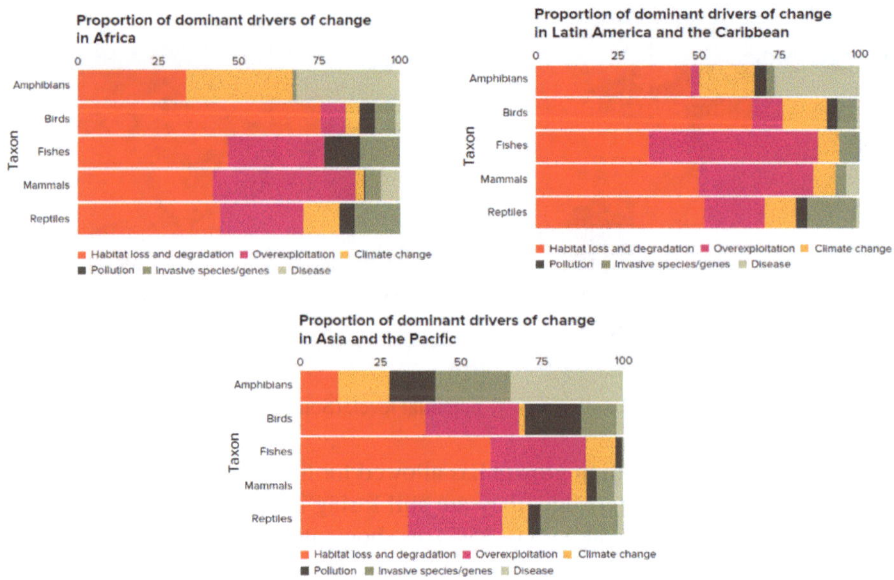

Fig. 11.4 Key factors in the degradation of terrestrial ecosystems in developing countries. (Source: WWF (2024))

particularly in areas where higher temperatures coincide with lower precipitation. Unlike other forms of life, in many cases the destruction of insects by humans has been deliberate, both to protect agricultural production and because some insects are vectors of pathogens (malaria, yellow fever, among others). Of course, when swamps are drained or pesticides are applied, individuals from many "innocent" species also die (Gebremariam 2024).

As referred to in the Living Planet Report, these are the "drivers" of biodiversity loss but underlying them are the market failures already analyzed in Part III. Land use change is largely because many of the services provided by forests and grasslands are unrecognized, non-appropriable, and therefore lack market prices that would signal their scarcity and the true costs of their exploitation. Added to this is human impatience, which leads to seeking short-term benefits at the expense of the loss of future services. A similar situation occurs with marine ecosystems, where overexploitation is because marine resources are common property, which increases the urgency to extract them now because of uncertainty about their future availability.

A factor influencing this outcome is the fact that poorer regions tend to depend more on natural capital compared to human and manufactured capital (Lange et al. 2018). At the same time, the poorest people tend to live in fragile natural ecosystems (Dasgupta 2021). Given the well-established positive correlation between fertility and poverty, demographic pressure on biodiversity tends to be greater in the world's poorest regions, many of which are currently especially rich in biodiversity. As Wilson (2002) noted, regions that have experienced major extinctions in the past tend to show slower subsequent losses. Thus, the remaining major reservoirs of biodiversity are found in remote areas under little human pressure. As populations grow in low-income areas, the pressure to use the main available form of capital—natural capital—increases.

Human action is particularly evident in the case of species of all kinds that provide provisioning services (such as fish for food), where markets fail to signal scarcity, leading to their overexploitation. This not only directly affects their future availability for humans, as previously discussed, but also depletes ecosystems and puts at risk other species that depend on them.

There are also losses of natural areas and biodiversity due to competition for the use of ecosystem services. The clearest example is that of freshwater bodies and wetlands, which, due to their abundance of water, offer especially favorable conditions for the development of all kinds of species, including humans. Competition for water and space has meant that nature has disappeared from these ecosystems, except for those species that have adapted to live in artificial environments. Furthermore, the expansion of crops and the massive use of fertilizers and pesticides to produce the food demanded by the population have caused losses of species of all kinds, particularly "nuisance" or dangerous insects, which has undermined the resource base for many other species that depend on them for food or for other services such as pollination.

Moreover, the analysis of biodiversity loss by region conceals the fact that these losses are not only due to local factors but also to demands from the rest of the world, particularly the consumption of goods by developed countries, which account for most of the global consumption. Thus, a significant part of deforestation in countries such as Brazil or Indonesia originates from the demand for food (meat, palm oil, coffee) from the rest of the world.

11.4 Actions to Contain Ecosystem Degradation and Protect Biodiversity

When it comes to protecting biodiversity and caring for nature, the essential task is to preserve habitats that are part of ecosystems capable of functioning autonomously, with minimal human intervention. This requires large, continuous areas, that is, areas not fragmented by human presence. Additionally, it is also possible to attempt to preserve specific species and limited but highly valuable natural areas, but this is very difficult to achieve because species are generally highly dependent on their habitat. Attempting to preserve species or individual specimens in altered habitats or those with a strong human presence can be very costly and may have low chances of long-term success. Nevertheless, there are also relatively small ecosystems that are essential for migratory species, as well as for amphibians, reptiles, and fish. For example, this would be the case for wetlands for certain bird species. The preservation of these areas depends on the actions of local authorities.

Since markets do not signal the value of preserving nature, state intervention is inevitable, both at the national and local levels. This is further justified by the fact that in many countries the state owns a large proportion of the land and water within its territory. In addition, the state has the capacity to regulate land use in particularly important areas and under strong human pressure.

At the international level, there is a wide variety of formally protected areas, mainly state-managed, but also under private or community management. These are often complemented by other areas under effective conservation management that differ from the former (OECM, for its acronym in English), including forest reserves and others that combine tourism and recreational use with a conservation objective. Figure 11.5 presents a schematic representation of the possible combinations of institutional arrangements used to classify different forms of protection.

Although the Kunming-Montreal agreement on a framework to protect global biodiversity, reached in 2022, sets the goal of protecting 30% of land and water by 2030, this target is far from being achieved. Currently, just over half of the global target has been met for terrestrial ecosystems (17.6%), and only 8.4% for inland waters. Moreover, the biodiversity coverage of these protected areas is low, representing about 25% of the most environmentally valuable ecosystems (UNEP-WCMC and IUCN 2024).

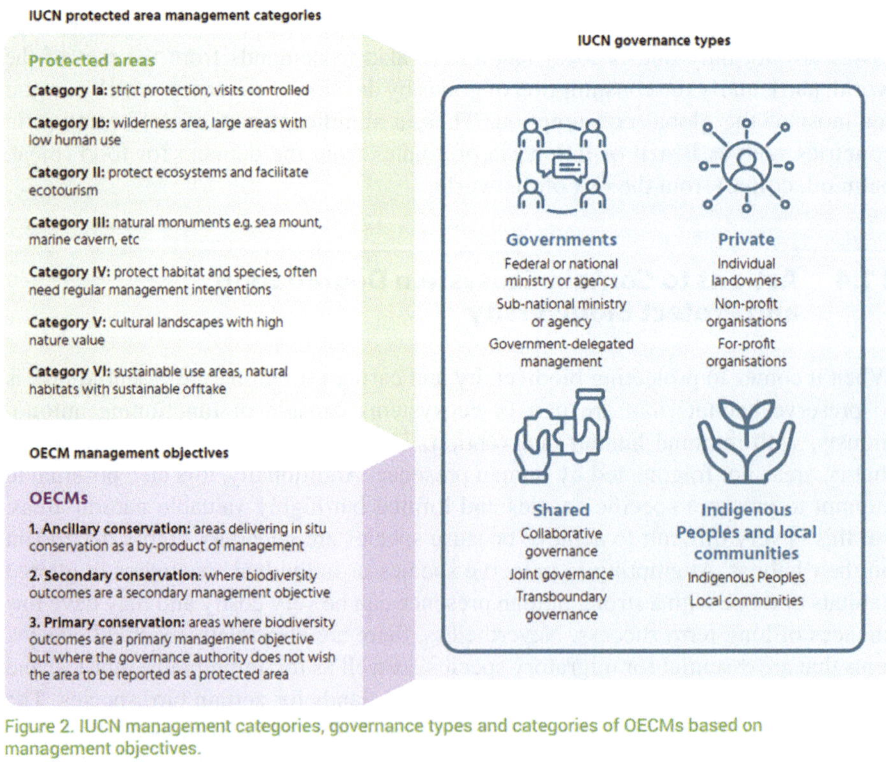

Fig. 11.5 Classification of different forms of nature protection. (Source: UNEP-WCMC and IUCN (2024))

While the current concept of protected areas emphasizes the goal of preserving ecologically valuable ecosystems, this was not their original purpose. In fact, an important motivation for the creation of national parks, for example, was the preservation of natural beauty for the enjoyment of future generations, which implies that at least parts of these parks are open to the public for recreational purposes. Paradoxically, their success has resulted in strong human pressure to visit and experience these places, altering the landscape, fragmenting some habitats, and disturbing wildlife. Congestion, waste disposal, noise, and, in some cases, fires, are serious problems for the management of the most attractive national parks.

Another threat to protected areas comes from residents of adjacent zones, often livestock farmers who see their activities threatened by predators, or poor families who have traditionally depended on extracting resources from now-protected forests or grasslands and exert pressure on these areas through legal means or simply without authorization. At the borders of these protected areas, there is also coexistence of native and exotic species of all kinds, raising the risk of invasive species entering and disrupting biological balances within protected areas.

In poor countries with large natural systems that persist, such as the forests of the Amazon, central-southern Africa, or countries in the East and South Pacific, many of these areas, with or without formal protection, are under strong pressure from land use change, whether to expand the agricultural frontier or to exploit mineral resources. In many of these cases, fire is still used for this purpose. In areas supposedly under protection, state capacities to safeguard these territories are very limited, both due to the lack of available resources and vulnerability to corruption when highly valuable resources are present in these areas.

Albagli and Vial (2022) attempt to distinguish between the effects of population growth and per capita income growth on land use changes, as a "proxy" for biodiversity loss. While both factors are found to influence such change, this relationship has shifted over time, with the effects of increases in per capita income losing importance, while population growth continues to exert pressure, especially in poorer countries that are further behind in the demographic transition.

In addition to the State, other actors play an important role in conservation. On the one hand, there are individuals and foundations funded by philanthropic contributions, specializing in the acquisition and/or management of relatively undisturbed lands for protection purposes. They are playing a very important role, and there are large areas in Patagonia and other regions that have been purchased by foreign investors for conservation. However, this type of intervention carries some risks that should be mitigated.

On the one hand, there is the risk of continuity, especially when it comes to individual investors or family businesses that, by definition, have a limited lifespan. This risk also exists in the case of foundations that may face insolvency and end up dissolving. In such cases, there may be strong public pressure, or pressure from the investors themselves, for the State to take over these lands, which forces the already scarce resources available for conservation to be spread over areas that may be of lower priority because they are already represented within state-protected zones.

Another source of private funding for nature protection originates from environmental damage compensation schemes, in countries that have incorporated conservation targets into their territories and allow for compensation.[4] While there are some relatively successful examples, these are more the exception than the rule. For such schemes to function well, adequate monitoring and control systems for protected areas are required, which must be certified by independent entities, and it may also be necessary to establish insurance and guarantee mechanisms to ensure the continuity of conservation efforts in the future. In any case, this would be a first step toward incorporating market-based instruments to increase resources for nature protection and improve the management of protected areas.

In many countries, the natural ecosystems that remain have a significant presence of indigenous or relatively poor peasant communities. These actors can be agents of destruction of protected areas, as mentioned earlier, or active participants in conservation, insofar as it helps them improve their incomes. This is an important factor to consider in the creation and management of protected areas, among other reasons, because these are their ancestral habitats and sources of livelihood for generations, and it is a matter of justice to do so, as well as to achieve cost-effective protection.

Given that states usually have better information than private actors about priority areas that require resources for their protection, and the fact that, in principle, they have continuity across generations, it is almost inevitable that they play a very important role in the protection of wilderness areas. It would be highly desirable to foster a climate of dialogue and cooperation among these actors.

References

Albagli E, Vial J (2022) Crecimiento económico y biodiversidad: Algo tiene que ceder (Documento de Política Económica No. 70), Banco Central de Chile

Brannen P (2017) The ends of the world: volcanic apocalypses, lethal oceans, and our quest to understand Earth's past mass extinctions. HarperCollins Publishers

Dasgupta P (2021) The economics of biodiversity: the Dasgupta review. HM Treasury

Gebremariam G (2024) A systematic review of insect decline and discovery: Trends, drivers, and conservation strategies over the past two decades. Psyche

IPBES (2019) Global assessment report on biodiversity and ecosystem services. IPBES Secretariat. https://doi.org/10.5281/zenodo.3831673

Johnson JA, Baldos UL, Cervigni R, Chonabayashi S, Corong E, Gavryliuk O, Hertel TW, Nootenboom C, Gerber JF, Ruta G, Polasky S (2021) The economic case for nature: a global earth-economy model to assess development policy pathways. World Bank. https://hdl.handle.net/10986/35882

Kolbert E (2014) The sixth extinction: an unnatural history. Henry Holt & Company

Lange GM, Wodon Q, Carey K (2018) The changing wealth of nations 2018: Building a sustainable future. World Bank Publications

Newbold T, Hudson LN, Arneth A, Contu S, De Palma A, Ferrier S et al (2016) Has land use pushed terrestrial biodiversity beyond the planetary boundary? A global assessment. Science 353(6296):288–291. https://doi.org/10.1126/science.aaf2201

[4] More on this topic in the next chapter, as there are strong synergies between ecosystem protection for biodiversity conservation and carbon capture, for which formal offset markets do exist.

Sachs JD (2008) Common wealth: economics for a crowded planet. The Penguin Press
Sachs JD (2015) The age of sustainable development. Columbia University Press
UNEP – WCMC, & IUCN (2024) Protected Planet Report 2024. UNEP – WCMC & IUCN
United Nations (2019) World population prospects: The 2019 revision
Wilson EO (2002) The bottleneck. Sci Am 286(2):82–91
WWF (2024) Living Planet Report 2024: a system in danger. WWF

Climate Change: The Most Manageable Component of the Global Environmental Crisis

12

Abstract

Climate change is a well-known dimension of the global environmental crisis: the causes have been identified, the impacts at the global and regional levels have been projected under different scenarios, targets have been set to limit emissions, and significant resources are devoted to research and development for technical solutions. However, we are still far from reaching those targets, and substantial changes must be made both in consumption patterns and in the efficiency in the use of Nature's services to contain catastrophic increases in global temperatures. The fast decline in fertility rates across the world and the pace of technical innovation are sources of hope, but they are not enough to remain within the emission limits. For many emerging countries the most efficient path is to concentrate mitigation efforts on the preservation of their forests and ecosystems, while devoting significant resources to protect their people from the consequences of climate change. Fast adoption of cleaner technologies for mobility and energy use might help them to sustain growth while leapfrogging over the dirty stages of industrialization and growth.

Keywords

Climate change · Greenhouse gases · Emissions · Externalities · Regulations · Carbon tax · Compensation mechanisms · Carbon captures · Economic impacts · Environmental impacts · Mitigation · Adaptation · Fiscal pressures · International aid

As Thomas Friedman (2016) noted a decade ago, in recent times, comments about powerful and unexpected climate events that have disrupted people's daily lives have become a recurring topic, gaining importance in conversations around the world. Of course, this is not scientific proof that we are experiencing a climate crisis, but there are changes perceptible to the population. On the other hand, the

J. Vial, *Sustainable Development and Natural Resources*, Integrated Science 39,
https://doi.org/10.1007/978-3-032-12263-6_12

scientific community has been warning for several decades about the excesses that are altering the climate and the potential risks this poses for humanity. As discussed in Part I, this is one of the systems that govern the proper functioning of the biosphere, in which the threshold for safe operation has already been exceeded.

12.1 Origin of Climate Change

The current risk of climate change[1] originates from the excessive concentration of gases in the atmosphere that block the dissipation of heat generated by the sun into and beyond the atmosphere (the so-called greenhouse effect), which threatens to dramatically alter living conditions on the planet within less than a century. The evidence for global temperature increases is now indisputable. While there are still questions about the influence of human activity on this phenomenon, the scientific community reached a consensus on this issue almost two decades ago (IPCC 2007), urging action. As noted in Part I, atmospheric timescales are not the same as human timescales, and with current concentrations of greenhouse gases (GHGs), significant temperature increases in the coming decades are already locked in.

When these effects are mentioned, ordinary people often wonder how severe their impacts will be. The figures discussed refer to increases in average temperatures of 2 or 3 degrees Celsius over decades and rises of several tens of centimeters in average sea level over these timeframes. Is this a relevant cause for concern?

On the other hand, there are also people—including U.S. President Trump—who deny the existence of this phenomenon, or at least that it is caused by human activity.

The answer is that unfortunately, the issue is real and could have serious consequences for the living conditions of current generations and, especially, for our children and grandchildren, essentially for three reasons[2]:

The use of fossil fuels for energy generation since the Industrial Revolution, as well as other sources of emissions of gases that block heat dissipation, has steadily increased since the early twentieth century, resulting in a rapid rise in the concentration of these gases in the atmosphere. Figure 12.1a, b shows the annual flow of GHG emissions, expressed in CO_2 equivalent, from the Industrial Revolution to the present (a), and the stock of GHGs that has accumulated to date, compared with previous periods for which data are available.

The greenhouse effect depends on the volume of gases that trap heat in the atmosphere, shown in Fig. 12.1b. Annual emissions correspond to the flow of gases entering the atmosphere each year. The concentration of gases also depends on the flow of GHGs that dissipate beyond the atmosphere and the gases that return to the surface and are "captured," either through photosynthesis, absorption by the soil,

[1] There have been many episodes of climate change in the 600-million-year period that encompasses the history of life on Earth. Several of these had catastrophic consequences that led to mass extinctions of species, and subsequently to the emergence of other species better adapted to the new conditions. See Kolbert (2014).

[2] For a more extensive and well-documented discussion, see Dasgupta (2021), especially Chap. 4.

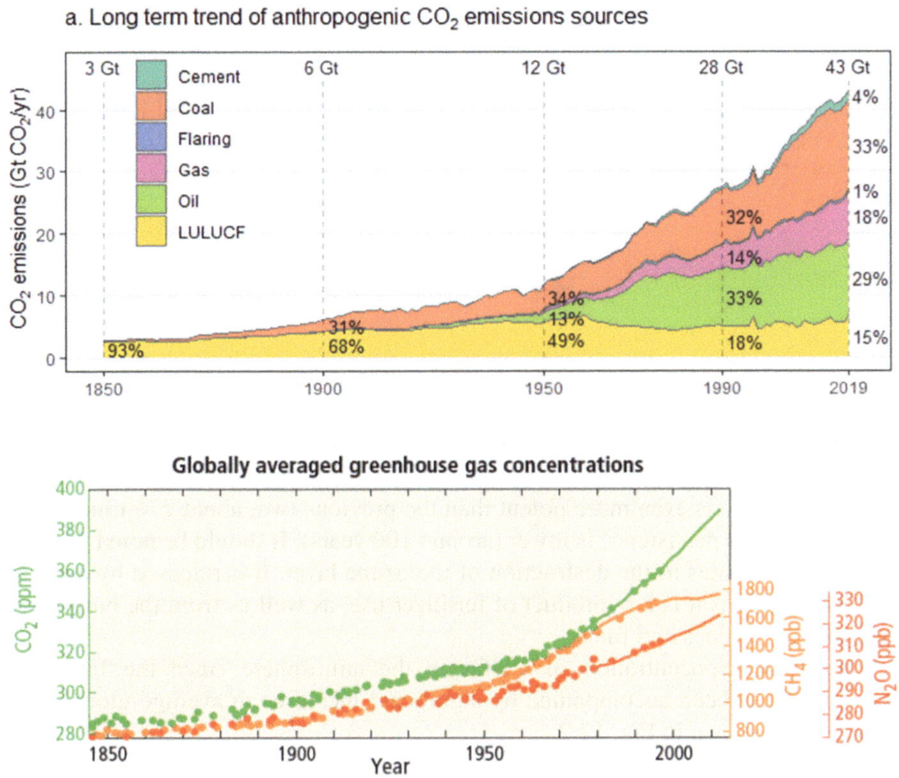

Fig. 12.1 (**a**) Annual global GHG emission flows. (Source: IPCC (2021)). (**b**) GHG concentration in the atmosphere. (Source: IPCC (2014))

and most importantly, due to the area they cover, by being absorbed by water, especially in the oceans.

Figure 12.1b also shows that several gases contribute to the greenhouse effect, with CO_2 (carbon dioxide), CH_4 (methane) and N_2O (nitrous oxide) are the most important. Each of these gases has a different origin and distinct impacts on the concentration of GHGs in the atmosphere.

Carbon dioxide is produced mainly from the combustion of fossil fuels and wood, and to a lesser extent, from the release of carbon stored in soil and vegetation during deforestation and land-use change. This gas is characterized by its high persistence in the atmosphere (between 300 and 1000 years).

Methane is found primarily in oil and natural gas deposits and is released into the atmosphere during the extraction of these fuels. Enteric fermentation processes, mainly in cattle, also contribute. Although methane represents a much smaller proportion than carbon dioxide among atmospheric GHGs, its ability to trap heat is between 25 and 80 times greater than that of CO_2. However, its atmospheric persistence is much lower, dissipating in about 10 years, so curbing these emissions could yield a relatively rapid reduction in the greenhouse effect.

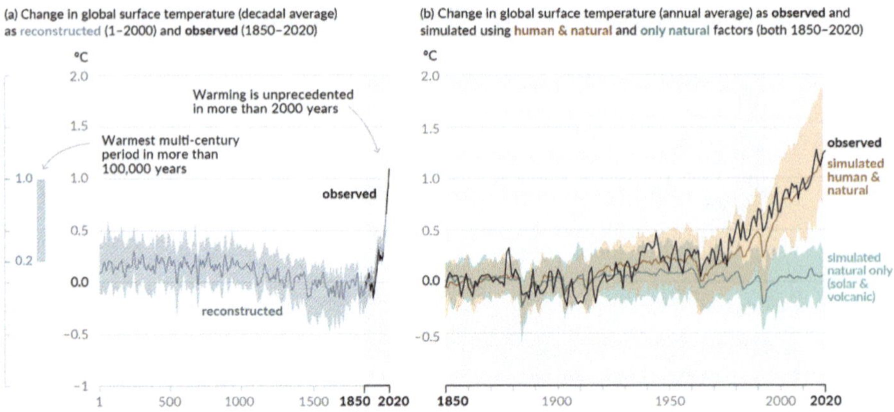

Fig. 12.2 Changes in global average temperature. (Source: IPCC (2021))

Nitrous oxide is even more potent than the previous two, about 270 times more than CO_2, but its persistence is lower (around 100 years). It should be noted that this gas also contributes to the destruction of the ozone layer. It is released by agricultural activities, as it is a byproduct of fertilizer use, as well as from the burning of agricultural residues and fuels.

The higher concentration of GHGs in the atmosphere since the Industrial Revolution has been accompanied by significant increases in average global temperatures, as shown in Fig. 12.2.

12.2 Impacts of Climate Change

Although the impacts of GHG emissions on global temperatures are already being observed, the most important effects are those in the future, because each year more GHGs are accumulating in the atmosphere, where they will remain for many years. As a result, future impacts on the climate will be greater unless these emissions are controlled and their concentration in the atmosphere decreases—very gradually.

To assess the effects of current and future GHG emissions on temperature changes, it is necessary to make long-term climate projections that take these factors into account. The IPCC brings together leading groups of scientists specializing in these topics in a forum under the auspices of the United Nations, who provide trend projections using their best models under different scenarios. These are then combined to obtain the most probable trajectories and confidence intervals.

Although climate science is far from exact, it has advanced significantly in recent decades, taking advantage of new information and data on GHG concentrations and ancient climate changes obtained with new technologies (for example, by analyzing information contained in deep Arctic ice cores). To make these projections, various scenarios (SSPC) are constructed based on assumptions about population trends, economic activity, and policy actions aimed at addressing the problem. These projections leave little room for optimism. The scenarios range from the most extreme, SSP5–8.5, in which current trends continue, to others that vary in the intensity of climate change mitigation actions.

Figure 12.3 is taken directly from the 2022 IPCC Report and shows the different levels of annual GHG emissions, separated into CO_2 and other gases, associated with the various projection scenarios.

In turn, Fig. 12.4 shows the contribution of emissions from different greenhouse gases to the increase in global average temperatures under the five main scenarios.

The SSP1–19 scenario represents the most drastic action (associated with the emission targets of the Paris Agreement) and allows the stabilization of GHG concentrations between 430 and 480 ppm, limiting the temperature increase relative to the pre-Industrial Revolution period to 1.5 °C[3]. At the other extreme, a scenario such as SSP5–8.5, which assumes that historical trends ("business as usual") continue, is associated with GHG concentration levels exceeding 1000 ppm and temperature increases between 4 and 5 °C.

An important effect of these changes is that they are reflected not only in annual averages, but also in the increased frequency and intensity of extreme weather events, something that has already begun to be observed, as seen in more frequent and intense heat waves during the summer, as well as in the frequency and intensity of torrential rains and hurricanes.

What do these scenarios imply for ecosystem functioning and for living conditions for humanity?

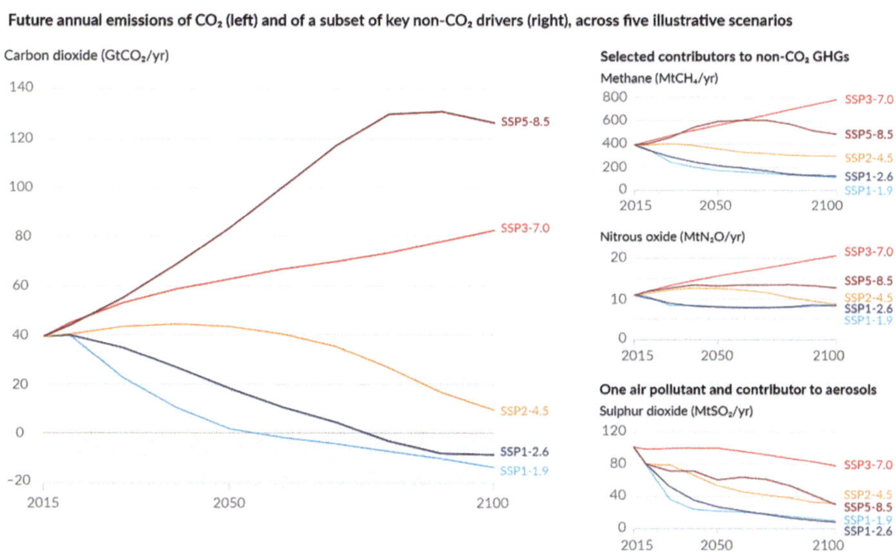

Fig. 12.3 Emission scenarios for different GHGs. (Source: IPCC (2021))

[3] The implications of this scenario in terms of annual GHG emission flows are examined below.

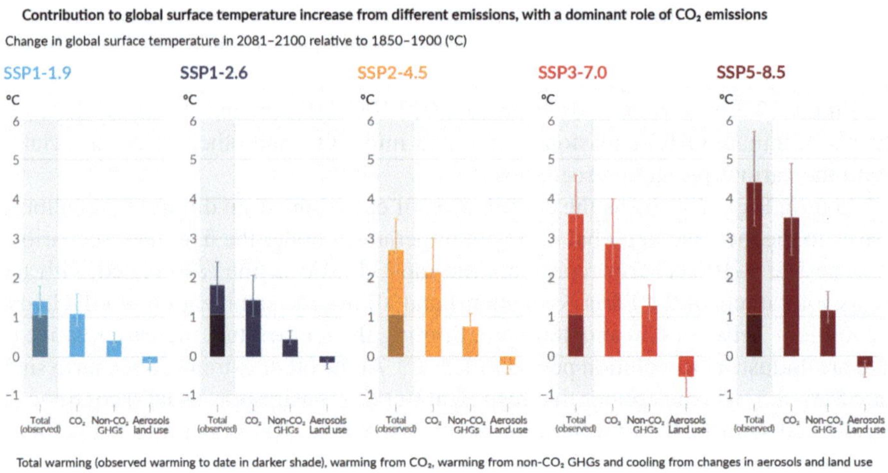

Fig. 12.4 Contributions of different GHGs to global temperature increase under different scenarios. (Source: IPCC (2021))

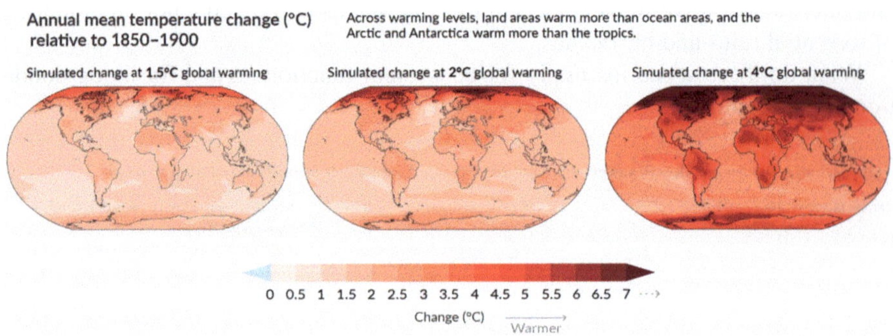

Fig. 12.5 Regional temperature increases at the end of the twenty-first century under different scenarios. (Source: IPCC (2021))

First, they will be reflected in higher average global temperatures, although with variations across different geographic regions and increasingly frequent episodes of extreme heat waves, something that has already begun to be felt over the past decade. These phenomena are neither linear nor continuous, and will manifest with varying intensity in different locations, as shown in Fig. 12.5, taken from the latest IPCC report.

The graphs show deviations in projected temperatures for the end of the twenty-first century relative to the late nineteenth-century average in different world regions, based on multiple simulations. These are not intended to be exact but provide an approximate idea of which regions would be most affected. While the poles will remain by far the coldest regions on the planet, they are expected to experience the greatest temperature increases, in contrast to tropical regions, which, although still the hottest, will see smaller average increases. A second conclusion is that the oceans tend to warm less than large land masses, for a very simple reason: the sea and its currents have a greater capacity to distribute and dissipate heat. This, in turn, means that the Southern Hemisphere tends to warm less than the Northern Hemisphere, which has much greater land coverage and concentrates more population and human activity. This same report shows that observed temperature changes have tended to match climate model projections so far.

Currently, the disruptive effects of heat waves that have affected the entire world in recent summers are already being observed, manifesting in poor agricultural yields, large-scale wildfires, and energy and health crises in urban areas.

A second dimension of the effects of climate change relates to changes in precipitation, since temperature changes affect water cycles. These are shown in Fig. 12.6, which presents the percentage changes between average annual precipitation by region and by scenario, comparing the end of the twenty-first century to observations from the late nineteenth century.

Unlike the case of temperatures, for precipitation there are areas where it increases (sometimes significantly), such as the polar regions and tropical oceanic latitudes, while, in subtropical regions, there are very significant decreases in precipitation, especially in the Caribbean and the Mediterranean in the Northern Hemisphere, the South Pacific coasts, the southern tip of Africa, and Western Australia.

The combination of rising temperatures and reduced precipitation significantly decreases atmospheric and soil moisture, which has major impacts on ecosystems as well as on living conditions and the capacity to produce food in affected areas.

Finally, a third dimension worth noting is the rise in sea level. This is shown in Fig. 12.7, which also represents the change—this time in absolute terms—in average sea level between the end of the twenty-first century and the end of the nineteenth century.

This graph differs significantly from the previous ones in that, under the extreme scenario, it incorporates the effect of a non linearity resulting from what is currently

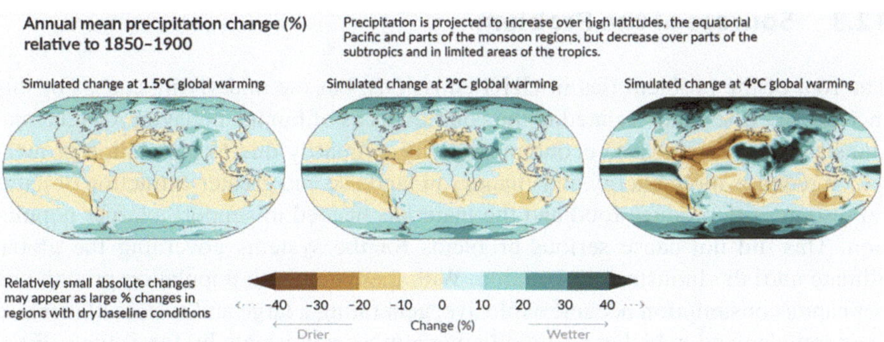

Fig. 12.6 Projected percentage changes in precipitation by region at the end of the twenty-first century under different scenarios. (Source: IPCC (2021))

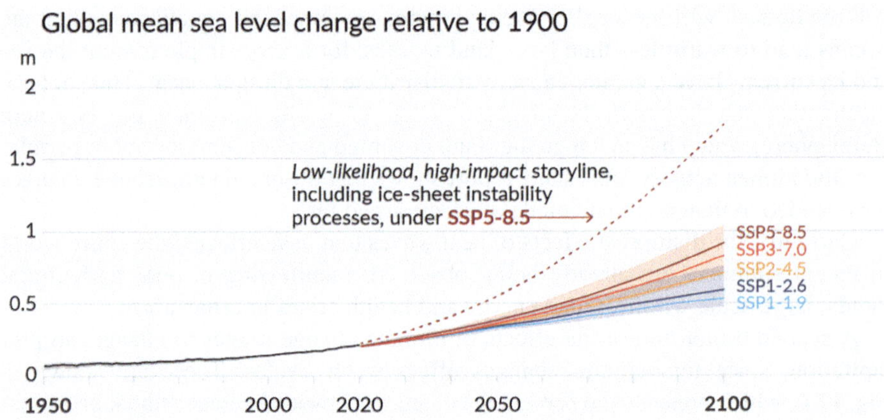

Fig. 12.7 Absolute changes in average sea level projected for the end of the twenty-first century and observed at the end of the nineteenth century. (Source: IPCC (2021))

perceived as a high-impact but very low-probability event, such as a drastic reduction in ice sheets over the sea level. Such an event would lead to an average sea level rise of more than 1.5 meters. This may seem like a modest increase that should not cause problems in coastal areas, but that is not the case: the sea is in constant motion due to currents and winds, the position of the Moon, and the Earth's rotation, among many other factors, so the average conceals a much more complex reality. Under adverse conditions—such as during a hurricane, for example—this could mean several meters of water height in the most affected locations. In cities like Miami, London, Shanghai, or Lima, for instance, this makes an enormous difference, and a single event of this kind every few years would be enough to radically change the landscape and habitability of those areas. The same would severely affect port infrastructure worldwide, with highly disruptive effects on the transportation of all types of goods.

12.3 Sources of the Problem

The root of the problem lies in GHG emissions, but the underlying causes of this increase are essentially related to the consequences of human behavior. GHG emissions arise from increased economic activity, primarily due to population growth, which requires more energy and changes in land use and greater extraction of natural resources to produce food and the materials needed to support a larger population. This did not cause serious problems for the systems governing the global climate until the Industrial Revolution. With the latter, both population growth and per capita consumption became explosive, generating a large and sustained increase in energy demand, which was initially met almost exclusively by fossil fuels. First, timber and coal were used to generate steam and electricity; then oil, due to its higher energy density and ease of storage, quickly replaced coal in all activities related to mobility and power; and more recently, natural gas, mainly for electricity generation. At the household level, heating during the cold winters of northern Europe and America also shifted from firewood to coal, which was later replaced by petroleum derivatives and, more recently, by natural gas and electricity. Fossil fuels have provided cheap and highly efficient energy for nearly two centuries, during which their carbon emissions have accumulated by more than 120 ppm, that is, over 40%. Thus, carbon that had been captured and stored in the ground for millions of years and transformed into coal, oil, and natural gas, has been released back into the atmosphere through combustion in just a couple of centuries.

The primary cause of the problem is ignorance: humanity simply did not know that burning fuel at the scale it was being done could harm the functioning of the atmosphere. There was concern about their potential depletion, as discussed in Part III, but concern for the state of the atmosphere did not go beyond local worries about particulate pollution resulting from coal burning in urban areas with high industrial concentration. At the beginning of the twentieth century, the planet was still not perceived as a closed and limited ecosystem. Although the era of exploration and discovery had already ended, there was still a perception of a vast world to populate and exploit for the benefit of humanity.

However, by the mid-twentieth century, the issue was already known, and warnings began to be raised about the state of global ecosystems and the risks that CO_2 emissions posed to the climate and the habitability of vulnerable areas.[4] By the late 1980s, there was sufficient scientific knowledge identifying GHGs as the source of the problem, although there were still doubts about how much it was due to human activity and how much to natural climate variability. At that time, the issue entered the international political agenda and became increasingly established as scientific uncertainty about the origin of the problem has dissipated. The creation of the IPCC (Intergovernmental Panel on Climate Change) by the United Nations was a decisive step in bringing together the global scientific community to work on diagnosing, projecting, and estimating the impacts of climate change.[5]

Solving the problem has proven to be much more complex, because even knowing the causes of the climate crisis and the risks it poses to the survival of civilization, there have been no significant behavioral changes and GHG emissions continue to rise as if nothing had happened. The major issue is that, since the atmosphere is free and common good, the market does not reflect the cost of its degradation.

The main reason for inaction in the face of the risk of a global climate catastrophe is that this is a classic case of externality, further aggravated by the fact that the costs will be observed in the distant future (several generations on a human scale) and are subject to considerable uncertainty. In this sense, the problem is

[4] In 1958, Frank Capra filmed a documentary for Bell Labs, presenting scientific advances, among which he highlighted the risks of climate change. See https://ossfoundation.org/projects/environment/global-warming/frank-capra-warns-of-global-warming-1958/

[5] The successive IPCC reports constitute the best and most up-to-date source of scientific knowledge on the subject. See: IPCC — Intergovernmental Panel on Climate Change

conceptually no different from the causes of deforestation: impatience to use cheap energy quickly and the undervaluation of the environmental costs of using fossil fuels, as well as of land use change.[6]

To put it in very concrete terms, when a person decides to commute to work in a private car, they do so because they can afford the fuel and the car, but also because they value comfort, privacy, and—though increasingly less so due to congestion—speed of travel. The carbon emissions from their trip simply do not factor into their decision. However, it has been observed that when a congestion charge is imposed, as has been done in some cities, behavioral changes occur, and more people switch from private to public transportation.

Many industries used firewood or coal simply because it was cheaper than alternatives or because the investment required to replace furnaces and boilers was too high. However, in cities where pollution was penalized and the replacement of these fuels with less polluting alternatives (gas or electricity) was supported, such a transition did occur.

Households and businesses respond to incentives, but when the market fails to provide the correct incentives due to externalities, there will be no behavioral change without government intervention. However, in the case of climate change, there is an additional difficulty: the problem is global and transcends national borders. Making the costs of GHG emissions transparent by reflecting them in market prices is costly for individuals and businesses and, obviously, politically unpopular. If it is a small country, the temptation to do nothing is overwhelming: Why impose a cost on voters if the country's actions will have no effect on global emissions and their impact on the climate? This is the classic case of the so-called "free rider."

But there is also a variant of the "free rider" argument based on ethics: if emissions were caused by countries that are now rich and developed, why should poor countries or those aspiring to become rich sacrifice their consumption to prevent a crisis created by others?

In practice, countries' priorities differ, especially when what is at stake is the possibility of growth and achieving levels of material well-being and consumption like those of countries that have already developed by taking advantage of fossil fuels. The fight against climate change became politicized at the global level, and there was very little practical progress.

This has only changed in the last couple of decades, since China and other emerging countries became major emitters of CO_2 and, at the same time, international studies revealed the vulnerability of their large urban centers and industrial development hubs to climate change.

[6] Figure 12.1a shows that until the mid-twentieth century, land use change and deforestation were the sources of more than half of GHG emissions.

12.4 Actions and Policies to Mitigate Climate Change

Analyses by IPCC experts for more than 30 years have indicated that the flow of emissions must decrease drastically and reach zero before the middle of this century to avoid a catastrophe of great proportions. However, despite growing awareness of the problem and the agreements adopted by countries on various occasions, GHG emissions continue to grow at an accelerated pace. In fact, since the Earth Summit in Rio de Janeiro in 1992, the annual flow of emissions has doubled.

Figure 12.8 highlights three very important issues: first, as global emissions of kept rising, exceeding 43 billion tons in 2024 it is clear that the agreements have not been effective so far. Second, the longer it takes, the more abrupt (and costly) the necessary corrections will have to be. Finally, it is also recognized that the problem has worsened and, instead of converging toward a target of zero emissions by 2050, the timeline has shortened, and it is now estimated that this must occur by 2040. The target of 400 ppm of GHG concentration in the atmosphere was recently surpassed, which in practice means that the 1.5 °C target will be exceeded, and in its most updated version, the scenario that allows reaching that target will involve a period of temperature "overshooting," followed by a return "from above" to the target, as already acknowledged in the 2022 IPCC Report.

Climate change mitigation has two fundamental components: actions aimed at reducing GHG emissions and, on the other hand, those aimed at capturing GHGs, especially CO_2, from the atmosphere.

Regarding emission reduction, the main focus of policies implemented by countries has been to accelerate the energy transition away from fossil fuels. To this end, there are various mechanisms:

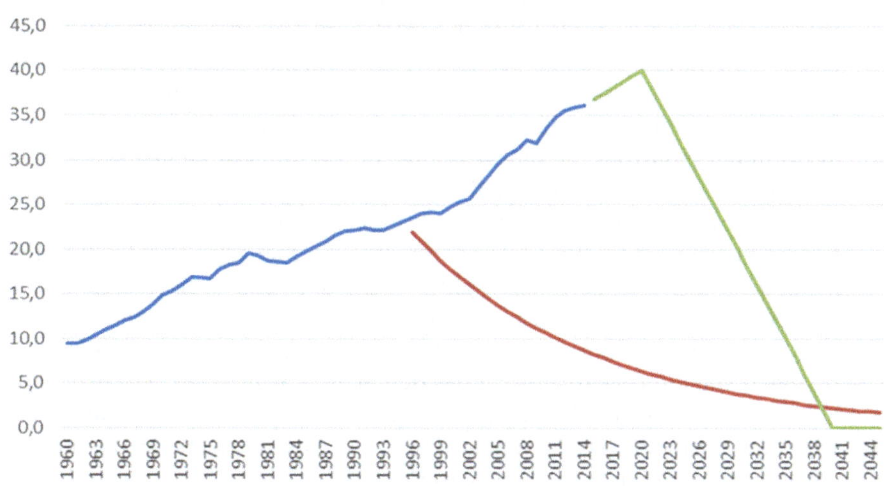

Fig. 12.8 Annual global GHG emissions, actual and agreed (million kt/year). (Source: World Bank, World Development Indicators)

(i) Administrative restrictions that set limits on emissions from certain industries, which, at the extreme, are leading to the establishment of targets for the closure of coal-fired power plants or the prohibition, at some future date, or the sale of gasoline-powered vehicles. An important case in this area is methane emission regulations. As previously discussed, methane is a gas with very potent effects but low persistence, so rapidly reducing these emissions creates room to move more slowly on other fronts. The main sources of methane emissions are leaks during hydrocarbon extraction, which can now be monitored globally via satellites. Unfortunately, progress on this field has been rather slow so far. The other source is emissions from livestock digestive processes, which are more difficult to control, but advances are being made in both feed technologies and in capture and reuse as an energy source.

(ii) A softer version of the above aims to facilitate the energy transition by converting emission limits into permits to emit within those limits, which can be traded in markets specifically developed for this purpose. This allows activities with technological alternatives to quickly reduce their emissions to finance those projects by selling permits to companies that face greater difficulties in doing so (not unlike transferable individual quotas in fisheries). This approach seeks to leverage market synergies to help address the underlying problem. In its early stages, the European Emissions Trading System (ETS) saw very little activity, partly because the emission limits, set at the national level, were very lax and the prices per ton of CO_2 emissions were very low (between 10 and 15 euros). More recently, emission limits have been set centrally by the EU, leading to more restrictive targets, which has resulted in a sharp increase in price, which after reaching levels close to 100 euros per ton, has stabilized between 60 and 70 euros.

(iii) Another option, often favored by economists (IMF, 2019), is the introduction of carbon taxes, which are relatively easy to monitor for large emitting industries and can also be extended to transportation. This category could include taxes on the various types of fuel used by different industries. In practice, however, these have encountered considerable resistance. On the one hand, in many countries fuel subsidies have been used to maintain the competitiveness of declining industries, as shown in Fig. 12.9. The most emblematic case is industrial fishing, whose profitability has declined globally due to the overexploitation of marine resources. Some developed countries with a strong fishing tradition have tried to offset higher costs through fuel subsidies. On the other hand, these types of taxes are highly visible and tend to have a significant impact on household purchasing power, where transportation costs are a relevant item. Shifting from subsidies to taxes is not easy for any political system. A similar situation occurs in countries rich in oil or natural gas. The nationalist logic of not passing on the (high) international fuel prices to the local market, when the population has the "natural right" to enjoy cheap energy provided by "their" natural resources, is very powerful.

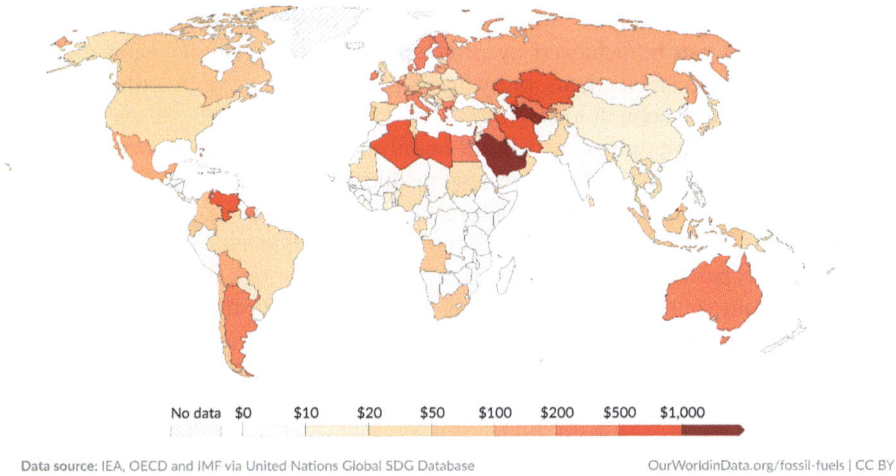

Fig. 12.9 Map of oil use subsidies per capita in 2021. (Source: Ritchie (2025))

Unfortunately, the revenue forgone by the state because of these pricing policies tends to lead to extravagant energy consumption (such as ski resorts in the middle of the desert, for example) or to a lack of funds for investment in the maintenance and development of the domestic oil industry, as has occurred in Venezuela and Mexico, for instance. A tax on the externalities of fossil fuels is a good option, but in many cases, the elimination of subsidies—which are estimated to reach the astronomical figure of 1.5% of global GDP—may be a better alternative in the short term.

(iv) There is also the possibility of subsidizing the use of alternative technologies that help reduce emissions. In this area, the range of options is very broad, but in general, policies tended to favor the energy transition, for example, by subsidizing the use of clean energy at the household level (tax exemptions or reductions for electric vehicles, or direct subsidies for the installation of solar panels in homes, for instance). Subsidies have also been used in this area in combination with industrial policy objectives, such as subsidizing the construction of solar panels or wind turbines, or the development of alternative sources of clean energy (green hydrogen). At one time, this argument was also used to subsidize agriculture to produce biofuels, but this option has fallen out of favor because the carbon footprint of that activity is very high.

(v) Nuclear energy deserves special mention. After a very promising start in the postwar period, fears of nuclear power plant accidents that emerged from the 1980s onward, as well as the problem of waste disposal—which remains radioactive for hundreds or thousands of years—led to a halt in new investments in the sector and, in some countries such as Germany, for example, to the implementation of a gradual phase-out of nuclear generation. However, concerns about energy security and the growing urgency to address climate change have led to a reversal in recent years, and several countries that had opted to reduce nuclear activity have changed course and are once again building nuclear plants.

The other approach to climate change mitigation involves increasing carbon capture capacity, which can be achieved by protecting and strengthening natural carbon capture and storage mechanisms, or by developing and investing in artificial carbon capture and storage technologies.

Natural carbon capture is carried out mainly by forests. Data shows that forest destruction and land-use change are responsible for approximately 20% of GHG emissions. In other words, halting deforestation and land-use change, particularly regarding forest destruction, could reduce current carbon emissions by almost a quarter. This is particularly important for developing countries, as for many of them, this is the main source of emissions, meaning they could offset higher emissions resulting from economic growth simply by limiting the destruction of their forests. Additionally, this has the added benefit that the forests of the Amazon, central Africa, and some countries in South Asia and the Pacific are the planet's most important biodiversity reserves. Protecting these ecosystems is a win-win proposition, and it only remains to assign them proper value. A well-functioning emissions offset market, complemented by financing for biodiversity preservation, can be a highly efficient tool for reconciling economic growth with emissions reduction in both economic and environmental terms.

Figure 12.10 shows the main sources of GHG emissions by region. For the developed world (Europe and North America), the most efficient path to carbon neutrality necessarily involves energy transition.

For the developing world, on the other hand, there is a dual opportunity to take advantage of technological advances that enable cleaner economic growth than that of the countries that industrialized in the twentieth century, while also advancing emissions offsetting by preventing the destruction of their natural ecosystems. This is possibly the best alternative, both environmentally and economically, available to many middle- and low-income countries to reconcile economic growth with emissions reduction, and it should be given high priority both at the national level in these countries and for development assistance, whether from multilateral organizations or donors.

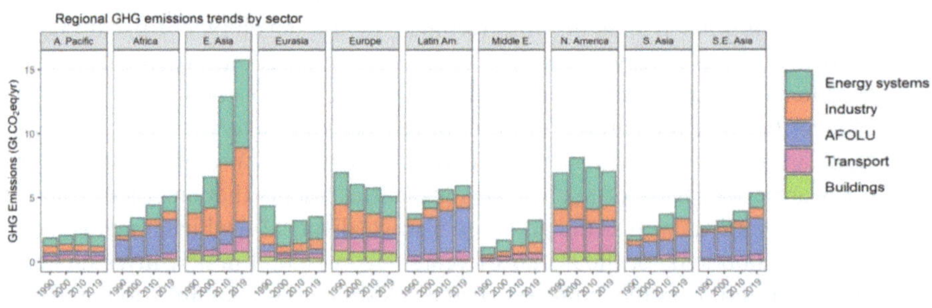

Fig. 12.10 Sources of GHG emissions by region. (Note: AFOLU Agriculture, Forestry and Other Land Use. Source: IPCC (2022b))

Unfortunately, this highly attractive option is under severe pressure, both from human activity and from climate change itself. In Chap. 7, in addition to highlighting the contribution of forests to carbon sequestration, it was shown how wildfires are increasing in frequency and intensity, driven by climate change but also by the pressure to "clear" forested land to expand areas for agriculture and livestock. The latter could be curbed through more effective state action, combined with policies aimed at increasing the productivity of small-scale agriculture in areas adjacent to forests.

Regarding climate change, the challenge is even greater, as higher temperatures and intense heat waves create favorable conditions for fires and their spread. Fire prevention efforts and the capacity for rapid response to contain fires in their initial stages are essential tools for success in this area, but so far, they have proven insufficient. Looking ahead, it is difficult to imagine that forest protection can reach its full potential without effective fire control mechanisms, as well as certification processes that ensure adequate forest protection.

Artificial carbon capture and storage is moving from an experimental phase toward commercial development. Currently, the first commercial carbon storage plants are still under construction. While industry sources present an optimistic outlook,[7] the most recent developments have been influenced by the sharp rise in carbon prices in the ETS market and by subsidies implemented by the Biden administration in the United States, which were eliminated by the Trump administration, a move that should at least delay the development of this option.

In summary, the means to achieve emission reductions that would mitigate climate change do exist. However, they are not being implemented due to a lack of public willingness to accept the loss of purchasing power required to internalize environmental costs, as well as governments with relatively short-term horizons, which tend to postpone decisions or even reverse unpopular actions. A positive trend worth highlighting is the fact that countries such as China and other major middle-income emitters are taking action to seize the opportunities offered by the energy transition and are even actively reforesting areas that lost their forests decades ago.

12.5 Actions and Policies for Adapting to Climate Change

The fact that the 400-ppm threshold of GHGs in the atmosphere has already been surpassed means that climate will inevitably change in the coming decades, as the effects are persistent and cumulative. Even if carbon neutrality were miraculously achieved tomorrow, it would take many decades, or even centuries, for atmospheric GHG concentrations to return to levels like those that prevailed before the Industrial Revolution. This means that humanity will inevitably have to face a significant

[7] See: https://www.globalccsinstitute.com/resources/global-status-report/

deterioration in the environmental conditions that favored the development of mammals and humans in particular and must therefore prepare for it.

The first and most easily identifiable aspect involves the risks and "defensive" investments that will be required to prevent the damage caused by extreme weather events. These range from severe droughts, with increased wildfires and water shortages affecting a wide range of economic activities, to landslides caused by torrential rain, sometimes even in the same locations. This is already being observed, and there is not a day when international news channels do not report such phenomena occurring somewhere in the world. Substantial investments will be required for this purpose, but these are "defensive" in nature: they will help maintain what already exists but not foster economic growth.

Unlike climate change mitigation, adaptation investments are specific to each region: a fishing village will need to reinforce its infrastructure to protect against storm surges; a drinking water supplier will have to invest to ensure resource availability, perhaps by building reservoirs, but will also need to invest in protection against landslides that could disrupt the normal operation of water treatment plants. A mining company will need to secure water availability, while ensuring that the relevant infrastructure is resilient to extreme weather events, and so on.

This will require substantial resources to finance costly investments that will not generate additional income to pay interest, thereby imposing a heavy burden on governments that are already facing strained public finances. This will be especially challenging for low-income countries, many of which are highly vulnerable to climate change, because being poorer, their populations depend more on the resources they can obtain from nature, and their survival is closely linked to environmental conditions. Here, multilateral organizations have an important role to play, and there is a need for concessional financing from wealthy countries. This is not only a matter of altruism, but also of preventing massive flows of migrants fleeing famines and other disasters associated with climate change, including resource wars.[8]

In this context, an essential tool is environmental risk mapping, which helps identify at-risk infrastructure and supports medium- and long-term urban and industrial planning. These maps should contribute to the planning and prioritization of public and private investments. Such tools should become increasingly important in medium- and long-term urban planning. They should also serve as useful instruments for financial systems to properly assess environmental risks, such as they are already being implemented by institutions that finance loans with mortgage guarantees in more advanced countries. In fact, these types of tools are already being adopted by insurance companies in the United States, in response to very high losses from hurricanes and wildfires.

In the coming decades and centuries, development will only be sustainable if it is resilient to the impacts of climate change and other environmental disasters. As time passes and the energy transition is further delayed and the destruction of natural areas persists, the investments required for adaptation will become larger and less effective (IPCC 2022c), which will exacerbate tensions both between and within countries.

[8] See Diamond (2005).

References

Dasgupta P (2021) The economics of biodiversity: the Dasgupta review. HM Treasury
Diamond J (2005) Collapse. How societies choose to fail or succeed. Viking
Friedman TL (2016) Thank you for being late: An optimist's guide to thriving in the age of accelerations. Farrar, Straus and Giroux
Intergovernmental Panel on Climate Change (2007) Climate change 2007: Synthesis report. Contribution of working groups I, II and III to the fourth assessment report of the intergovernmental panel on climate change (RK Pachauri, A Reisinger, eds)
Intergovernmental Panel on Climate Change (2014) Climate change 2007: Synthesis report. Contribution of working groups I, II and III to the fifth assessment report of the intergovernmental panel on climate change (RK Pachauri, A Reisinger, eds)
Intergovernmental Panel on Climate Change (2021) Summary for policymakers. In: Masson-Delmotte V et al (eds) Climate change 2021: The physical science basis. Cambridge University Press. https://doi.org/10.1017/9781009157896.001
Intergovernmental Panel on Climate Change (2022a) Agriculture, forestry and other land uses (AFOLU). In: Climate change 2022: Mitigation of climate change. Cambridge University Press, pp 747–860
Intergovernmental Panel on Climate Change (2022b) Summary for policymakers. In: Shukla PR et al (eds) Climate Change 2022: Mitigation of Climate Change. Cambridge University Press. https://doi.org/10.1017/9781009157926.001
Intergovernmental Panel on Climate Change (2022c) Summary for policymakers. In: Pörtner H-O et al (eds) Climate change 2022: impacts, adaptation and vulnerability. Cambridge University Press. https://doi.org/10.1017/9781009325844.001
Ritchie H (2025) How much in subsidies do fossil fuels receive? Our world in data. https://ourworldindata.org/how-much-subsidies-fossil-fuels

Risks of the Global Environmental Crisis 13

Abstract

The last chapter is devoted to the analysis of risks coming from the global environmental crisis in different dimensions. First is the increasing probability of provoking non-linear catastrophic reactions in the environment, triggered by interactions between different phenomenon, such as warming and losses of biodiversity and carbon release from forest fires. A second area of concern is the impact of the environmental crisis on the economic value of real assets, such as houses, ports, energy and water infrastructures, among others. Natural disasters linked to climate are already affecting insurance costs and availability in some areas of the world. A third issue are the tensions between economic development and protection of the environment: they are on the rise as environmental awareness increases and the political pressures mount to pursue progress "as usual", something happening all over the world. Finally, there is a brief discussion about the tensions arising from social changes coming from latest technologies affecting all spheres of society, combined with concerns about natural disasters and fears about the state of nature in the future.

Keywords

Environmental feedback · Tipping points · Economic impacts · Financial impacts · Technological changes · Social tensions

Partha Dasgupta (2021) argues that the management of natural assets must be examined from a risk management perspective, which is inevitable given the complexity of natural phenomena and our limited understanding of the impacts of interactions between human actions and nature. The previous chapters have analyzed the specific risks associated with the degradation of natural areas, biodiversity loss, and climate change. These have been treated as if they were isolated and independent

J. Vial, *Sustainable Development and Natural Resources*, Integrated Science 39,
https://doi.org/10.1007/978-3-032-12263-6_13

phenomena. The reality is far more complex. These and other critical phenomena occurring in parallel interact with each other, generally amplifying their negative effects: rising temperatures and other climate changes are altering ecosystem functioning and, in many cases, contributing to habitat destruction and biodiversity loss. For example, rising temperatures and changes in ocean acidity due to increased CO_2 in the seas is contributing to the rapid deterioration of coral reefs, which in turn results in habitat loss in marine ecosystems that are extraordinarily rich in biodiversity.

Moreover, one of the major concerns regarding the environmental crisis is the existence of tipping points, beyond which the process of environmental degradation accelerates explosively, giving rise to what are known as high-impact, low-probability events.

These tipping points can occur at the "micro" level of a particular ecosystem, or at the level of the large "macro" systems that regulate the climate, ocean circulation, and the biosphere's capacity to absorb damage and regenerate. In both cases, it is crucial to identify what is happening before a crisis occurs. At the "micro" level, responses may involve targeted interventions to mitigate damage, such as limiting construction permits in areas adjacent to wetlands, or definitively designating the area as a protected zone for conservation purposes (Dasgupta 2021, Section 3.4). This section will address the "macro" issue, where what is at stake is the planet's ability to maintain conditions favorable for the development of life as known today.

When thresholds are crossed in any global dimension, cascading effects begin to occur in multiple domains, turning what was a gradual process of deteriorating living conditions into a catastrophe, likely irreversible, leaving very little time to react. Mitigating these types of risks must be prioritized.

13.1 Tipping Points in the Environmental Crisis

The identification of tipping points has evolved over time, and today there is even talk of "positive tipping points" to refer to events that can trigger a positive chain of interactions that help establish a sustainable balance in nature. The major development in this area is that 20 years ago it was considered highly unlikely that these thresholds would be crossed near the 1.5 °C target set by the Paris Agreement. However, over time, more scientific evidence has accumulated indicating that at temperature levels between 1.5 and 2 °C, it is likely that some thresholds will be crossed, as reaffirmed in the 2022 IPCC Report, albeit in somewhat oblique language[1].

That report highlights the increased likelihood of more frequent, intense, and prolonged heatwave extremes, as well as a faster rate of Antarctic ice sheet melting,

[1] "C.3. Low-likelihood outcomes, such as ice sheet collapse, abrupt ocean circulation changes, some compound extreme events and warming substantially larger than the assessed very likely range of future warming cannot be ruled out and are part of risk assessment" IPCC AR6 WGI S (2022a).

higher forest mortality[2], and a weakening of the North Atlantic Current (AMOC). Other risks of this type are also seen as more probable, such as the collapse of permafrost in the Siberian tundra and the melting of the Greenland ice sheet. The melting of Arctic ice, meanwhile, is already a reality, to the extent that navigation through the legendary "Northwest Passage" is now commonplace and neighboring powers are competing for its control.

Since these events take time to manifest their effects and persist over time, it is very likely that some thresholds have already been crossed, so processes such as polar ice melt will likely continue for centuries and millennia, even if emissions are reduced to zero in the coming decades. This must be incorporated into risk assessments and add greater urgency to the planning and implementation of climate change adaptation policies.

The OECD, in a relatively recent study (OECD 2022), analyzes the risks of tipping points in three major subsystems of the global climate: ice cover, water and atmospheric circulation, and the biosphere. Each of these subsystems, in turn, has components distributed across various regions of the world. One aspect under investigation is the potential for cascading effects when any of these components undergo severe degradation by crossing the threshold that marks the onset of inflection in the deterioration of their respective ecosystems.

Figure 13.1, adapted from OECD (2022), schematically illustrates the components of the three subsystems that are very close to their respective tipping points under current climate conditions.

Below, the potential impacts of major disruptions in three of these subsystems are described and analyzed: one related to ocean circulation (AMOC disruption), one to the biosphere (destruction of the Amazon rainforest), and a third concerning ice systems (collapse of the West Antarctic ice shelf). These three subsystems, highlighted with red circles in Fig. 13.2, share the characteristic that their disruption could occur abruptly and that there is a relatively high probability of being close to the threshold of a tipping point.

(i) Disruption of ocean currents, particularly the North Atlantic (AMOC): Ocean currents are essential for regulating sea temperatures and the transport of nutrients across the oceans. They play a decisive role in the climate of various terrestrial regions, not necessarily adjacent, affecting both temperatures and precipitation patterns. Rising atmospheric temperatures are also causing ocean surface warming, which has accelerated in recent years, exceeding 1.5 °C above late nineteenth-century sea temperatures[3]. In the case of the AMOC, this temperature increase accelerates the melting of Arctic and Greenland ice, which contains freshwater and gradually reduces ocean salinity, thereby slowing down water circulation and potentially even reversing it at some point.

[2] In the years following the report, the occurrence of all the phenomena listed has become increasingly frequent.

[3] EPA. See: https://www.epa.gov/climate-indicators/climate-change-indicators-sea-surface-temperature

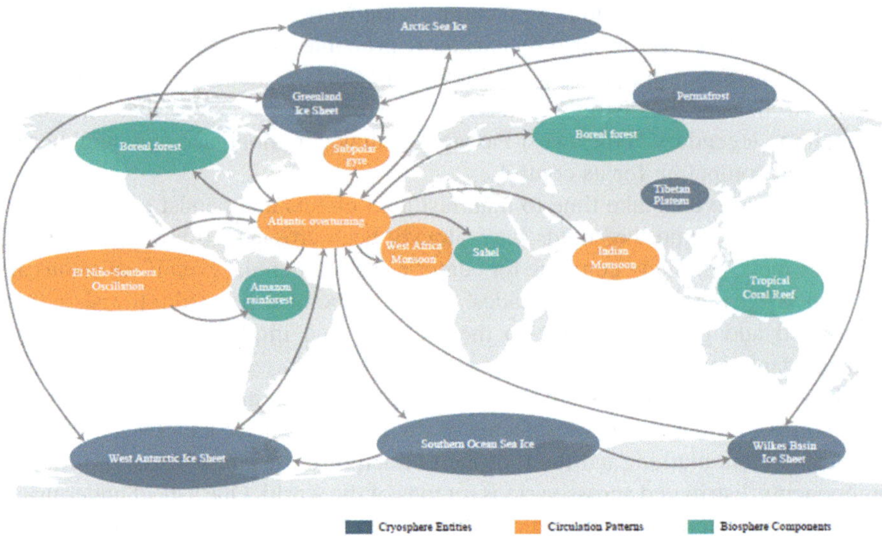

Fig. 13.1 Potential components of climate systems near tipping points. (Source: OECD (2022). Own translation)

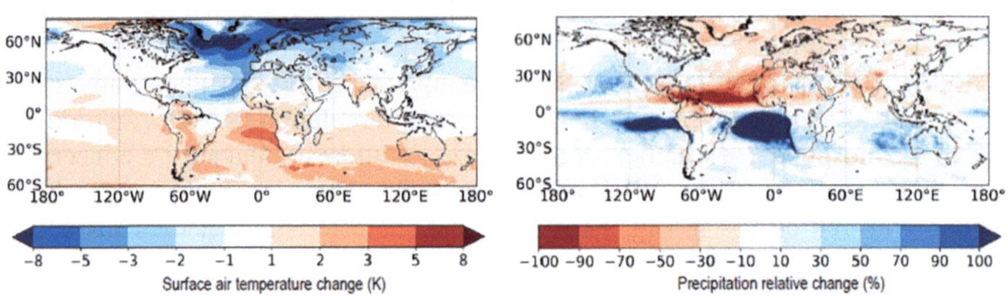

Fig. 13.2 Effects of an AMOC collapse on temperatures and precipitation. (Source: OECD (2022))

Historical evidence shows that changes in the AMOC, and its collapse, have occurred in the past and have coincided with major climate shifts in various regions of the world. The latest IPCC projections indicate that this system will continue to weaken throughout the century, but that a collapse within this century appears unlikely, although this probability may be underestimated (OECD 2022). Such a change in the AMOC could trigger cascading effects on other fragile systems, such as the Amazon rainforest, which would experience reduced precipitation, thereby increasing the likelihood of crossing a tipping point, as well as on global monsoon patterns.

(ii) The mass dieback of the Amazon rainforest is the result of direct deforestation, which is already affecting regional temperatures and precipitation, and has led to a prolonged drought that in turn has increased the frequency and intensity of forest fires. Even assuming deforestation were to cease, it is estimated that if temperatures continue to rise as a consequence of climate change, reaching levels 1.5–2 °C above those prevailing at the end of the nineteenth century, it is highly likely that the tipping point will be exceeded within 20–30 years, and this would occur sooner if deforestation continues (OECD 2022). Beyond this point, the Amazon rainforest would irreversibly transition to savanna, with an incalculable loss of biodiversity. From a climate impact perspective, it is worth noting that the Amazon stores around 200 gigatons of CO_2, which would be released into the atmosphere over a period of decades, undermining the possibility of stabilizing global warming at safe levels. Figure 13.2 shows that drastic changes in the AMOC would exacerbate temperature increases and drought in the western Amazon.

(iii) The collapse of land-based ice sheets would cause a significant rise in sea level, although it is highly unlikely that this would occur abruptly. The largest terrestrial ice masses are in Antarctica (14 million km^2) and Greenland (1.7 million km^2), and both regions are experiencing a significant acceleration in the rate of ice mass loss (OECD 2022). In both cases, it is estimated that the tipping point beyond which a process of continuous and irreversible melting is triggered is very close to, or slightly above, a 1.5 °C increase relative to the pre-industrial period. So, the situation has reached the verge of surpassing that threshold, if it has not already done so. The complete melting of the ice mass in both locations (something that no one is currently predicting) would result in a sea level rise of approximately 65 m above the current level. In this case, the increase in air and ocean temperatures are combining to melt the ice sheets at a faster rate than previously projected, a process that will continue for a long time, even if GHG emissions were to cease now. IPCC projections indicate that sea level rise for the remainder of this century will be on the order of 1 m, which would already create major problems due to the expected increase in extreme weather events such as hurricanes and torrential rains. These losses of ice reduce the capacity to reflect sunlight, so a greater proportion would be absorbed by land, thus contributing to global warming. Furthermore, the reduction in ice-covered surface area threatens the survival of species that depend on it.

Figure 13.2 shows the potential marginal effects of an AMOC collapse on global temperatures and precipitation. As can be seen, the changes are substantial, even when compared to those projected in IPCC emissions scenarios.

What is evident in these three examples is that not only are there thresholds beyond which affected systems enter a process of exponential deterioration, but they also interact, triggering cascading effects in other systems, thereby mutually reinforcing each other. Another important issue is that the window of opportunity to prevent crises in some of these systems is rapidly closing, making it inevitable that humanity will have to adapt to living in an environment very different from the one that has prevailed throughout the history of our species. This is possible, but the question that arises given the magnitude of the changes is whether human societies and their institutions will be able to withstand the stresses these phenomena will cause and, at the same time, act to mitigate the risk of more extreme scenarios.

13.2 Environmental Risks and Their Economic and Financial Impacts

Another important dimension that is beginning to be studied is the impact of climate change on the value of assets held by families, businesses, and the state, and its potential effects on financial activities, either because resources are being lent for investments subject to these types of risks, or because the value of assets used as collateral for loans could be affected by climate change.

In this context, it is common to distinguish between direct and indirect risks of climate change to these assets. The former are clear and can be easily illustrated with some examples: a coastal real estate development, with very little elevation above average sea level, must factor in the risk of more intense and frequent storm surges. An orchard in an area highly exposed to episodes of low water availability and/or extreme temperatures must take these risks into account. A mining operation that requires water to process ore is exposed to the effects of climate change. All these examples have become evident in recent years due to various extreme events such as droughts, floods, storm surges, and torrential rains, among others, which are occurring worldwide and are already leading insurance companies to adjust their policies. In fact, the US press has reported that increases in homeowners' insurance premiums have left many owners without coverage, and some have chosen not to cover certain risks. Since these recent climate anomalies may well be a preview of what could become a "new normal" in 20 or 30 years, this gives an idea of the potential direct economic impacts of climate change.

Indirect risks are more varied and are usually classified into at least two categories: so-called "transition risks" and "liability risks."

Transition risks are defined as those arising because of actions to mitigate climate change, which are largely associated with public policy. For example, a coal-fired power plant could become uncompetitive if a carbon emissions tax is implemented. Similarly, the value of a property near the sea could be affected if building codes change in areas at risk of storm surges.

Liability risks refer to those arising from perceptions of environmental harm caused by companies or activities. These may materialize through compensation claims or court-ordered halts to operations, or even through a drop in sales due to consumer backlash.

As environmental problems, and global warming in particular, gain more prominence in public opinion and on the political agenda, it becomes increasingly necessary to incorporate these three types of risks into credit risk assessments. In recent years, it has become more common for financial regulators to require banks to ask their clients for reports on these types of risks as a condition for granting loans. Natural disasters, such as wildfires that have affected many countries across all continents in recent years, as well as floods and storms in coastal areas, have heightened the need to include these risks in financial analyses.

For developing countries that depend on exports based on natural advantages, it is very likely that the impact of transition risks will be as significant as, or even more important than, the direct risks caused by climate change, since export activities such as mining or fishing are highly capital- and energy-intensive. Furthermore, the distance from export destination markets (and from the origin of imports) is also an obstacle, as maritime and air transport are significant sources of emissions that, if penalized, would reduce the competitiveness of exports from emerging countries that are farther from markets such as China and Europe.

A fundamental tool for addressing these risks is to have good information on the carbon footprint of the most important economic activities, as well as complementary instruments such as a taxonomy of polluting activities that allows for the effective targeting of economic instruments for climate change mitigation, such as emissions taxes, as well as facilitate the operation of emission offset markets. This is an area where some progress has already been made in some upper-middle-income countries, but there is still a long way to go. The same can be said for emission monitoring capabilities, as well as for legislating emission taxes and ensuring their effective enforcement.

In recent years, there have been increasing calls for a proactive role of financial systems in climate change mitigation and adaptation efforts, including from key figures within the sector, such as the presidents of major central banks in France and the United Kingdom, as well as the former Managing Director of the IMF and current President of the European Central Bank, Christine Lagarde.

In 2017, at the initiative of the President of the Bank of France, a global coalition of central banks and financial supervisors was established to promote the "greening" of financial markets (NGFS, Network for the Greening of the Financial System). An increasing number of Central Banks and financial supervisors from emerging economies are participating in this initiative. While the original focus was on climate change issues, it soon became clear that the environmental crisis is multidimensional, and the scope of analysis has since expanded to include ecosystem protection and the risks associated with biodiversity loss.

What explains this particular concern within financial systems, and why might Central Banks and regulators require special attention to this issue? There are a wide range of perspectives: some argue that this is simply a result of the need to broaden risk management frameworks, while others contend that the climate problem would be solved if the financial system stopped providing resources to oil companies and the entire fossil fuel production and distribution chain.

The NGFS has had a dual focus: on the one hand, to raise awareness of the potential risks to financial systems arising from both the direct effects of climate change and the measures needed to facilitate the transition to clean energy. On the other hand, it seeks to identify and promote best practices for both regulators and financial intermediaries in the disclosure of information, as well as in the management of environmental issues, and climate change.

At the root of this concern is the sense that the various actors within the financial systems are lagging behind events. Until a few years ago, the prevailing view was that this problem would have consequences beyond the normal time horizon of financial sector activities, except perhaps for institutions specializing in long-term savings management, such as pension funds and life insurance companies. Moreover, there was considerable reluctance to make decisions based on scientific projections with wide uncertainty ranges and with concrete impacts on economic activities that were difficult to specify.

In recent years, progress on two fronts has heightened the urgency of this issue. On the one hand, there have been significant advances in identifying specific risks associated with climate change. On the other, the succession of climate-related natural disasters—consistent with projections previously made for decades into the future—has created a widespread sense that the problem is much closer and far more serious than previously thought.

NGFS recommendations primarily call for financial institutions to identify and report the risks in their loan and investment portfolios that are associated with environmental and climate-related issues.

From the regulator's perspective, this means establishing rules for reporting these risks and their impacts on the value and solvency of companies that raise funds in capital and credit markets, since there is no possibility that banks and other financial institutions could become a kind of "environmental police": that is not their role, nor do they have the means to do so. Even the task of having companies, banks, and investors identify environmental risks, estimate their potential economic impacts, and then report them is complex and requires a learning process. The same applies to regulators, so these will be slow processes that are likely to take years rather than months to implement.

Environmental impact assessments, required in many countries for investment project approval since the last decade of the twentieth century, are an important tool for identifying potential risks associated with companies' direct interaction with the environment. This is particularly relevant in resource-dependent economies, as is the case in many emerging countries. Today, after more than 30 years of applying these regulations in developed countries, it is highly likely that multinational companies and some medium and large domestic firms that export to developed markets are fully aware of their direct environmental risks and, in fact, some are already regularly reporting them in their annual reports and investor information brochures.

However, the reporting of environmental risks by firms often requires information that goes beyond their own operational scope, which is particularly true in the case of physical risks associated with global phenomena such as climate change. This information has a public good component, and it is necessary for the governments to assume, both some responsibility for its production and availability, as well as its funding. This requires collaboration between the public sector and the scientific community, at least for the identification of potential events and their consequences at the national and local levels. In poorer countries or those lacking technical capacity, multilateral organizations and the development aid donor community have an important role to play.

On the other hand, information such as that provided by carbon footprint measurements enables governments to assess the economic impacts of mitigation measures and allows companies to evaluate the risks arising from them, both nationally and globally. While it is true that these are aggregate measurements for different sectors of economic activity, they provide a reference so that companies and the financial system itself can estimate these impacts at the sectoral level, and even for larger companies.

At the international level, some propose that investors and banks adopt an aggressive stance by restricting financing to polluting activities. This carries several risks. The most significant thing is the belief that by quickly ending pollution, the problem can be solved.

In practice, the opposite may occur: the costs of a disorderly transition can be so high that they exceed the political capacity of governments to implement them. For example, if fuel prices rise so much, governments may end up reversing measures that could have enabled a more orderly and less costly transition.

In this regard, it is very risky to demand that financial intermediaries take responsibility for implementing a transition strategy, as this falls squarely within the political sphere and for which they have no legitimacy. Often, it seems that environmental groups push for these demands because they know they lack the necessary power to influence policy changes by governments or parliament, the institutions mandated to implement public policies, to put their agenda into practice.

Moreover, imposing such restrictions would create a vacuum that could easily be filled by financial institutions based in countries that do not enforce these standards, which in the long run could lead to greater "informalization" of global financial markets, with potentially serious consequences for global financial stability, as well as for the environment.

In fact, in some cases, the activism of pressure groups that tends to discourage financing for energy projects can have adverse effects on the coherence of public policies. There is no doubt that an orderly and efficient transition will require a flow of financial resources toward clean energy, but it is also necessary to allow for the continuity of backup systems that do not jeopardize the ongoing provision of energy or cause a loss of purchasing power among the population that could ultimately erode political support for the energy transition. A relatively orderly energy transition at both global and national levels is a complex exercise that requires balancing the increased costs associated with the transition with the need to preserve its political viability. This calls for careful design, appropriate institutional frameworks, and a continuous and highly effective communication effort.

13.3 Tensions Between Development and Environmental Protection

Global warming and the broader environmental crisis gained significant prominence on political agendas up until the Covid-19 pandemic. However, subsequently, the issue became politicized in developed countries, especially in the United States, and there have been reversals that could delay the environmental agenda, just as the scientific community is increasingly convinced that time is running out to act and mitigate the risk of catastrophic scenarios.

Even in countries that had begun to respond with large-scale programs to facilitate the energy transition, such as the United States, resistance quickly emerged from those whose purchasing power and usual consumption patterns were affected, as well as from business owners who saw the value of their investments threatened, with dramatic effects that led the new government in that country to undo much of the previous progress and significantly backtrack on the preservation and care of natural areas.

The tension between economic growth and progress in environmental protection and climate disaster prevention is not easy to manage anywhere. Much has been said about developed countries, but this tension is even more pronounced in middle- and low-income countries, which face the imperative to grow to reduce poverty. The effort must be shared everywhere, although it is also important for countries to contribute to solving problems in the areas where they are most efficient, or, in economic terms, where they have comparative advantages.

A country like Brazil, for example, with the enormous responsibility of preserving the biodiversity of much of the Amazon, as well as preventing GHG emissions from land-use change, should focus its mitigation efforts in this area, dedicating fewer resources to changing the energy matrix in those components that would entail higher costs for the population and businesses.

On the other hand, in countries like Japan, which have ample resources, emissions stem from industrial activity and transportation, and they face significant risks from the direct impacts of climate change. These countries should leverage all their resources to accelerate and facilitate the energy transition, advancing the development of technologies and products that support it. This, in turn, could be utilized by poorer economies to accelerate their development, skipping the stage of heavy fossil fuel consumption.

The first to feel the pressure to change will be business, whether due to increased restrictions, emissions taxes, or simply the risk of losing customers or facing civil lawsuits for damages. In recent years, companies have made efforts to modify technologies, replace equipment, change products, and adapt business models to this new reality. Furthermore, as these practices become widespread, new business opportunities will arise for companies that are more attuned to changes in the environment. Concern for the climate will cease to be a matter for the "Corporate Social Responsibility Department" and will begin to play a central role in business strategies.

But it is not only production techniques that must change. Achieving the goal of net zero emissions by 2045 will require a profound shift in lifestyles and consumption patterns. The aspiration for car ownership among emerging middle classes, especially in the densely populated developing world, may not be compatible with twenty-first-century realities, at least if these vehicles use fossil fuels. Giving up this dream without losing the hope of achieving greater mobility in a comfortable way is not a utopia. It is very interesting to note how China, in just a few years, has come to lead the penetration of electric cars in its local market, thereby generating a competitive capability that, in a very short time, is making automobiles available to the world's middle classes at prices competitive with those of fossil fuel vehicles. Of course, the fact that twenty-first-century cars are "cleaner" does not necessarily mean that part of the climate crisis is resolved: they will still require steel and paved roads to operate, so if motorization continues to increase, emissions will also continue to rise, although now originating from sources other than the combustion of fuels by automobiles. Advances in public transportation, including digital platforms such as Uber, as well as the shared use of individual means of transport, from cars to bicycles, are demonstrating possible alternative pathways. Emissions taxes and other regulations requiring disclosure of the carbon footprint of products and services will become increasingly important.

In summary, addressing climate change will not be easy, because as a global problem, it necessarily requires collaboration among countries whose governments must respond to the demands of their citizens, who often perceive the costs of such actions without a clear understanding of the risks of inaction. This is the difficult dilemma posed by externalities with global impact. National governments have experience dealing with externalities at the local or national level, and those harmed by them can organize to counterbalance the pressures from those who benefit from such externalities. As has been seen many times, governments are not always successful in regulating these externalities, especially when the harm they cause are difficult to perceive and are distributed among many dispersed actors. When externalities have supranational effects, the difficulty of regulating them and enforcing such regulations increases exponentially.

13.4 Technological Revolution, Environmental Crisis, and Social Tensions

As discussed in the preceding pages, the tremendous advances in scientific knowledge, as well as their practical applications across all human activities, have radically changed the material quality of life, the physical environment, and the ways in which human societies are organized.

However, little has been said about what can be expected from the new advances currently underway, whose effects will be seen in the coming decades. It is always difficult to predict the future, but something can be said with certainty, given recent advances and trends: the world in which our children live is radically different from that experienced by the "baby boomer" generation, and it is hard to imagine the world our grandchildren will inhabit. The advances of recent decades in digitalization, biotechnologies, and, in recent years, in the development of artificial intelligence are radically altering ways of life, opening unforeseen opportunities, and generating risks that are difficult to assess, all within a span of just a few decades.

To begin with, perhaps the most important point: the increase in the capacity to produce new knowledge and applications is directly related to the number of educated people dedicated to this task, and to the resources available for developing new ideas. With the inclusion of large numbers of educated people from emerging countries and the income growth of recent decades, the leap in innovation capacity currently underway is monumental and unprecedented. Just a few decades ago, the world of scientists conducting research and connected with one another comprised only a few hundred thousand people. Today, it is tens or perhaps hundreds of millions. This is without yet mentioning the information processing and analytical capabilities, and all kinds of equipment now available for their work, which is no longer limited to isolated laboratories in various Western cities, but operates on a global scale, with interconnected centers around the world, and even with experimentation and data collection platforms beyond the planet's boundaries. To this has been added the nascent development of artificial intelligence, which puts this vast base of accumulated global knowledge within reach of anyone with a mobile phone and sufficient skill to formulate the right questions and verify the accuracy of the answers.

The impacts of this increase in capabilities are evident in all fields: from the production of synthetic foods to the application of genomics in health, or the development of methods to capture solar energy for electricity generation, to mention just a few advances with clearly positive impacts for the human population and which make it feasible to envision sustainable development.

It should not be forgotten, however, that for each of these advances, there are also dangerous applications for humanity and society: biotechnology opens the door to new forms of warfare and bioterrorism. Advances in connectivity and three-dimensional printing allow new weapons to be accessible to small groups with limited means. Cyberattacks can cause enormous damage to all activities that depend on interconnection, both in communications and in physical operations, etc. Artificial intelligence can be used to enhance criminal activities and provide unprecedented destructive capabilities.

It does not take much imagination to realize that every new scientific discovery brings with it both opportunities for progress and threats. To reach a society where science and engineering enable us to travel to other worlds and eventually overcome physical limitations, we must traverse a path fraught with dangers, many of which will stem from the shortcomings of human organizations in adapting to and overcoming the conflicts arising from change.

One aspect that cannot be overlooked is the role of incentives in determining the direction of technological advances. In the nineteenth and twentieth centuries, nation-states were the only entities with sufficient resources to drive large-scale scientific and technological research, and they played a central role in the development of such advances.

The nationalisms of the late nineteenth and early twentieth centuries were decisive for the development of transportation and communications infrastructures[4]. The terrible wars that marked that era, reinforcing these trends and serving as a major catalyst for industrial development, culminated in the advent of nuclear energy.

The globalization of economic relations, communications, and culture, combined with the exhaustion of the fiscal capacities of nation-states, has led to the emergence of mega-corporations and, consequently, to a growing predominance of private profit expectations in directing resources toward scientific research and

[4] MacMillan (2020) suggests that the emergence of social protection systems in Europe is one side of a coin, on the other of which lies the extreme nationalism that ultimately led to the major wars of the early 20th century.

technological development. By the mid-twentieth century, these large multinationals emerged in sectors such as energy, particularly oil extraction, which operated globally and generated substantial rents. Paradoxically, they were not major promoters of innovation[5], and in many cases even became obstacles to the development of alternatives to fossil fuels. However, from the mid-twentieth century onward, new multinationals began to emerge, driven by scientific advances in fields such as computing (IBM) and the pharmaceutical industry, among others.

Unlike the major oil companies, which were based in the most powerful Western countries and could pressure their governments to defend their interests in remote parts of the world, the new knowledge-based multinationals used their power and influence to ensure that legal systems protected their market power through the enforcement of intellectual property rights.

This is changing: the globalization of knowledge and production chains, together with the rapid development of Asian countries, has enabled large companies based in these regions to challenge the hegemony of Western-based multinationals. This has become a source of tension between rival powers (such as China and the United States in the field of information technology) and even among close allies (such as the United Kingdom and Europe, regarding the conduct of major pharmaceutical companies in the commercialization of vaccines). This shift in the control of scientific research from states to private companies is most evident in the field of space exploration, where private enterprises backed by vast fortunes are setting priorities and developing new technologies. In the United States, NASA has gone from holding an absolute monopoly in this area to having to contract private companies to transport cargo and people into space.

In practice, this means that the direction of research and the development of new goods and services for the future world is no longer a monopoly of national governments, at least in the West. For now, they still retain a monopoly over research in strict military applications, but they can no longer dictate the direction of future advances, which have become so decentralized that they are increasingly beyond the control of states, at least in the West.

In this context, profit opportunities are driving an increasing share of resources allocated to innovation, and markets are playing a more significant role in this allocation. This is evident in the contrast between the speed of development of inexpensive drugs to treat male erectile dysfunction, compared, for example, to treatments for infectious diseases prevalent in poor African countries.

However, not all outcomes are negative: the Human Genome Project was launched in 1990 under the auspices of U.S. government agencies and with the collaboration of research centers from several developed Western countries. By the end of that decade, private competitors and developers emerged, building on previous advances and accelerating and expanding progress to the study of multiple genomes. The effect was explosive: in 2003, the goal of completing the human genome

[5] Where they have made significant contributions, it is in those areas necessary to reduce costs and develop alternative sources of supply, something that has benefited other mineral extractive activities.

sequence was finally achieved. Less than 15 years later, it became possible to identify genes that make individuals more susceptible to cancer, and to develop vaccines against Covid-19 by manipulating viral genes, largely because the promise of profits for those who discover treatments for these diseases has motivated numerous private research projects in these areas.

This is not limited to biology and medicine. It is also occurring in the energy sector: after many years of discussion about new technologies to harness scientific advances for renewable energy production, just a few years of oil prices above 100 USD per barrel were enough to spur the development of new applications that made wind and solar energy profitable options for electricity generation. Growing concern over global warming has generated worldwide pressure to develop initiatives that make investment in alternative energy development more attractive, as well as new products to replace those with a large carbon footprint, ranging from the development of electric vehicles to the production of "artificial meat."

A key variable for aligning the objectives of companies and consumers with environmental protection is the use of market instruments: taxes and subsidies. Given the fiscal pressures most countries are facing after the pandemic, it seems reasonable to assume that taxes should be the first line of defense. This is why many economists, and some opinion leaders are advocating for the introduction of carbon emission taxes.

Regarding biodiversity protection, where causal links are more diffuse and direct impacts are first felt at the local level, a more active role for subsidies or fiscal compensation for restricting certain activities in areas critical to the provision of ecosystem services may be required.

Of course, this does not mean that nation-states have lost all relevance in these matters. It should not be forgotten that the governments of the United States, Japan, China, Russia, and the European Community have allocated significant resources to scientific research and have been fundamental to the explosion of technological advances currently underway. While much of this has been motivated by the desire to maintain or gain political, military, or economic power, applications in the civilian sphere have been at the heart of today's transformations.

Looking ahead, the outlook is more complicated. Virtually all these countries are increasingly pressured to allocate their resources to more immediate needs, such as providing social protection in old age and security against new threats. With societies that are ever more informed and connected, both democratic societies and autocratic regimes, including dictatorships, are compelled to be more responsive to the most urgent priorities of their populations, making it more difficult to dedicate resources to more diffuse objectives whose benefits will only be realized in the longer term. This increases the risk that technological progress will become increasingly concentrated in mega corporations with the resources to finance it.

Another aspect to highlight in this section is the speed at which technological capabilities are spreading. It is often noted how stark the differences are between lifestyles in the cities of developed countries and the villages of poor countries in Africa and Asia. In the former, children have nearly all the knowledge they can absorb at their fingertips on their phones, can travel between home and school in minutes, and, in medical emergencies, have access to diagnostic equipment, specialists, and treatments at the cutting edge of human knowledge. All of this is unimaginable for children without access to electricity or the internet, who must walk for hours to attend school, and in health emergencies, can rely only on their natural defenses to survive.

These differences, however, conceal a more complex reality: children living in the cities of those same poor countries do have access to electricity and the internet, to formal education, to some forms of public transportation, and to modern medicine. This demonstrates that technological advances can spread rapidly once a minimum level of infrastructure and income is achieved.

When poor countries begin to make progress, elites gain access to services very similar to those available in developed countries. Soon after, as access to information and political power spreads, states begin to allocate more resources to provide services for the broader population. All of this occurs amid tensions, political and social conflicts, but ultimately, progress is seen almost everywhere in the world. In these same countries, access to the internet via mobile phones is accelerating the access of middle and lower-income groups to sophisticated services, such as financial transactions, for example. This would have been unthinkable 50 years ago, when most of the population was illiterate and had no access to electricity. The widespread expansion of education and the dramatic reduction in the cost of acquiring accumulated knowledge are opening unprecedented opportunities to democratize knowledge and technological innovation on a global scale.

How does all this connect with the prevention and mitigation of the environmental crisis?

First, as mentioned earlier, technological progress is enabling a much faster energy transition in developing countries, which could make it possible to leapfrog the most polluting stage of the economic development model that dominated the twentieth century. The development of new materials and the application of Artificial Intelligence should accelerate these processes even further.

Second, it is important to mention the enormous opportunities emerging to better monitor and control environmental threats: from satellite tracking of methane emissions as well as the evolution of major ecosystems, to active surveillance using remote technologies such as drones, which have proven so effective in warfare, can play a decisive role in protecting natural areas.

A third aspect to consider is that advances in robotics, which are currently perceived as a major threat to traditional jobs, also offer the possibility of alleviating the labor shortages associated with the rapid demographic transition underway in almost every part of the world. This phenomenon is already generating significant tensions and fears, but it is an essential component for a transition toward a sustainable development path. This is particularly important for achieving continued income growth, independent of population growth, which will be fundamental to financing the higher costs associated with increased human longevity. Of course, this will also require redesigning tax mechanisms, especially those that increase the cost of human labor, which will become scarcer and, possibly, less valuable, at least in relative terms.

At the dawn of the third millennium, we are witnessing prodigious advances in science and technology, which have led various authors to suggest that the storage of personal memory, intelligence, and consciousness by artificial means could become a reality within decades rather than millennia, fundamentally changing the concept of "life" as we have understood it until now[6]. What was once a fantasy for the "baby boomers" now appears to be a real, if still future, possibility.

However, this aspiration collides with the harsh reality that a large proportion of the world's population is still far from even considering this dream, and their aspirations are much more prosaic: to have enough to eat, access to minimally adequate health services, and basic education. But this is not only a problem for the poor: the large middle classes in emerging countries aspire to access quality services and greater security in their current and future incomes. On the other hand, workers in developed countries who have lost stable jobs due to competition from industries that have lifted workers in poorer countries out of poverty also aspire to do the same. This is a complex environment, in which populists are already emerging with promises of easy—but ineffective—solutions to problems that are nearly insoluble in the short term.

These major global trends will be the driving forces of foreseeable change in the twenty-first century. They will place societies under strain, testing the limits of the capacity of organizations and political systems in the decades to come. Differences in the pace of population growth, the accelerated loss of biodiversity, the impacts of climate on the most vulnerable regions, especially those suffering from high levels of poverty and hunger, as well as the effects of the changes we are witnessing in the energy sector and the rapid pace of technological response, will profoundly reshape the global geopolitical landscape.

In a world where economic activity is truly globalized, with trade, production, and finance integrated into networks operating in real time and with a physical presence across various hemispheres, the issue of political and social tensions will become increasingly significant in the decision-making processes of families, businesses, and governments. They must be prepared to respond to profound changes in this area. At the end of the last decade, the resurgence of authoritarian regimes, and the weakening of democratic institutions, even in the most developed countries, are manifestations of the impact these tensions can generate. The recent COVID-19 pandemic has added an additional layer of pressure to an already highly strained world. Although the official death toll was relatively low compared to previous pandemics, the fear of a much higher mortality rate, combined with the speed of its spread, was enough to challenge the healthcare systems of the wealthiest countries, trigger the most severe global economic recession in a long time, and leave nearly all states around the world far more indebted than before. All this, not to mention that future generations, who are already present in the world, will have to face the enormous costs of the environmental crisis and the adjustments required by the new demographic reality.

[6] See Kurzweil (2005) and Harari (2015).

In the coming decades, the world, countries, businesses, and individuals will face enormous pressures arising from the trends mentioned above. Success in overcoming these challenges will depend critically on the flexibility to adapt and respond to them promptly and effectively. This is easier at the individual level and becomes increasingly difficult as organizational complexity grows. A concerning issue is that political systems, based on the governance of nation-states, face serious difficulties in addressing global problems and are struggling to cope with the "denationalization" of culture and economic activities.

References

Dasgupta P (2021) The economics of biodiversity: the Dasgupta review. HM Treasury
Harari YN (2015) Homo Deus: A brief history of the future. Penguin Random House
Kurzweil R (2005) The singularity is near: When humans transcend biology. Viking
MacMillan M (2020) War: How conflict shaped us. Random House, New York
OECD (2022) Climate tipping points: Insights for effective policy action. OECD Publishing. https://doi.org/10.1787/abc5a69e-en

Final Reflections

To conclude, I would like to briefly summarize personal messages that I believe are important to record. These are lessons learned over more than 45 years of professional life and may prove useful to some readers.

To begin with, I believe it is necessary to highlight the tremendous progress achieved by humanity, which has materialized in a minuscule period. Of course, we should not underestimate the tens of thousands of years that preceded the Industrial Revolution: this time was essential for accumulating the knowledge and building the institutions that made it possible.

This is an inescapable conclusion and lesson at a time when the human species faces unprecedented challenges to survive and keep its civilizations alive. My grandparents witnessed the arrival of electricity and electric lighting in Santiago, Chile, when public transportation in the city was provided by "blood cars" (horse-drawn carriages). I have no doubt that they never imagined they would witness humankind's journey into space, or that their great-granddaughters would travel to Calcutta, Morocco, London, Bangkok, Miami, or Berlin without being millionaires, while speaking with their families, on camera, from their phones. Not to mention that they could pay their bills online and manage their savings and payments in Chile, also from their phones.

Whatever metric is used to measure progress or success as a species, all show the same: an explosive increase in the human population occupying all habitable spaces on the planet; unprecedented improvements in nutrition and longevity; exponential growth in "leisure time" devoted to activities other than survival; the ability to manipulate the environment and nature to suit human needs; artistic creation in its most diverse forms and with global reach, and so on.

It is tempting to fall into exuberant arrogance when admiring these achievements and to become enthusiastic about the idea that virtually anything we set out to do is possible, and that we are leaving behind all our material limitations. This is a temptation we must avoid.

A second message concerns some of the factors that have made this progress possible. I would like to mention at least three: science and knowledge, institutions, and exchange and specialization. All are interrelated and could hardly have developed without one another. The fact that they have flourished independently in different parts of the world suggests that human ingenuity, curiosity, and memory, as well

J. Vial, *Sustainable Development and Natural Resources*, Integrated Science 39,
https://doi.org/10.1007/978-3-032-12263-6

as the ability to learn from one's own and others' experiences, are inherent characteristics of the species that have underpinned its success. At the same time, they are the foundation of their capacity to adapt to the most diverse environments and the most adverse contingencies.

The enormous increase in the scale and complexity of human relationships is testing the limits of these capacities, and it cannot be overlooked that progress has also been marked by the bloodiest and most destructive conflicts in history, some even on a global scale. Most notably, regions that have been the cradle of scientific and artistic progress in Europe and Asia have led the most terrible massacres of the twentieth century. Those who have followed in their footsteps in the Americas and Eurasia are not doing much better today.

Scientific knowledge and institutions are extremely important topics, with multiple determinants. What has been emphasized in this book is the importance of economic incentives in guiding innovation, especially in what is one of the main focuses of this work: alleviating the scarcity of natural resources in a closed global ecosystem. When resource scarcity has been reflected in market prices, research and investment have eased these constraints. The Green Revolution and the nascent energy transition are proof evident of this. On the contrary, when scarcity is not reflected in markets, whether due to externalities or failures in the exercise of property rights, innovation not only helps but can worsen the problem. The overexploitation of marine species by industrial fishing fleets, the destruction of ecosystems with the consequent loss of biodiversity, and the uncontrolled emission of greenhouse gases associated with the use of fossil fuels and land use changes are also evidence of the limitations of technology to solve, on its own, the problems associated with the rapid development of humanity.

Exchange and specialization are indeed at the core of economic analysis and play a central role in our discussion. Without exchange, there can be no specialization, and without this, progress would be much slower. This holds true for both individuals and nations, as not all have the size and resources to be self-sufficient. Indeed, Adam Smith[1] stated this long before and far better than this author ever could. It only remains to add that the experiences of the twentieth and twenty-first centuries show that those societies or countries that have rapidly closed the gap with global leaders have done so only after opening to trade and exchange, specializing in activities where they held relative advantages. This condition was not sufficient, but all the evidence suggests it is necessary. This approach is especially important for small countries far from global economic centers, which may possess great natural wealth but lack domestic markets large enough to exploit the economies of scale essential for growth and for competing with others in manufacturing.

A corollary of the above is that if a country's natural resources provide it with comparative advantages in activities related to their exploitation, these should be leveraged, but without forgetting that such resources are limited and that sustainable use over time requires respecting the conditions imposed by nature for their regeneration. In the case of exhaustible resources, it is necessary to allocate

[1] Smith (1776).

resources to replace the natural capital consumed with other forms of capital. Otherwise, the country's Net Wealth will decline, resulting in lower income flows in the future.

A third message is the antidote to the temptation mentioned at the outset: the spectacular progress of humanity is generating adverse impacts on nature that are compromising its regenerative capacities, on the one hand, and its ability to support the diverse forms of life inhabiting the planet, on the other. Part of this deterioration stems from the use of resources to meet humanity's consumption needs, most notably the destruction of forests and grasslands that support a great diversity of life forms, for agriculture and livestock, and to a lesser extent, urban development. However, the most severe effects are related to the generation of various types of waste from human activity: fertilizers that pollute soils and water, gases, mainly from the use of fossil fuels, released into the atmosphere, threatening its integrity and impacting the climate (GHG), pollution of the oceans (CO_2, industrial waste, organic matter, etc.). All of this has resulted in massive losses of wildlife, even leading to the extinction of entire species, and in environmental crises on a genuinely global scale: the planet is becoming too small for humanity, even as a large proportion of the world's population remains far from achieving minimally acceptable levels of consumption of basic goods and services by current standards.

This tremendous advance, both in the number of inhabitants and in per capita consumption levels, is reaching a point where, even with advances in scientific knowledge and technology, we cannot ignore that increased consumption in the most disadvantaged regions is becoming a zero-sum game at the global level. In other words, until technology and human behavior provide us with the means to reduce our impact on nature, progress in improving the living conditions of the poorest will not be sustainable without setbacks in the consumption of the rest of humanity. Something must give, but if what continues to give way is the state of nature, this will result in catastrophic losses for all humanity and, in turn, for other forms of life on the planet.

One way to view this dilemma is to break down the impact of human activity on the biosphere into three factors: population size, per capita consumption, and "productivity" in the use of the services provided by nature.

Population is key because every living person occupies space, energy, and resources, so even if per capita consumption does not increase, this translates into greater pressure on the environment. It should be remembered that the population has multiplied tenfold in just over a century. In this regard, there is news that offers hope, but at the same time poses serious challenges that must be addressed: if current trends continue, the world population will begin to decline rapidly in the last quarter of this century, especially in developed countries and in some emerging countries, as large as China or as small as Chile, where birth rates have fallen particularly sharply. Unfortunately, without an adjustment in population size, it is impossible to achieve an orderly adjustment of humanity's footprint on the planet that would allow a return to a sustainable development path. The good news is that this adjustment is occurring endogenously and voluntarily, without the need for repressive policies.

Per capita consumption is also essential, as it too demands space, energy, and resources, and the species has proven to be insatiable. Even in the wealthiest democracies, governments fall, and politicians are ousted if employment, incomes, and consumption do not increase steadily. This adjustment must be focused on the most prosperous societies, which are already meeting their resource requirements from the environment, because there are still nearly one billion people who today are not able to finance a level of consumption that would allow them to overcome poverty.

Efficiency in the use of Nature's services is the third tool for achieving an adjustment to the global environmental crisis: this requires, for example, the replacement of fossil fuels, the reduction of land and water use for food production, and the reduction of pressure on key ecosystems to preserve the diversity of human, animal, and plant life. This also implies decreasing the intensity of all types of materials whose extraction and disposal as waste leave a significant footprint on the various ecosystems that support life.[2] In this area, the availability of natural resources has not been as significant a limitation as the impact on ecosystems from the production systems of goods, as well as from the disposal and treatment of waste generated by human activity. For this change to happen, economic incentives should be mobilized in areas in which markets do not work properly, either penalizing negative externalities, or favoring positive externalities, as well as by creating institutions to compensate for the absence of property rights. Command and control measures might help in some cases, but seldom are the most efficient way to achieve the goals.

The final global message concerns the likelihood of success I assign to achieving the goal of sustainable improvement in the conditions of human life on the planet. That is, to achieve the necessary progress on the three fronts mentioned above.

On the one hand, I believe we must keep in mind tremendous capacity for innovation and adaptation that humanity has demonstrated throughout its history. In the face of the magnitude of these challenges, it is easy to become discouraged, to see everything in a negative light, and to fall into fatalism, but there are many examples that show how easy it is to underestimate our capabilities. Today, it seems natural to think that Ireland, South Korea, or Taiwan, societies that emerged amid poverty and bloody conflicts, have, in less than half a century, become advanced, modern, prosperous societies and leaders in the development of cutting-edge technologies in various fields. On a very different note, it is almost incredible that more than half of the fish we consume worldwide now comes from aquaculture, an activity that was practically insignificant just 40 years ago.

On the other hand, we must also highlight that important changes are already taking place on the three fronts we have identified:

- The growth of the world population is rapidly slowing, especially in wealthy countries, but also in middle-income countries: progress in demographic transition has gone hand in hand with economic development. It is no coincidence that declines in birth rates are lagging most among the poorest populations. A necessary condition for accelerating the global demographic transition appears to be

[2] See Smil (2022).

the creation of new opportunities for economic and social advancement in the poorest societies, and for women in those societies. Here lies an opportunity for a global positive-sum game.
- Per capita consumption is possibly the most difficult variable to alter, especially because the prices of goods and services do not incorporate many of the environmental costs associated with their production and consumption. In the absence of economic incentives, expecting cultural changes to modify mass consumption behaviors seems very unlikely complicated. It is true that there has been progress, for example, in areas such as waste recycling, reduced preference for meat consumption, and, in general, simpler lifestyles, but these are more niche behaviors than widespread changes. Nevertheless, pressure on decision-makers in democratic political systems has begun to be felt. For example, in Europe, a "cap and trade" system for greenhouse gas emissions has been consolidated, which is a good substitute for carbon emission taxes. The "caps," which were quite lax at first, have become more stringent, and today we see carbon prices that are three to four times higher than a decade ago. Spain has enacted a law to reduce food waste in restaurants, and thus, changes are being observed—still very slow but moving in the right direction.
- Finally, we are also seeing significant progress in the better use of Nature's services: the development of low-impact renewable energies is the clearest example, although there is still a long way to go. The widespread adoption of hydroponic crops, the search for new substitutes for animal meat, and the replacement of traditional livestock farming with lower-impact alternatives are all advances pointing in the right direction. However, there remains tremendous pressure to extract resources from forests for energy, as well as to replace them with crops or livestock, both generally of very low productivity. The greatest losses of tropical forests today occur in regions of deep poverty, where rapid population growth continues to drive the encroachment into natural areas. Here, too, there is an opportunity for positive-sum outcomes: better opportunities for progress for these poor rural populations should help reduce both population growth and the pressure to expand the agricultural frontier.

All of the above considered, I believe that a non-traumatic solution to the global environmental crisis is possible, and although the steps we are seeing are late and still insufficient, except with regard to the evolution of the world population, we still have time to make adjustments before the accumulation of GHGs in the atmosphere causes catastrophic effects for the survival of our species. However, everything suggests that current political will and institutional frameworks are still not up to the challenges we face.

"Fortunately," the effects of the environmental crisis are already being felt, and this should begin to influence decision-making at the national level, although the politicization of these issues has led to an incomprehensible rejection of scientific evidence. Hopefully, this will change soon, because otherwise, an orderly adjustment will no longer be viable and the impacts on human societies could be catastrophic.

In any case, this means that the most vulnerable countries to climate change climate will have to make significant efforts and investments to mitigate the impact of what now seems inevitable: a stabilization of GHGs in the atmosphere leading to increases in average temperature of 2 or 2.5 degrees C above pre-industrial levels. This would be enough to render some regions of the planet almost uninhabitable, especially in Asia and North Africa.

To conclude, I will briefly address some specific messages for small countries that are rich in natural resources.

Small, middle- or low-income countries with a small domestic market and abundant natural resources will inevitably continue to depend on these resources for some time to come. Comparative advantages are not created overnight, and developing previously nonexistent capabilities require sustained, long-term efforts, beginning with the development of human capital.

This necessitates great care in designing the rules that govern the exploitation of these resources, which must be based on a deep understanding of their characteristics and the associated economic activities. The most challenging aspect is achieving an appropriate balance between state capture of rents, while maintaining incentives for investment, and protecting the environment. This is not easy, because carrying out large-scale, long-maturing investment projects, as required by some of these activities, demands credible long-term commitments. In practice, this means immobilizing large amounts of capital for extended periods in sectors that, as we have seen, are vulnerable to political risks. For the same reason, the cost of financing plays a fundamental role in making these projects viable. This cost is sensitive to the country's overall macroeconomic and political situation, leaving investment highly vulnerable to changes in risk perception.

A second point to highlight is that this specialization faces major challenges in expanding production capacity, precisely because of the environmental impacts of natural resource extraction. The increase in regulations on mining, oil extraction, and energy generation in general, and, increasingly, on agriculture itself, is no coincidence. These are activities that, due to their direct contact with nature and their immersion in ecosystems that are ever more interconnected with populations demanding their protection and preservation, face growing difficulties in making the investments needed to maintain or increase production. Complaints about the costs and delays caused by environmental regulations are increasingly common and frequent in all types of countries. This represents an ever more significant challenge for these countries and is a factor to consider for its impact on medium- and long-term growth.

A third message I wish to emphasize is that contrary to what many believe, the trend toward stabilization of population size should not be seen as a threat. On the contrary, considering the fragility and endemic biodiversity of territories that still have low human population density, this is very good news. Moreover, a lower proportion of children and adolescents, and the possibility of scarcity of workers in the future, presents a great opportunity to reduce poverty and inequality, if access to and the ability to acquire human capital by new generations is improved.

Social tensions have risen and will continue to play a very important role in the design and functioning of the institutions and public policies needed to address the challenges we have identified in these pages. This issue has gained enormous importance in recent years, amid a resurgence of isolationist and populist trends in many countries, some of them extraordinarily important. When evidence is denied and data that do not fit ideological, racial, gender, or other prejudices are dismissed, it becomes clear that much remains to be done to achieve more cohesive societies, with institutions and policies that are coherent and effective in facing the complex challenges ahead. This cannot be neglected; the world is undergoing a massive communicational and social transition, just beginning to adapt to the decentralization and massification of content provision that is threatening the viability of traditional forms of democracy. These forms created the conditions for innovation and exchange underpinning humanity's enormous success in recent centuries, but today they are under tremendous pressure, and it is by no means certain that they will continue to play that role in the future.

It is too early to advance proposals that would reconcile sustainable material progress with respect for personal freedoms and the protection of human rights, but amid the global crises we are experiencing, finding effective responses to these dilemmas is the most important task for current generations.

Appendices

Appendix 1 Emissions Reduction Scenario Simulation Exercise

Inertia Scenario			
Population (thousands of people)	2022	2040	2100
World	8,021,407	9,177,190	10,180,161
High-income countries	1,254,317	1,290,599	1,252,364
Middle-income countries	6,020,971	6,758,785	6,746,443
Upper-middle-income countries	2,804,740	2,846,409	2,026,151
Lower-middle-income countries	3,216,231	3,912,375	4,720,292
Low-income countries	716,866	1,097,776	2,151,930
Per capita GDP (constant 2015 US$)			
World	11,315	20,031	73,019
Per capita GDP (constant 2015 US$)			
World	11,315	20,031	73,019
High-Income Countries	43,791	54,406	98,838
Middle-Income Countries	5,796	12,673	46,961
Upper-Middle-Income Countries	9,747	22,340	73,299
Lower-Middle-Income Countries	2,347	4,215	29,650
Low-Income Countries	757	831	1,133
CO2 Emissions (million kt/year)			
World	33.6	47.7	65.2
High-Income Countries	10.9	0.2	−15.8
Middle-Income Countries	21.2	46.7	80.6
Upper-Middle-Income Countries	16.4	33.0	11.3
Lower-Middle-Income Countries	4.8	13.6	69.3
Low-Income Countries	0.2	0.4	0.7

(continued)

J. Vial, *Sustainable Development and Natural Resources*, Integrated Science 39,
https://doi.org/10.1007/978-3-032-12263-6

Rapid demographic transition			
Population (thousands of people)	2022	2040	2100
World	8,021,407	8,784,545	6,986,814
High-Income Countries	1,254,317	1,240,396	900,686
Middle-Income Countries	6,020,971	6,462,103	4,530,591
Upper-Middle-Income Countries	2,804,740	2,727,519	1,334,229
Lower-Middle-Income Countries	3,216,231	3,734,584	3,196,362
Low-Income Countries	716,866	1,051,903	1,536,868
Per capita GDP (constant 2015 US$)			
World	11,315	20,031	73,019
High-Income Countries	43,791	54,406	98,838
Middle-Income Countries	5,796	12,673	46,961
Upper-Middle-Income Countries	9,747	22,340	73,299
Lower-Middle-Income Countries	2,347	4,215	29,650
Low-Income Countries	757	831	1,133
CO2 Emissions (million kt/year)			
World	33.6	42.2	26.6
High-Income Countries	10.9	3.9	−8.4
Middle-Income Countries	21.2	49.9	99.6
Upper-Middle-Income Countries	16.4	36.8	33.4
Lower-Middle-Income Countries	4.8	13.0	66.2
Low-Income Countries	0.2	0.3	1.0

Both rapid transitions			
Population (thousands of people)	2022	2040	2100
World	8,021,407	8,784,545	6,986,814
High-income countries	1,254,317	1,240,396	900,686
Middle-income countries	6,020,971	6,462,103	4,530,591
Upper-middle-income countries	2,804,740	2,727,519	1,334,229
Lower-middle-income countries	3,216,231	3,734,584	3,196,362
Low-income countries	716,866	1,051,903	1,536,868
Per capita GDP (constant 2015 US$)			
World	11,315	20,031	73,019
High-income countries	43,791	54,406	98,838
Middle-income countries	5,796	12,673	46,961
Upper-middle-income countries	9,747	22,340	73,299
Lower-middle-income countries	2,347	4,215	29,650
Low-income countries	757	831	1,133
CO2 emissions (million kt/year)			
World	33.6	31.7	−16.2
High-income countries	10.9	−0.7	−18.1
Middle-income countries	21.2	31.6	1.5
Upper-middle-income countries	16.4	32.3	1.4
Lower-middle-income countries	4.8	−0.7	0.1
Low-income countries	0.2	0.3	0.4

Accelerated adjustment across all dimensions			
Population (thousands of people)	2022	2040	2100
World	8,021,407	8,784,545	6,986,814
High-income countries	1,254,317	1,240,396	900,686
Middle-income countries	6,020,971	6,462,103	4,530,591
Upper-middle-income countries	2,804,740	2,727,519	1,334,229
Lower-middle-income countries	3,216,231	3,734,584	3,196,362
Low-income countries	716,866	1,051,903	1,536,868
Per capita GDP (constant 2015 US$)			
World	11,315	20,031	73,019
High-income countries	43,791	54,406	98,838
Middle-income countries	5,796	12,673	46,961
Upper-middle-income countries	9,747	22,340	73,299
Lower-middle-income countries	2,347	4,215	29,650
Low-income countries	757	831	1,133
CO2 emissions (million kt/year)			
World	33.6	13.1	−11.2
High-Income Countries	10.9	−0.7	−18.1
Middle-Income Countries	21.2	13.6	7.3
Upper-Middle-Income Countries	16.4	9.6	1.4
Lower-Middle-Income Countries	4.8	4.0	5.9
Low-Income Countries	0.2	0.2	−0.3

Accelerated adjustment with progress in poor countries			
Population (thousands of people)	2022	2040	2100
World	8,021,407	8,784,545	6,986,814
High-Income Countries	1,254,317	1,240,396	900,686
Middle-Income Countries	6,020,971	6,462,103	4,530,591
Upper-Middle-Income Countries	2,804,740	2,727,519	1,334,229
Lower-Middle-Income Countries	3,216,231	3,734,584	3,196,362
Low-Income Countries	716,866	1,051,903	1,536,868
Per capita GDP (constant 2015 US$)			
World	11,315	21,143	99,458
High-Income Countries	43,791	54,406	98,838
Middle-Income Countries	5,796	12,673	46,961
Upper-middle-income countries	9,747	22,340	73,299
Lower-middle-income countries	2,347	4,215	29,650
Low-income countries	757	1,289	13,559
CO2 emissions (million kt/year)			
World	33.6	13.4	−10.5
High-income countries	10.9	−0.7	−18.1
Middle-income countries	21.2	13.6	7.3
Upper-middle-income countries	16.4	9.6	1.4
Lower-middle-income countries	4.8	4.0	5.9
Low-income countries	0.2	0.5	0.4

Appendix 2 Categories of Goods Included in the Definition of International Specialization

New aggregates	Leamer aggregates	#	SITC	Description
I. Primary products (PP)	1. Petroleum (PETRO)	1	33	Petroleum and petroleum products
	2. Raw materials (MAT)	2	27	Crude fertilizers and crude minerals, nes
		3	28	Metalliferous ores and metal scrap
		4	32	Coal, coke and briquettes
		5	34	Non ferrous metals
		6	35	Gas, natural and manufactured
		7	36	Electric energy
II. Crops	3. Forest products (FOR)	8	24	Wood, lumber and cork
		9	25	Pulp and paper
		10	63	Wood and cork manufactures excluding furniture
		11	64	Paper, paperboard and manufactures thereof
	4. Tropical products (TROP)	12	5	Fruit and vegetables
		13	6	Sugar, sugar preparations and honey
		14	7	Coffee, tea, cocoa, spices & manufs. Thereof
		15	11	Beverages
		16	23	Crude rubber including synthetic and reclaimed
	5. Animal products (ANL)	17	0	Live animals
		18	1	Meat and meat preparations
		19	2	Dairy products and eggs
		20	3	Fish and fish preparations
		21	21	Hides, skins and fur skins, undressed
		22	29	Crude animal and vegetable materials,
		23	43	Animal and vegetable oils and fats, processed
		24	94	Animals, incl. Zoo animals, dogs and cats
	6. Cereals, etc. (CER)	25	4	Cereals and cereal preparations
		26	8	Feed. Stuff for animals excl. Unmilled cereals
		27	9	Miscellaneous food preparations
		28	12	Tobacco and tobacco manufactures
		29	22	Oil seeds, oil nuts and oil kernels
		30	26	Textile fibres, not manufactured, and waste
		31	41	Animal oils and fats
		32	42	Fixed vegetable oils and fats
III. Labor intensive (LAB)	7. Labor intensive (LAB)	33	66	Non metallic mineral manufactures,
		34	82	Furniture
		35	83	Travel goods, handbags and similar articles
		36	84	Clothing
		37	85	Footwear
		38	89	Miscellaneous manufactured articles,
		39	91	Postal packages not class. According to kind
		40	93	Special transact. Not class. According to kind
		41	96	Coin, other than gold coin, not legal tender
IV. Capital Intensive (CAP)	8. Capital intensive (CAP)	42	61	Leather, lthr. Manufs., nes, & dressed fur skins
		43	62	Rubber manufactures,
		44	65	Textile yarn, fabrics, made up articles, etc.
		45	67	Iron and steel
		46	69	Manufactures of metal,
		47	81	Sanitary, plumbing, heating and lighting fixt.

Appendix 3 Evolution of Fertility Rates in Selected Emerging Countries

	Fertility rate		Population 2020
País	2000	2023	(Millions)
Algeria	2.6	2.8	44
Argentina	2.6	1.5	45
Bangladesh	3.2	2.2	166
Brazil	2.3	1.6	208
Chile	2.1	1.2	19
China	1.6	1	1425
Colombia	2.6	1.7	50
Corea	1.5	0.7	52
Costa Rica	2.4	1.3	5
Cuba	1.6	1.4	11
Dem. Rep. Congo	6.7	6.1	94
Dominican Rep.	2.9	2.2	11
Ecuador	3.1	1.8	17
Egypt	3.5	28	109
El Salvador	3.1	1.8	6
Guatemala	4.6	2.6	17
India	3.4	2	1,396
Indonesia	2.5	2.1	274
Iran	2	1.7	88
Kenya	5.1	3.2	52
Malaysia	2.9	1.6	34
Mexico	2.7	1.9	126
Myanmar	2.8	2.1	53
Nigeria	6.1	4.5	212
Pakistan	5.4	3.6	233
Peru	2.9	2	33
Philippines	3.8	1.9	112
Russia	1.2	1.5	147
South Africa	2.4	2.2	60
Thailand	1.7	1.2	72
Taiwan	1.5	0.9	24
Turkey	2.5	1.6	86
Venezuela	2.8	2.1	29
Vietnam	2	1.9	99
Average Fertility Ratio	**2.86**	**1.94**	
Total Population			**5,409**

References

Albagli E, Vial J (2022) Crecimiento económico y biodiversidad: Algo tiene que ceder (Documento de Política Económica No. 70), Banco Central de Chile
Alesina A, Perotti R (1995) The political economy of budget deficits. IMF Staff Pap 42(1):1–31
Alesina A, Hommes R, Hausmann R, Stein E (1999) Budget institutions and fiscal performance in Latin America. J Dev Econ 59(2):253–273
Alexander P, Brown C, Arneth A, Finnigan J, Rounsevell MDA (2016) Human appropriation of land for food: the role of diet. Glob Environ Chang 41:88–98
Arellano JP (1980) Sistemas alternativos de seguridad social: Un análisis de la experiencia chilena. Colección Estudios CIEPLAN N°4
Arellano JP (2015) ¿Qué hacemos ahora que terminó el super ciclo del cobre? Programa CIEPLAN—Universidad de Talca
Barma N, Kaiser K, Le TM, Viñuela L (2012) Rents to riches? The political economy of natural resource-led development. World Bank
Bar-On YM, Phillips R, Milo R (2018) The biomass distribution on Earth. Proc Natl Acad Sci 115(25):6506–6511
Barry JM (2021) The great influenza: the deadliest pandemic in history. Capitán Swing Libros
Bosch M, Melguizo Á, Pages C (2013) Better pensions, better jobs: towards universal coverage in Latin America and the Caribbean. IDB
Brannen P (2017) The ends of the world: volcanic apocalypses, lethal oceans, and our quest to understand Earth's past mass extinctions. HarperCollins Publishers
Bravo-Ortega C, De Gregorio J (2005) The relative richness of the poor? Natural resources, human capital, and economic growth, Policy Research Working Paper Series 3489. World Bank
Brooke J (2014) Climate change and the course of global history. Cambridge University Press
Brundtland GH (1987) Our common future: Report of the World Commission on Environment and Development. United Nations
Caballero RJ, Panageas S (2008) Hedging sudden stops and precautionary contractions. J Dev Econ 85(1–2):28–57
Centro de Estudios del Cobre y la Minería (2023) Las interrogantes del presente y futuro de CODELCO. https://cesco.cl/wp-content/uploads/2024/10/CESCO_Las-interrogantes-del-presente-y-futuro-de-CODELCO.pdf
Collier P (2014) Resource revenue management: Three policy clocks. In: Caputo R, Chang R (eds) Commodity prices and macroeconomic policy. Central Bank of Chile
Cornelius P, Vial J (eds) (2002) The Latin American competitiveness report 2001–2002. CID – WEF
Costanza R, d'Arge R, de Groot R et al (1997) The value of the world's ecosystem services and natural capital. Nature 387:253–260. https://doi.org/10.1038/387253a0
Crutzen PJ, Stoermer EF (2000) The "Anthropocene". Glob Change Newsl 41:17–18
Daily G (1997) Nature's services: Societal dependence on natural ecosystems. Island Press
Dasgupta P (2021) The economics of biodiversity: the Dasgupta review. HM Treasury

J. Vial, *Sustainable Development and Natural Resources*, Integrated Science 39,
https://doi.org/10.1007/978-3-032-12263-6

Deaton A (2013) The great escape: health, wealth, and the origins of inequality. Princeton University Press
Diamond J (1997) Guns, germs, and steel: the fates of human societies. W. W. Norton
Diamond J (2019) Upheaval: How nations cope with crisis and change. Penguin UK
Diodato D, Hausmann R, Schetter U (2022) A simple theory of economic development at the extensive industry margin (CID Faculty Working Paper No. 416). Harvard Kennedy School. https://growthlab.hks.harvard.edu/files/growthlab/files/2022-09-cid-wp-416-extensive-industry-margin.pdf
DIPRES (2022) Indicador del balance cíclicamente ajustado: Metodología y resultados 2021
El Serafy S (1989) The proper calculation of income from depletable natural resources. In: Yusuf A, El Serafy S, Lutz E (eds) Environmental accounting for sustainable development. World Bank
European Commission (2021) The 2021 ageing report: Economic & budgetary projections for the EU member states (2019–2070) (Institutional Paper No. 148)
Fagan B (2000) La pequeña edad del hielo. Gedisa
Ferguson N (2006) The War of the World: Twentieth century conflict and the decline of the West. Penguin Press
Ferguson N (2008) The ascent of money: A financial history of the world. The Penguin Press
Fogel RW (2004) The escape from hunger and premature death, 1700–2100: Europe, America, and the Third World. Cambridge University Press
Food and Agriculture Organization of the United Nations (2021) Global Forest Resources Assessment 2020: Main report. https://doi.org/10.4060/ca9825es
Food and Agriculture Organization of the United Nations (2022) The state of world fisheries and aquaculture 2022: Towards blue transformation. FAO
Food and Agriculture Organization of the United Nations (2024) The state of world fisheries and aquaculture 2024. FAO
Frankel JA, Végh CA, Vuletin G (2013) On graduation from fiscal procyclicality. J Dev Econ 100(1):32–47
Friedman TL (2016) Thank you for being late: An optimist's guide to thriving in the age of accelerations. Farrar, Straus and Giroux
Gazzani A, Herrera V, Vicondoa A (2025) The asymmetric effects of commodity price shocks in emerging economies [Unpublished manuscript]. https://drive.google.com/file/d/13DdP7pVRSrYwyfWH7WEyrfEPsZm0XdYs/view
Gebremariam G (2024) A systematic review of insect decline and discovery: Trends, drivers, and conservation strategies over the past two decades. Psyche
González A, Sanchez F, Castillo E (2023) The nationalization of the large-scale copper mines in Chile: successful investment or financial failure? Mineral Economics. Advance online publication. https://doi.org/10.1007/s13563-023-00412-z
Gordon HS (1954) The economic theory of a common-property resource: the fishery. J Polit Econ 62(2):124–142
Gordon RJ (2016) The rise and fall of American growth. Princeton University Press
Goulder LH, Kennedy D (1997) Ch. 3: Valuing ecosystem services: Philosophical bases and empirical methods. In: Daily GH (ed) Nature's services: Societal dependence on natural ecosystems. Island Press
Gruss B, Kebhaj S (2019) Commodity terms of trade: a new database. International Monetary Fund
Harari YN (2015) Homo Deus: A brief history of the future. Penguin Random House
Hardin G (1968) The tragedy of the commons: The population problem has no technical solution; it requires a fundamental extension in morality. Science 162(3859):1243–1248
Hasell J, Roser M, Ortiz-Ospina E, Arriagada P (2022) Poverty. Our world in data. https://ourworldindata.org/poverty
Hausmann R, Hidalgo CA (2011) The network structure of economic output. J Econ Growth 16:309–342
He K, Gan Y, Wu H (2024) Natural resources, decentralized system, financial inclusion and sustainable development: Evidence from top emerging economies with resources abundance. Resour Policy 89:104674. https://doi.org/10.1016/j.resourpol.2023.104674

Herre B, Samborska V, Hasell J, Roser M (2017) Famines. Our world in data. https://ourworldindata.org/famines
Holzman R, Palmer R (eds) (2006) Pension reform: Issues and prospects for non-financial defined contribution (NDC) schemes. World Bank
Holzman R, Palmer E, Robalino D (eds) (2012) Non-financial defined contribution pension schemes in a changing pension world. World Bank
Intergovernmental Panel on Climate Change (2007) Climate change 2007: Synthesis report. Contribution of working groups I, II and III to the fourth assessment report of the intergovernmental panel on climate change (RK Pachauri, A Reisinger, eds)
Intergovernmental Panel on Climate Change (2014) Climate change 2007: Synthesis report. Contribution of working groups I, II and III to the fifth assessment report of the intergovernmental panel on climate change (RK Pachauri, A Reisinger, eds)
Intergovernmental Panel on Climate Change (2021) Summary for policymakers. In: Masson-Delmotte V et al (eds) Climate change 2021: The physical science basis. Cambridge University Press. https://doi.org/10.1017/9781009157896.001
Intergovernmental Panel on Climate Change (2022a) Agriculture, forestry and other land uses (AFOLU). In: Climate change 2022: Mitigation of climate change. Cambridge University Press, pp 747–860
Intergovernmental Panel on Climate Change (2022b) Summary for policymakers. In: Shukla PR et al (eds) Climate Change 2022: Mitigation of Climate Change. Cambridge University Press. https://doi.org/10.1017/9781009157926.001
Intergovernmental Panel on Climate Change (2022c) Summary for policymakers. In: Pörtner H-O et al (eds) Climate change 2022: impacts, adaptation and vulnerability. Cambridge University Press. https://doi.org/10.1017/9781009325844.001
International Monetary Fund (2012) Macroeconomic policy frameworks for resource-rich developing countries (Policy paper). https://www.imf.org/external/np/pp/eng/2012/082412.pdf
International Monetary Fund (2019) Fiscal policies for Paris climate strategies: from principles to practice (IMF Policy Paper). https://www.imf.org/en/Publications/Policy-Papers/Issues/2019/05/01/Fiscal-Policies-for-Paris-Climate-Strategies-from-Principle-to-Practice-46826
International Monetary Fund (2020) State-owned enterprises: the other government. Chapter 3, Fiscal monitor. April. https://www.imf.org/en/Publications/FM/Issues/2020/04/06/fiscal-monitor-april-2020#:~:text=Chapter%203%20:%20State%2DOwned%20Enterprises,of%20transparency%20of%20their%20activities
International Monetary Fund (2023) Chile: Technical Assistance Report—Fiscal considerations in managing stabilization funds. Country Report. https://www.imf.org/en/Publications/CR/Issues/2023/07/07/Chile-Technical-Assistance-Report-Fiscal-Considerations-in-Managing-Stabilization-Funds-535833
IPBES (2019) Global assessment report on biodiversity and ecosystem services. IPBES Secretariat. https://doi.org/10.5281/zenodo.3831673
Johnson JA, Baldos UL, Cervigni R, Chonabayashi S, Corong E, Gavryliuk O, Hertel TW, Nootenboom C, Gerber JF, Ruta G, Polasky S (2021) The economic case for nature: a global earth-economy model to assess development policy pathways. World Bank. https://hdl.handle.net/10986/35882
Kim J (2016) Female education and its impact on fertility. IZA World of Labor
Kolbert E (2014) The sixth extinction: an unnatural history. Henry Holt & Company
Krautkraemer JA (2005) Economics of natural resource scarcity: The state of the debate (Discussion Paper 05-14). Resources for the Future. https://media.rff.org/documents/RFF-DP-05-14.pdf
Kurzweil R (2005) The singularity is near: When humans transcend biology. Viking
Lange GM, Wodon Q, Carey K (2018) The changing wealth of nations 2018: Building a sustainable future. World Bank Publications
Leamer EE (1985) Sources of comparative advantage: Theory and evidence. MIT Press
Leamer EE, Maul H, Rodríguez S, Schott PK (1999) Does natural resource abundance increase Latin American income inequality? J Dev Econ 59:3–42

Lederman D, Maloney WF (2007) Neither curse nor destiny: Introduction to natural resources and development. In: Lederman D, Maloney WF (eds) Natural resources: Neither curse nor destiny. World Bank, pp 1–14

Livi Bacci M (2012) A concise history of world population, 5th edn. John Wiley & Sons

Loayza N, Fajnzylber P (2005) Economic growth in Latin America and the Caribbean: Stylized facts, explanations, and forecasts. World Bank Publications

MacMillan M (2020) War: How conflict shaped us. Random House, New York

Maddison A (2000) The world economy: A millennial perspective. OECD Development Centre Studies

Maddison A (2006) The world economy: Volume 1: A millennial perspective and volume 2: Historical statistics. OECD Development Centre Studies. https://doi.org/10.1787/9789264022621-en

Malthus TR (1798) An essay on the principle of population

Markham B (1942) West with the night. Open Road

Marshall A (1890) Principles of economics. Palgrave Macmillan

Meadows D, Meadows D, Randers J (1972) The limits to growth. Potomac Associates

Miller T, Mason C, Holz M (2011) Impact of demographic change in ten Latin American countries: Projecting public expenditures in education, health and pensions. In: Cotlear D (ed) Population aging: Is Latin America ready? World Bank, pp 233–272. https://documents1.worldbank.org/curated/en/117291468299147625/pdf/588420PUB0Popu11public10BOX353816B0.pdf

Naudon A, Vial J (2016) The evolution of inflation in Chile since 2000. BIS Papers, 89g

Newbold T, Hudson LN, Arneth A, Contu S, De Palma A, Ferrier S et al (2016) Has land use pushed terrestrial biodiversity beyond the planetary boundary? A global assessment. Science 353(6296):288–291. https://doi.org/10.1126/science.aaf2201

Nordhaus W (1992) Lethal model 2: The limits to growth revisited. Brookings Pap Econ Act 1992(2):1–59

OECD (2022) Climate tipping points: Insights for effective policy action. OECD Publishing. https://doi.org/10.1787/abc5a69e-en

OECD (2023) Pensions at a glance 2023: OECD and G20 indicators. OECD Publishing. https://doi.org/10.1787/678055dd-en

Otto J (2017) The taxation of extractive industries: Mining (WIDER Working Paper 2017/75). United Nations University WIDER. https://www.wider.unu.edu/sites/default/files/wp2017-75.pdf

Otto J, Andrews C, Cawood F, Doggett M, Guj P, Stermole F, Stermole J, Tilton J (2006) Mining royalties: a global study of their impact on investors, government and civil society. World Bank

Pearce DW, Turner RK (1990) Economics of natural resources and the environment. Johns Hopkins University Press

Peña-Torres J, Muñoz R, Quezada F, Kaufmann C (2022) Entry deterrence and collusion at repeated multi-unit auctions of ITQs. Mar Resour Econ 37(4):417–442

Pérez Rosales V (1882) Recuerdos del pasado (Ed. 1970). Biblioteca Francisco de Aguirre

Peskin HM (1976) A national accounting framework for environmental assets. J Environ Econ Manag 2:255–262

Pomeranz K (2000) The great divergence: China, Europe, and the making of the modern world economy. Princeton University Press

Prebisch R (1959) Commercial policy in underdeveloped countries. Am Econ Rev 49(2):251–273

Pritchett L (1997) Divergence, big time. J Econ Perspect 11(3):3–17

Repetto R (1991) La erosión en el balance general: Cómo contabilizar la pérdida en recursos naturales. In: Vial J (ed) Desarrollo y medio ambiente: Hacia un enfoque integrador. CIEPLAN

Ritchie H (2021) The world has lost one-third of its forest, but an end of deforestation is possible. Our world in data. https://ourworldindata.org/world-lost-one-third-forests

Ritchie H (2025) How much in subsidies do fossil fuels receive? Our world in data. https://ourworldindata.org/how-much-subsidies-fossil-fuels

Rockström J, Steffen W, Noone K, Persson Å, Chapin FS III, Lambin E et al (2009) Planetary boundaries: Exploring the safe operating space for humanity. Ecol Soc 14(2)

Rouzet D et al (2019) Fiscal challenges and inclusive growth in ageing societies. OECD Economic Policy Papers, No 27. https://doi.org/10.1787/c553d8d2-en
Sachs JD (2008) Common wealth: economics for a crowded planet. The Penguin Press
Sachs JD (2015) The age of sustainable development. Columbia University Press
Sachs J, Vial J (2002) Can Latin America compete? In: Vial J, Cornelius P (eds) The Latin American Competitiveness Report 2001–2002. Oxford University Press
Sachs J, Warner AM (2001) The curse of natural resources. Eur Econ Rev 45(4–6):827–838
Scartascini, C., Spiller, P., Stein, E., & Tommasi, M. (Eds.). (2010). The political game in Latin America: How are public policies decided? Inter-American Development Bank
Smil V (2022) How the world really works: the science behind how we got here and where we're going. Viking
Smith A (1776) The Wealth of Nations. Reprinted by Penguin Books, 1982
Solow RM (1956) A contribution to the theory of economic growth. Q J Econ 70(1):65–94
Stein, E., Tommasi, M., Echeverría, K., Lora, E., & Payne, M. (Eds.). (2005). The politics of policies: economic and social progress in Latin America (2006 Report). Inter-American Development Bank
Swan TW (1956) Economic growth and capital accumulation. Econ Rec 32(2):334–361
Tietenberg TH, Lewis L (2015) Environmental & natural resource economics, 10th edn. Pearson
Toye J, Toye R (2003) The origins and interpretation of the Prebisch–Singer thesis. Hist Polit Econ 35(3):437–467
UN Environment Programme (2019) Global Environmental Outlook—GEO-6: Healthy planet, healthy people. https://www.unep.org/resources/global-environment-outlook-6
UNEP – WCMC, & IUCN (2024) Protected Planet Report 2024. UNEP – WCMC & IUCN
United Nations (2019) World population prospects: The 2019 revision
United Nations (2024) World population prospects 2024: Online edition
United Nations Environment Programme, International Resource Panel (2024) Bend the trend: Pathways to a livable planet as resource use spikes (Global Resources Outlook 2024). UNEP. https://wedocs.unep.org/20.500.11822/44901
Vial J (2019) Reserves management and FX intervention: Chile's experience. BIS Papers, 104f
Vial J (2023a) Desafíos para un Desarrollo sostenible de Chile en el Siglo XXI. Ediciones UC
Vial J (2023b) Desarrollo sostenible y capital natural. https://scioteca.caf.com/handle/123456789/2030
Vial J (2024) Capital forestal y sus desafíos en Chile y el mundo. (Working Paper No. 133). Clapes UC
Vial J, Isea M (2025) Errores en el cálculo de la brecha del PIB y sus impactos en la aplicación de la regla fiscal en Chile. Clapes UC
Wildlife Conservation Society (n.d.) Human footprint. https://wcshumanfootprint.org/
Williamson J (1990) Latin American adjustment: How much has happened? Institute of International Economics
Wilson EO (2002) The bottleneck. Sci Am 286(2):82–91
World Bank (1997) Expanding the measure of wealth: Indicators of environmentally sustainable development. World Bank. https://elibrary.worldbank.org/doi/epdf/10.1596/0-8213-3956-7
World Bank (2011) Overview of state ownership in the global minerals industry: long-term trends and future (Extractive Industries for Development Series No. 20). https://hdl.handle.net/10986/16968
World Bank (2024) The changing wealth of nations: Revisiting the measurement of comprehensive wealth. World Bank
WWF (2024) Living Planet Report 2024: a system in danger. WWF

MIX
Papier aus verantwortungsvollen Quellen
Paper from responsible sources
FSC® C105338

If you have any concerns about our products,
you can contact us on
ProductSafety@springernature.com

In case Publisher is established outside the EU,
the EU authorized representative is:
Springer Nature Customer Service Center GmbH
Europaplatz 3, 69115 Heidelberg, Germany

Printed by Libri Plureos GmbH
in Hamburg, Germany